The Federal Plan for Meteorological Services and Supporting Research

FISCAL YEAR 2015

50th Anniversary Edition

FEDERAL COORDINATOR FOR METEOROLOGICAL
SERVICES AND SUPPORTING RESEARCH

Silver Spring Metro Center, Building 2 (SSMC 2)
1325 East West Highway, Suite 7130
Silver Spring, MD 20910
301.628.0112

FCM-P1-2014
September 2014
WASHINGTON, D.C.

Editor: Mr. Anthony Ramirez
Assistant: Ms. Erin McNamara
Assistant: Mr. Kenneth Barnett

PREFACE

This edition of the *Federal Plan for Meteorological Services and Supporting Research* marks the 50th anniversary of the annual plan. The plan today articulates the provision of meteorological services and supporting research by agencies of the Federal Government just as it did 50 years ago. Dr. Robert M. White, the first Federal Coordinator for Meteorology, noted in the preface to the first plan:

> "The best guarantee of achieving an economical use of Federal meteorological resources is to have all agencies working together toward a common goal. The many similarities in agency requirements and programs present both a challenge and an opportunity for the Federal Government to achieve a higher return on its investment in meteorological programs, better use of scientific manpower, and improved meteorological services."

These words still ring true today, reverberating across decades that have seen interagency coordination and collaboration deliver better weather reconnaissance aircraft systems, numerical weather prediction, and research strategies to improve hurricane forecasting and warning, a national network of highly-capable Doppler weather radars, significant reductions in weather-related aircraft accidents, and successful advocacy for the high performance computing that powers today's weather forecasts and atmospheric dispersion models critical to responding effectively to major emergencies.

I am pleased to report that today our agencies are continuing to work together toward a common goal in a number of important areas. In 2014, we reactivated the Committee for Operational Environmental Satellites to improve interagency communication and coordination of matters regarding the use of environmental satellites. In 2014 we began and in 2015 will complete a mid-course assessment of progress in achieving our strategic research plan for tropical cyclones. We will continue to plan, organize, and host the Interdepartmental Hurricane Conference to share information and prepare for next year's hurricane season, including publishing the 2015 *National Hurricane Operations Plan*. I am also eager to see the proof of concept system under development for dual-polarization phased array radar and the possibility to move forward with a full-scale Advanced Technology Demonstrator for Multifunction Phased Array Radar (MPAR). MPAR represents a potential breakthrough in economy across a number of radar applications. This is also an exciting time for discovering, acquiring, and employing new data streams from satellites, networks of observing networks across the country, and unmanned aerial systems—all of which our interagency coordinating groups will be addressing in fiscal year (FY) 2015. In data-related work, we will also seek to reactivate the Satellite Telemetry Interagency Working Group (STIWG) in FY 2015. Reactivating STIWG will heighten agency awareness of the Geostationary Operational Environmental Satellite (GOES) Data Collection System (DCS) and opportunities for data sharing and collaboration. GOES DCS has grown to approximately 26,000 stations sending almost 700,000 transmissions containing an estimated 6 million observations per day.

This edition of the *Federal Plan* provides additional information on the topics I've just described as well as much more information about activities across the agencies. It provides Congress and the Executive Branch with a review of agency programs and activities in FY 2014 and a comprehensive account of proposed programs for FY 2015. The *Federal Plan*'s narratives, timelines, and schedules are current as of September 2014.

I extend my thanks to our agency partners and their staffs for their dedicated efforts to contribute to this important document.

//Signed//
Samuel P. Williamson
Federal Coordinator for Meteorological
Services and Supporting Research

Table of Contents

SECTION 1
AGENCY FUNDING FOR METEOROLOGICAL OPERATIONS
AND SUPPORTING RESEARCH

RESOURCE INFORMATION AND AGENCY PROGRAM UPDATES

The narratives and tables in this section summarize the budgetary information for the Federal government for fiscal years (FY) 2014 and 2015. The funds shown are used to provide meteorological services and associated supporting research with service improvements as their immediate objectives. Fiscal data are current as of the end of September 2014 and are subject to later changes. The data for FY 2015 do not have legislative approval and do not constitute a commitment by the United States Government. The budget data are prepared in compliance with Section 304 of Public Law 87-843, in which Congress directed that an annual horizontal budget be prepared for meteorological programs conducted by the Federal agencies.

AGENCY BUDGET SUMMARIES

DEPARTMENT OF AGRICULTURE

The Department of Agriculture's (USDA) budget request for FY 2015 is $70.0 million for operations and supporting research, a 2.2% decrease from the FY 2014 funding level. The decrease in funding is mainly in the areas of supporting research and costs, although a decline in meteorological operations is also anticipated. USDA staffing levels remain unchanged at 103 full-time equivalent (FTE) employees. USDA has requested $49.9 million for research and development programs, a decrease of $1.2 million (2.4%) from 2014. Most of the decrease is for the Agricultural Research Service (ARS) in the area of Agricultural and Land Management. ARS is the USDA's principal in-house scientific research agency, conducting research on how to cope with annual variations of weather on crop and animal production, ecosystem services, and the environmental and economic sustainability of agricultural enterprises. Funding is also decreased for the Forest Service (FS) in the area of Wildland Fire Weather. The research and development mission of the Forest Service is to develop and deliver knowledge and innovative technology to improve the health and use of the Nation's forests and grasslands. Research at the Forest Service includes studies of the long-term effects of air pollution on forests and water resources. The Forest Service is also the world leader in developing emissions factors from fires and modeling its dispersion. For the National Institute for Food and Agriculture (NIFA), there is a slight increase in funding. NIFA funding supports research projects that collect, analyze, and utilize short and long-term weather and climate data as a base of information for the projection and prediction of climatic trends related to environmental impacts on agro-ecosystems, forests, and rangelands and the development of adaptation and mitigation strategies for natural resources and production management.

The FY 2015 amount requested for meteorological operations is $20.0 million, down (1.5%) from the $20.3 million funding level in FY 2014. Operational activities include specialized weather observing networks such as the SNOTEL (SNOw pack TELemetry) network operated by the Natural Resources Conservation Service (NRCS) Snow Survey and Water Supply Forecasting program (SSWSF) and the Remote Automated Weather Stations (RAWS) network managed by the Forest Service. The SNOTEL and RAWS networks provide cooperative data for NOAA's river forecast activities, irrigation water supply estimates, and Bureau of Land Management operations. The SSWSF program, managed by the NRCS National Water and Climate Center, provides western states and Alaska with information on future water supplies. The Forest Service uses meteorological data and interpretation skills data for decision making regarding wildland fire management. The meteorological staff of the Office of the Chief Economist's World Agricultural Outlook Board (OCE/WAOB) routinely collects global weather data and agricultural information to assess the impact of growing season weather conditions on crops and livestock production prospects, keeping USDA commodity analysts, the Chief Economist, and the Secretary of Agriculture and top staff well informed of weather impacts on crops and livestock worldwide. The Risk Management Agency (RMA) provides administration and oversight of programs authorized under the Federal Crop Insurance Act.

DEPARTMENT OF COMMERCE/NATIONAL OCEANIC AND ATMOSPHERIC ADMINISTRATION

National Weather Service

The National Weather Service (NWS) funding request for the FY 2015 President's Budget totals $1.06 billion and 4,617 full-time equivalent (FTE) employees. This is a 0.1% increase over the FY 2014 Spend Plan. In 2015, NOAA proposes to restructure NWS Operations, Research, and Facilities (ORF) and Procurement, Acquisition and Construction (PAC) account Programs, Projects, and Activities (PPA) as part of a broader effort to evolve the NWS to deliver more efficient, responsive, and advanced operations to the Nation. This proposal, which will align the NWS budget to both function and performance, is critical to advancing the Department's Weather-Ready Nation goal. Significant requested increases and decreases in funding over the FY 2014 program include the following:

- *Responding to Congressionally-Requested Studies of NWS.* NOAA requests $3.0 million to support effective response to recommendations of two recently-completed independent studies of the National Weather Service. The requested increase will support analysis of workforce and infrastructure; enhanced demonstration and testing capabilities; improved stakeholder outreach; and improved evaluation capacity.

- *Centralized Water Forecasting Demonstration.* NOAA requests $4.0 million to design, develop, and test a new national hydrologic modeling and forecast system to be deployed at the National Water Center in Tuscaloosa, Alabama. This centralized approach enables both the implementation of scientific advances and national consistency.

- *Next Generation Weather Radar (NEXRAD) Service Life Extension Program.* NOAA requests $9.3 million to extend the useful life of the aging NEXRAD infrastructure that underpins severe weather forecast and warning services for high-impact events. A Service Life Extension Program is required to sustain current weather forecast and warning services until the next generation of weather radars are identified, developed, and deployed. Without this investment, NEXRAD availability will degrade beginning in 2020, resulting in radar outages and gaps and negative impact to tornado and flash flood warnings.

- *Ground Readiness Project.* NOAA requests $6.0 million to ensure utilization of the substantial increase in environmental satellite, radar, and model data that will improve weather warnings and forecasts. NWS's ground readiness activities are a multi-year effort that was initiated in the FY 2013 President's Budget and began work with funding from the Disaster Relief Appropriations Act, 2013.

- *NWS Telecommunications Gateway (NWSTG) Legacy Replacement.* NOAA requests $5.0 million to continue re-architecture of its Telecommunications Gateway and its backup into a modern, scalable, extensible, and reliable system. The Telecommunications Gateway is the hub for collection and transmission of hydrometeorological observations and products for NOAA's customers.

- *Relocation of the National Logistics Supply Center/National Reconditioning Center (NLSC/NRC).* NOAA requests $8.1 million for the relocation of NLSC/NRC from the Bannister Federal Complex in Kansas City, Missouri (MO) as required by the General Services Administration (GSA), who plans to sell the property.

- *Regional Enterprise Application Development and Integration (READI) Teams.* NOAA requests a $10.0 million reduction to reflect the significant efficiencies that can be achieved by transition to a new information technology service delivery model for the NWS forecast offices. NWS will create regionalized IT collaboration teams enabling higher consistency of service delivery with minimal impacts to NWS' mission.

- *Hurricane Forecast Improvement Project (HFIP).* NOAA requests an $8.2 million decrease to delay on-going research toward improving hurricane track and intensity prediction within the HFIP. NWS anticipates meeting HFIP goals of 20 percent improvement for both track and intensity in a demonstration mode by the end of the 2015 hurricane season. The remaining funding will provide for maintenance of already developed hurricane models within a research environment.

- *Next Generation Air Transportation System Weather Program (NextGen).* NOAA requests a $9.0 million dollar decrease to maintain continuity of NOAA support for the NextGen initiative and implement key IT infrastructure efforts, while conducting NextGen science and application development and implementation at a slower pace to reduce annual costs. This allows for a re-evaluation of NextGen goals and scope with NextGen partners.

National Environmental Satellite, Data, and Information Services

The National Environmental Satellite, Data, and Information Services (NESDIS) funding request for the FY 2015 President's Budget totals $2.25 billion and 909 FTE. This is a 7.7% increase over the FY 2014 Spend Plan.

NESDIS is responsible for managing all aspects of remotely gathered environmental data. This includes procurement, launch, operation, product development, and product distribution of the Nation's civil operational environmental satellites and corresponding data. In addition, NESDIS manages the NOAA environmental data collections, provides assessments that describe climate, and disseminates data and information to meet the needs of users in commerce, industry, agriculture, science, and engineering, as well as federal, state, and local governments.

Budget Restructure:

In FY 2015, NOAA is proposing to restructure NESDIS Operations, Research, and Facilities (ORF) and Procurement, Acquisition, and Construction (PAC) appropriation account projects, programs, and activities (PPAs) as part of a broader effort to reorganize several of its components to best fulfill its critically important mission and to examine more cost-effective means of providing its products and services. The new budget structure strengthens both the satellite and data management sides of NESDIS and better coordinates systems engineering and common ground services.

Modifications to major systems acquisition programs for FY 2015 include the following:

- *Joint Polar Satellite System (JPSS).* NOAA requests an increase of $95.4 million for a total of $916.3 million. These funds are necessary to continue the build of the JPSS-1 and JPSS-2 satellite missions. Additionally, these funds enable JPSS to pursue the procurement of ATMS and CrIS spares to reduce schedule risk. The FY 2015 priority is to maintain the critical polar weather satellite observations in the United States' afternoon orbit including the following activities: operate and sustain the Suomi-National Polar-orbiting Partnership (S-NPP) satellite, which was launched October 28, 2011; maintain the planned launch of JPSS-1 by no later than Q2 FY 2017 and launch JPSS-2 by Q1 FY 2022; add robustness to the JPSS flight segment; and continue re-development of the common ground system for the JPSS missions (S-NPP, JPSS-1, JPSS—2). The requested increase in funds are necessary to continue to maintain S-NPP operations, complete development of the JPSS-1 spacecraft, and support the integration of the Visible/Infrared Imager/Radiometer Suite (VIIRS), Cross-track Infrared Sounder (CrIS), Advanced Technology Microwave Sounder (ATMS), Clouds and Earth's Radiant Energy System (CERES), and Ozone Mapping Profiler Suite-Nadir (OMPS-N).

- *Jason-3.* NOAA requests an increase of $7.2 million for a total of $25.7 million to continue the development of the Jason-3 satellite in partnership with European Organisation for the Exploitation of Meteorological Satellites (EUMETSAT) and Centre National d'Etudes Spatiales (CNES). NOAA's European partners have provided the spacecraft, altimeter, and precision orbit components, and NASA, on behalf of NOAA, has completed the build of mission instruments and has delivered the completed instrument to CNES for satellite integration. NASA also completed spacecraft integration and testing in FY 2013. The Administration has received approval for an FY 2014 reprogramming package that maintains a Q2 FY 2015 Jason-3 launch.

- *GOES-R.* NOAA requests an increase of $38.9 million for a total of $980.8 million to continue satellite engineering development, production, integration, testing, and launch activities for the four-satellite GOES-R Series Program. NOAA needs the requested increase in funding to maintain instruments, satellite, and ground system developments that are all currently under contract in order to meet the planned launch dates of 2Q FY 2016 for GOES-R and 3Q FY 2017 for GOES-S. The funds will also be used to continue the development activities for GOES-T and GOES-U to maintain their launch schedules.

- *Deep Space Climate Observatory (DSCOVR).* NOAA requests a decrease of $2.6 million in FY 2015 for a total of $21.1 million to complete the refurbishment of the DSCOVR satellite and sensors for solar wind observations and to deliver the spacecraft for a U.S. Air Force (USAF) launch. NASA/Goddard Space Flight Center (GSFC), under a reimbursable agreement, will complete the refurbishment of DSCOVR, which is currently housed at GSFC in Greenbelt, MD.

- ***Solar Irradiance, Data and Rescue (SIDAR).*** NOAA requests an increase of $15.0 million and 2 FTE for a total of $15.0 million and 2 FTE in FY 2015 to implement the acquisition strategy for hosting the already purchased and built Total Solar Irradiance Sensor (TSIS). TSIS will provide measurements of the variability in the Sun's total output. As part of the SIDAR project, NOAA will also continue to support two international partnerships: satellite search and rescue via the Search And Rescue Satellite Aided Tracking (SARSAT) system and environmental data collection and relay via the Advanced Data Collection System (ADCS).

- ***COSMIC-2/Global Navigation Satellite System Radio Occultation (GNSS RO).*** NOAA requests an increase of $4.8 million and 0 FTE for a total of $6.8 million and 1 FTE for ground reception and processing of GNSS RO satellite data from GNSS RO satellites provided by Taiwan and USAF. The GNSS RO ground system is part of an international partnership between Taiwan's National Space Organization (NSPO), USAF, and Brazil, for a Constellation Observing System for Meteorology, Ionosphere, and Climate (COSMIC) follow-on. In this COSMIC follow-on partnership, NSPO has agreed to procure 12 satellite buses and integration of instruments, and USAF will procure the instruments and provide launch services. Together, NOAA, Taiwan, and Brazil will provide the ground system. In addition to the COSMIC follow-on partnership, NOAA will continue to support the reception and processing of COSMIC and foreign satellite radio occultation data.

Office of Oceanic and Atmospheric Research

The Office of Oceanic and Atmospheric Research (OAR) request for funding related to meteorological supporting research in the FY 2015 President's Budget totals $160.3 million and 408 FTE. This is a 14.6% increase over the FY 2014 Spend Plan.

Requested FY 2015 funding for Weather and Air Chemistry Research (W&ACR), including Laboratories and Cooperative Institutes and Weather and Air Chemistry Research Programs is $84.9 million. This amount represents a net increase of $3.8 million or 4.7% from the FY 2014 Spend Plan. Increases would go to improve the readiness of those weather and related research projects associated with critical technologies, model improvements, and service applications to a stage of development that will enable a successful future transition to operations for deployment by NOAA's operational entities. OAR's Weather and Air Chemistry Research supports R&D that provides the Nation with more accurate and timely warnings and forecasts of high impact weather events and their broader impact on issues of societal concern such as weather and air quality; and supports research that provides the scientific basis for informed management decisions about weather, water, and air quality.

Requested FY 2015 funding for Climate Labs and Cooperative Institutes is $75.5 million. This amount represents a net increase of $ 16.6 million, or 28% from the FY 2014 Spend Plan. These funds will go toward research activities that will help to gain a comprehensive understanding of the physical, chemical, and dynamical processes that shape our climate including maintaining and enhancing NOAA's Atmospheric Baseline Observatories to deliver policy-relevant information on greenhouse gas emissions, supporting northward development of NOAA's Arctic Observing Network, providing focused drought impacts research and applications development to underserved regions of the country and supporting of the U.S. Global Change Research Program's priority research areas. OAR's Laboratories and Cooperative Institutes are central to the effort of the climate research community to improve that understanding, to test our understanding through the development of state-of-the-art Earth System Models, and then to use those models to predict the future state of the climate. Observations of the Earth system and their analysis underpin the efforts that form the scientific basis for Climate Research.

National Ocean Service

The National Ocean Service (NOS) request for funding related to meteorological supporting research in the FY 2015 President's Budget totals $30.1 million and 117 FTE. This is a 0.8% increase over the FY 2014 Spend Plan. These funds allow for expanded operation of the National Water Level Observation Network (NWLON), and continued operation of the Physical Oceanographic Real-Time System (PORTS®) program, the data quality control program known as the Continuous Operational Real-time Monitoring System (CORMS), and the Ocean Systems Test and Evaluation Program (OSTEP), which is a development program for bringing new sensor technology into operations. Both the NWLON and PORTS® programs include subsets of operational water level stations with meteorological sensors installed for various partners and users, including the NWS. This is level funding with the FY 2014 Spend Plan.

Though traditionally an oceanographic observing system, NWLON/PORTS® technology allows multiple other sensors to be added, including meteorological sensors such as wind speed/direction/gusts, air temperature, relative humidity and barometric pressure. Additionally, NOS has eight PORTS® stations with operational visibility sensors, in Mobile Bay, AL, San Francisco Bay, CA and Jacksonville, FL. These meteorological observations provide important data for improving and verifying marine weather forecasts and warnings. NOS has upgraded and enhanced the majority of its NWLON stations with new meteorological sensors. This increase in meteorological observations has led to a 10% increase in the probability of detection of marine weather events and a ten minute increase in lead times for marine warnings, according to actual verification data for special marine warnings (WFO Sterling).

Navigation data users require a complete picture of their operating environment, which includes local meteorological data, to make safe and cost-effective decisions. Leveraging existing observing infrastructure is a cost-effective alternative to establishing new platforms to collect these data. The additional meteorological data improves the accuracy of NWS forecasts for storm surge, marine wind speed, and marine wave heights, used by both marine navigation and coastal communities when extreme weather events occur. The real-time data can be used by emergency responders to make sound decisions based upon the most up to date and accurate information. For example, when coastal areas are flooding, emergency responders must know which evacuation routes are still viable and other related information that most accurately reflects the current state of the physical environment.

In FY 2013, the *Disaster Relief Appropriations Act* provided $3.0 million to NOS' Center for Operational Oceanographic Products and Services in order to repair and rebuild NWLON and PORTS® stations which were damaged by the storm.

Office of Marine and Aviation Operations

The FY 2015 President's Budget requests $31.6 million and 121 FTE for the Office of Marine and Aviation Operations (OMAO) Aircraft Operations. This is a 1.4% increase over the FY 2014 Spend Plan. OMAO supports meteorological operations and research activities by collection of related data from ships and aircraft. The fleet of aircraft support NOAA's mission of science, service, and stewardship. The aircraft operate throughout the United States and around the world; over open oceans, mountains, coastal wetlands, and the Arctic. OMAO provides capable, mission-ready aircraft and professional crews to study global climate change and air quality, assess marine mammal populations, survey coastal erosion, investigate oil spills, conduct coastal mapping, survey snowpack levels for flood prediction, and improve hurricane prediction models. AOC flight crews operate in some of the world's most demanding flight regimes including flying into the eye of a hurricane.

DEPARTMENT OF DEFENSE

U.S. Air Force

The Air Force (AF) resources for meteorological support fall into two primary categories: general operations and investment and research. The total AF weather operations and investment funding requested for FY 2015 is $135.5 million, a one percent decrease from FY2014, 2948 active duty military and 440 civilian personnel authorizations. Additionally, the AF provides resources for space-based environmental monitoring development and operations managed by Air Force Space Command, including the Defense Meteorological Satellite Program (DMSP) and the Space Situational Awareness Environmental Monitoring (SSAEM).

Operations

The operations support portion of the AF weather FY 2015 budget request is $110.4 million to fund day-to-day environmental support to the Department of Defense (DoD), the Active and Reserve Components of the Air Force and Army, 10 unified commands, and other agencies as directed by the Chief of Staff of the Air Force. The AF employs over 3,388 military and civilian personnel to conduct these activities at more than 250 locations worldwide. Approximately 85 percent of personnel specialize in weather; the remainder includes communications, computer, information technology, program management, program analysis, administrative, and logistics specialists.

Investment (Acquisition and Research)

The total Air Force budget request for meteorological investment (acquisition and research) for FY 2015 is $25.1 million. As part of the Air Force Strategic Weather Modernization Plan, the AF continues investing in modernized environmental prediction and commercial-off-the-shelf technologies that enhance automation and optimize resources. The AF plans to invest in the following efforts in FY 2015 and beyond:

- *Weather Data Analysis (WDA).* WDA is a large-scale processing, services-oriented architecture system providing the capability to ingest, process, store, access/retrieve, and disseminate meteorological and oceanographic (METOC) data. WDA will provide Open Geospatial Consortium, Intelligent Data Services, and Web Coverage Processing Service for weather products across multiple security enclaves. These services include both machine-to-machine and machine-to-human capabilities. WDA will provide a centralized web service capability across multiple security enclaves employing a services oriented architecture via a common URL. WDA will use commercial-off-the-shelf software as the primary component for visualization of weather data for web browser-based presentation for weather, command and control, mission operations, and combat support personnel. WDA will add new data for presentation to web users and provide weather forces applications which enhance weather operations effectiveness and integration in joint/AF operations. WDA also will provide enterprise storage, cross domain capabilities, and network hardware and software for the AF's centralized meteorological production center.

- *Numerical Weather Modeling (NWM).* NWM includes numerical weather prediction (NWP) models; cloud analysis and forecasting models; land surface characterization models; aerosol, atmospheric constituent, and point analysis models/applications; and both global and mesoscale ensembles. Weather Research and Forecast Model (WRF) will develop advances in parameterization via science algorithms, atmospheric chemical constituent specification, and model coupling to improve forecasting performance in all levels of the atmosphere with an emphasis on the lower atmosphere. NWM will integrate new satellite data and assimilation schema into cloud analysis/forecasting products to improve our high cadence products. Ensemble capabilities, both global and regional scale, will provide operationally-focused meteorological intelligence for military operators/warfighters by objectively quantifying the forecast certainty of mission-impacting meteorological parameters to optimize operational risk management for decision making at all echelons. These ensembles will provide probabilistic algorithms for high-impact variables.

- *Space Weather Analysis and Forecast System (SWAFS).* SWAFS is the primary operational DoD source for space weather data, models, and products. SWAFS ingest and databases ground- and space-based space weather data and employs models and applications to create and disseminate specified space weather analysis, warning, and forecast products for weather and space operations personnel. SWAFS will perform integration work for the Global Assimilation of Ionospheric Measurements (GAIM) – Full Physics (FP) model to improve specification and forecasts of the ionosphere; verification and validation of D-region addition to GAIM-FP; incorporation of data sources from new satellite systems; and net-centric data availability of magnetospheric model output and data visualization for web browser and User Defined Operational Picture users.

- *Environmental Data Cube Support System (EDCSS).* EDCSS provides Air and Space Natural Environment data for DoD Modeling and Simulation (M&S) applications. EDCSS supports Live Virtual Constructive (LVC) Operational Training and provides DoD M&S users a correlated and realistic natural environment with tailorable scenarios for specific warfighter effects. EDCSS will integrate real-time meteorological data; service-oriented weather decision aids; and improved scenario management capabilities for LVC users.

The goals of these efforts are to provide accurate, relevant, and timely meteorological intelligence to operators and warfighters at all levels of operations, within their decision cycle(s), and in a manner that facilitates exploitation of the current and forecasted environmental conditions.

United States Army

The U.S. Army (USA) estimates a $27.1 million requirement and 46 civilian personnel authorizations for FY 2015 consisting of $14.1 million for operational support and $13.1 million for research and development. Funding for operational support decreased by 10 percent, and research and development decreased by 4 percent for FY 2015. Staffing levels decreased slightly.

Army monies for meteorology are spent on research and development related to the Army mission; the development, production, and maintenance of Army meteorological systems; staff meteorological functions at selected headquarters; and weather-related training at the Training and Doctrine Command (TRADOC) schools and centers.

Headquarters, Department of the Army, Deputy Chief of Staff, G-2, employs two full-time meteorologists for the coordination of meteorological support within the Department of the Army and with other DoD and Federal Agencies and organizations, and the development of Department of the Army policy concerning weather, environmental services, and oceanographic support to the Army (not to include those environmental services functions assigned to the Corps of Engineers). The United States Air Force provides one full-time staff weather officer to serve as a liaison between the AF and the Army Staff. Headquarters, U.S. Army North employs one civilian meteorologist to address meteorological issues at its headquarters. Forces Command, U.S. Army Europe, Eighth U.S. Army, U.S. Army Pacific, and U.S. Army South have Air Force Active Component, Reserve Component, civilian, or contract meteorologists who conduct meteorological staff services at these locations.

TRADOC employs both Army and Air Force personnel to manage its meteorological-related activities. As part of the TRADOC weather support structure, the U.S. Army Intelligence Center of Excellence (USAICoE) employs one full time meteorologist to manage weather capabilities development efforts. The two AF specialists also oversee USAICoE's combined Army/Air Force Battlefield Weather Course (3 weeks of instruction) held at Ft. Huachuca, AZ. TRADOC employs one meteorologist to oversee the Distributed Common Ground System-Army (DCGS-A) weather initiatives and objectives. DCGS-A is the Army's intelligence, surveillance, and reconnaissance enterprise for the analysis and processing, exploitation, and dissemination of information and intelligence data, across all echelons. DCGS-A leverages also Air Force Weather Capabilities, integrates tailored weather information and products provided by the Air Force and disseminates to Army command and control systems.

Weather equipment maintenance and training costs at the Artillery school accounted for the majority of weather-related expenditures within TRADOC. Funds were programmed for operations support related to training development, instructor/support personnel, logistics (expendable supplies), and repair costs for artillery meteorological systems at the U.S. Army Field Artillery School (USAFAS).

The Army requested 2014 procurement funds for the completion of Profiler Block III fielding that was initiated in 2013. The Army requested 2014 RDT&E funds for the Profiler Virtual Module (PVM). PVM is a product improvement to the Profiler Block III. The product improvement includes changing the weather model and changing the software to be Common Operating Environment compliant. FY 2015 procurement funds are required for procurement and fielding of PVM. FY 2015 RDT&E funds are required for PVM Common Operating Environment interoperability development with Advanced Field Artillery Tactical Data Systems (AFATDS) Version 2.

In its civil operational activities, the Corps of Engineers (COE) uses a network of about 10,850 land-based gages. About 55 percent of these sites collect meteorological data, 35 percent collect a combination of hydrologic and meteorological data, and 10 percent collect hydrologic or water quality data. The meteorological gages commonly measure precipitation and temperature. All data are used in the regulation of COE dams and other water projects, for flood control, navigation, hydroelectric power, irrigation, water supply, water quality, and recreation. The COE funds or partially funds nearly half of all the gages it uses.

The COE funds NOAA/National Weather Service (NWS) to collect and maintain precipitation information from 876 meteorological sites. The COE funds the NWS for hydro-meteorological studies and funds the U.S. Geological Survey (USGS) for maintaining hydro-meteorological data collection services for 2479 sites. The rest of the sites are maintained by the COE. Services performed by USGS vary by site and by year, and can include site visits, maintenance of equipment, replacement of damaged equipment, field measurements for verification of data and continuous monitoring of data results. About 90 percent of all COE sites provide real-time data via satellite, microwaves, meterbursts, landlines, or radio. Data from COE gage sites are available to NWS, and to other federal, state and local agencies.

U.S. Army Europe and U.S. Army Pacific maintain a small budget to fund Army-owned automated weather-sensing systems within their respective areas of responsibility.

Budgets for meteorological research decrease 10 percent from FY 2014 to FY 2015 as a result of fiscal constraints affecting all Department of Defense organizations. Within the Army Materiel Command/Research Development & Engineering Command will continue research and development efforts in basic and applied atmospheric science. The Army Test and Evaluation Command's meteorology program budget in FY 2015 will be nearly unchanged from the original FY 2014 budget. The High Performance Computing Modernization Office awarded ATEC a 2014 Designated Support Partition, a large allocation of computer resources at the Navy's DoD Supercomputing Research Center at Stennis Space Center, for extension of the ensemble version of the Four-Dimensional Weather (4DWX) weather modeling system currently in operations at Dugway Proving Ground, to all the ATEC test ranges. Development of this capability will continue in 2015.

U.S. Navy

The U.S. Navy FY 2015 budget request for meteorological programs is $ 113.3 million, made up of $ 96.3 million for operations and $ 17.0 million to support enabling research. These numbers reflect a slight decrease from FY 2014. Requested manpower authorizations for FY 2015 include 415 Navy active duty, 370 Marine Corps active duty, and 75 civilian personnel authorizations.

Naval Oceanography Program (NOP)

The Chief of Naval Operations, through the Oceanographer of the Navy, sponsors operational Navy Meteorology and Oceanography (METOC) services and related research and development. In 2012, the Oceanographer of the Navy acquired responsibility for funding the Navy's meteorology and oceanography Operations & Maintenance (O&M, N) funding from the Chief of Naval Operations (N43). Since August 2012, the Oceanographer of the Navy has served as the Director of Oceanography, Space, and Maritime Domain Awareness for the Navy, within the Information

Dominance Corps. In March 2014, the Chief Of Naval Operations signed the formal approval for an Information Dominance Type Command. Discussions are ongoing as to how this new development will shape Navy Weather activities.

The Navy provides meteorological services for Navy and joint forces, meteorological products to the Marine Corps, and oceanographic support to all elements of the Department of Defense, as well as to allied and coalition partners. The Navy sponsors programs in five closely related disciplines: meteorology, oceanography, space, maritime domain awareness, and positioning, navigation, and timing (PNT). All are used to protect ships, aircraft, fighting personnel, other platforms (manned and unmanned), and shore establishments from adverse ocean and weather conditions, and to provide a decisive tactical or strategic edge by exploiting the physical environment to optimize the performance and efficiency of platforms, sensors, and weapons. Naval METOC personnel (Navy and Marine Corps) are required to provide intelligence preparation of the operational environment (IPOE) for decision makers by assessing the impact of atmospheric and ocean phenomena on platforms, sensors, and weapon systems. Navy and Marine Corps METOC personnel provide for safe space, aviation, surface, and submerged movement, maneuver, and navigation in support of naval, joint, and combined forces, operating around the globe. This is done with a cadre of highly trained military and civilian personnel, educated in both sciences and warfighting services. By teaming with and leveraging the efforts of other agencies and activities, the NOP meets these challenges in a cost-effective manner, providing a full spectrum of products and services to provide decision makers in the field with environmental decision superiority while using only a small percentage of the Federal weather budget. Two high-profile elements of the NOP, Earth System Prediction Capability (ESPC) and Littoral Battlespace Sensing, are discussed in Section 2, in the subsection of Military Services on Navy Products and Services.

Operational Support

Naval METOC provides a wide array of essential tactical, operational strategic METOC products and services to operating forces afloat and ashore. These services include collecting and processing environmental data using resources such as oceanographic ships, aircraft, satellites, and computing systems. These products and services enhance the performance of active and passive sensor and weapon systems; optimize the effectiveness of the sea control mission for mine counter-measures; and identify the environmental effects that influence the performance of fixed and mobile warfare systems and tactics. General and tailored oceanographic, acoustic, and meteorological forecasts are provided daily to fleet commanders and individual operating units from the Meteorology and Oceanography Command's numerical modeling and forecasting centers and from forecasting support activities located worldwide. Funding primarily supports national security interests and also benefits maritime commerce. Operational support for the Navy and Marine Corps includes the day-to-day provision of METOC products and services. As naval operations in the littoral increase, Naval METOC support is directed towards providing on-scene capabilities to personnel that directly furnish environmental data for sensor, weapon system, and personnel planning and employment. These on-scene capabilities are key elements for enabling the war-fighters to take advantage of the natural environment as part of battlespace management. Owing to the crucial interrelationship of the ocean and the atmosphere, Naval METOC requires various oceanographic products to provide the requisite meteorological services. In addition to aviation and maritime METOC support, Navy and Marine Corps METOC teams provide a variety of unique services on demand, such as electro-optical, electro-magnetic, and acoustic propagation models and products, METOC-sensitive tactical decision aids, and global sea ice analyses and forecasts.

Systems Acquisition. Naval METOC systems acquisition is accomplished through the Program Executive Office for Command, Control, Communication, Computers and Intelligence and Space (C4I and Space) in San Diego, California. This funds new and replacement meteorological equipment for all Navy and Marine Corps Air Stations, all Navy ships, USMC Operational Forces units and other activities required to provide weather observations and provides safety of flight capabilities. The procurement has been thoroughly coordinated with other DOD and civilian agencies. Program also funds replacement of Survey Vessel shipboard mission equipment, deep and multibeam SONARs, Side Scan SONARs, Hydrographic Survey Launches, Ship Moving Vessel Profilers, Unmanned Under Water Gliders, and Autonomous Underwater Vehicles. The Oceanographer of the Navy also funded through its SCN, two new research vessels (one in FY11; one in FY12), to be operated by civilian research institutions in coordination with the University National Oceanographic Laboratory System (UNOLS). An additional Auxillary General Survey vessel (T-AGS) will be delivered in FY 15.

Navy's Earth System Prediction Capability (ESPC). The Navy's Earth System Prediction Capability (ESPC) is a program to develop next generation environmental prediction employing emerging technologies in physical coupled prediction across the air-sea interface as well as to land, near space, sea ice, and other physical interactions. Additionally, ensemble-based probabilistic prediction for informed decision making is being addressed and collaborations with other operational prediction agencies through common technology standards is being implemented to develop an improved coupled weather and ocean prediction across multiple time scales from tactical to strategic.

The Navy's ESPC will provide a more accurate, longer range, global ocean, atmosphere, and sea ice forecast system for decision support to safety of flight, safety of navigation, sensor and weapon performance, and mission planning, mitigation and effectiveness decisions. Development of global coupled ensemble technologies will provide increased accuracy for lead times of 1-10 days as well as a new capability for accurate forecasts from the Tropics to the Arctic at tactical, operational, and strategic lead times. It will develop a Navy interface to NOAA's products for seasonal to multi-annual lead times for deliberate planning through integrating atmosphere, ocean, ice, land and near-space forecast models into a seamless prediction system. This effort is the Navy contribution to a National ESPC for improved cross-Agency Research to Operations collaboration, and the development of more accurate forecast systems and more efficient computational architectures to allow for improved real-time operational prediction.

The Navy's ESPC directly enables achievement of warfighter decision superiority through improved knowledge of the physical environment. Such knowledge is key to the realization of Navy operational concepts critical to future warfighting capabilities such as the *Undersea Dominance Operating Concept* (UDOC) and *Electromagnetic Maneuver Warfare* (EMW). In both cases, knowledge of the current and future physical environment and its impact on the acoustic and electromagnetic spectrums directly supports the warfighter's ability to optimize courses of action and mitigate environmental operational impacts.

Through-the-Sensor (TTS) Capabilities. The Hazardous Weather Detection and Display Capability (HWDDC) and Tactical Environmental Processor (TEP) are TTS technologies which will passively tap Navy air-search radars to obtain and display hazardous weather information. The HWDDC and TEP systems will be based off common modular weather processing algorithms and will have similar data product and display capabilities. Essentially, they represent one common set of processing algorithms applied to two different radars. The differences in capabilities supported by the systems are driven by the differences in the individual radars.

The HWDDC addresses a fleet requirement for real-time hazardous weather detection/display to support safety of flight and operations planning within Carrier and Expeditionary Strike Groups. The knowledge of hazardous weather conditions afloat greatly enhances readiness and combat posture.

Research and Development (R&D)

Naval METOC R&D is cooperatively sponsored by the Oceanographer/Navigator of the Navy and the Chief of Naval Research. This program enables the warfighter of the future to effectively carry out their mission by transitioning to operational use research performed by the Office of Naval Research. The Space and Naval Warfare Systems Command is the primary office responsible for transitioning Naval research to operational use. All research and development funded by the Oceanographer of the Navy is in direct support of the Naval mission or selected missions.

Naval R&D efforts typically have applications to meteorological, oceanographic, and/or tactical systems. The Navy's tabulation of budget data includes R&D funding for basic research, applied research, demonstration and validation, and engineering and manufacturing development. Projects initiated by the Navy and Marine Corps, under sponsorship of the Oceanographer/Navigator of the Navy, transition from engineering development to operational naval systems. Such efforts include advances in Naval METOC forecasting capabilities, enhancements to communications and data compression techniques, further development and improvement of models to better predict METOC parameters in littoral regions, and an improved understanding of the impact these parameters have on sensors, weapons systems, and platform performance.

The Naval METOC community works closely with research developers and operational forces to ensure that naval and joint force commanders will always have the most accurate, timely, and geo-referenced METOC information available for successful operations.

DEPARTMENT OF HOMELAND SECURITY

U.S. Coast Guard

All of the U.S. Coast Guard's (USCG) funding for meteorological programs is for operations support. For FY 2015, the requested funding level is $29.9 million and 108 personnel authorizations (107 military and 1 Federal). The Coast Guard does not have a specific program and budget for meteorology—all meteorological activities are accomplished as part of general operations. The USCG does not track meteorological costs at an organizational level, so the funding level is an estimate. The Coast Guard's activities include the collection and dissemination of meteorological and iceberg warning information for the benefit of the marine community. The Coast Guard also collects coastal and marine observations from its shore stations and cutters and transmits these observations daily to the Navy's Fleet Numerical Meteorology and Oceanography Center and NOAA's National Weather Service. These observations are used by both the Navy and NOAA in generating weather forecasts.

The Coast Guard also disseminates a variety of weather forecast products and warnings to the marine community via radio transmissions. Coast Guard shore stations often serve as sites for NWS automated coastal weather stations, and the National Data Buoy Center provides logistics support in deploying and maintaining NOAA offshore weather buoys from Coast Guard cutters. The International Ice Patrol conducts iceberg surveillance operations and provides warnings to mariners on the presence of icebergs in the North Atlantic shipping lanes. Coast Guard efforts in meteorological operations and services have not changed significantly during recent years

DEPARTMENT OF THE INTERIOR

Bureau of Land Management

The Bureau of Land Management (BLM) requested funding for FY 2015 is $4.66 million and 38 FTE which is a 2.3% increase over the total allocated from FY 2014. The BLM funds two principal programs—the soil, water, and air (SWA) program and the fire weather activities of the Office of Fire and Aviation (OFA).

Soil Water and Air Program (SWA). The FY 2015 budget request for SWA meteorological operations is estimated at $1.55 million and 10 FTE, which is a 6.9% increase over the amount allocated in FY 2014. The FY 2015 estimate includes $600,000 for labor and logistics to support climate and weather data collection efforts by BLM resource management staff, and approximately $965,000 for project implementation. Projects include 1) collection of meteorological data and operation of RAWS stations equipped with additional instruments to measure soil moisture and winter precipitation not required for fire monitoring, 2) maintenance of eight stations in the NRCS SCAN network, 3) operation of six NADP sites, and 4) operation of a newly installed meteorological monitoring site at Inigok Station within the National Petroleum Reserve – Alaska.. The SWA program initiates efforts to collect additional data through cooperative efforts with other agencies or with resource management staff in state and field offices when existing monitoring networks are not sufficient to meet the needs for air-resource-related information.

Fire and Aviation Directorate (FA). The FY 2015 FA budget request for meteorological operations is estimated at $3.1 million and 28 FTE, no change from FY2014. This represents BLM support for meteorologists at the National Interagency Coordination Center (NICC) and Geographic Area Coordination Centers (GACCs) and BLM support of the Interagency Remote Automatic Weather Station (RAWS) network. An additional $920,000 is generally recovered through reimbursable accounts with non-Department of Interior (DOI) agencies for RAWS support. Funded activities related to the RAWS network include maintenance, travel, transportation, services, supplies, and equipment. Some agencies incur additional costs in support of the RAWS network through commercially contracted maintenance services.

The interagency RAWS network is an important tool for wildland fire management which directly supports the protection of life and property. All affected Federal agencies within DOI participate in its acquisition, operation, and support. The BLM, in particular, has a lead role in the maintenance of the RAWS network, providing both data distribution services and equipment support. Participating agencies address common issues and coordinate efforts to ensure the collection of accurate and useful fire weather data.

Under the Predictive Services Program, meteorologists who specialize in fire weather services team with intelligence specialists and wildland fire analysts at the GACCs and the NICC to form Predictive

Services units. The Predictive Services units act as centers of expertise to produce integrated planning and decision-support tools that enable more proactive, safe, and cost-effective fire management.

National Park Service

The National Park Service (NPS) FY 2015 budget request is approximately $2.8 million, essentially flat funding from FY 2014. The NPS expends about $0.9 million on atmospheric research with a focus on measurements of all forms of atmospheric reactive nitrogen and on aerosol science. The goal of this research is to identify the sources of air pollution that are affecting park ecosystems and visibility and to quantify their impacts. The NPS also expends approximately $1.9 million in routine air quality, visibility, and meteorological monitoring networks.

U.S. Geological Survey

The U.S. Geological Survey requests and expends approximately $66,000 per year conducting post-wildfire debris flow warning operations.

DEPARTMENT OF TRANSPORTATION

Federal Aviation Administration (FAA)

For FY 2015, the FAA is requesting a total of $398 million, 149 FTE and 964 Contractor Authorizations for Aviation Weather related, Operations Support, Major Systems Acquisitions, Recurring Research and Development Costs, and Systems Development; an approximate 6% increase from FY 2014 actual funding received. The actual funding for the Aviation Weather Programs in FY 2014 was $375 million. The changes are comprised of:

- No significant change in reported numbers for Operations Support from an actually received $296.3 million in FY 2014 to a requested $296.4 million in FY 2015.

- An increase in reported numbers for Major Systems Acquisitions from an actually received $55.4 million in FY 2014 to a requested $74.2 million in FY 2015.

- An increase in reported numbers for recurring Research and Development Costs from an actually received $18.2 million in FY 2014 to a requested $21.8 million in FY 2015.

- No significant change in reported numbers for Systems Development from an actually received $4.8 million in FY 2014 to a requested $5.2 million in FY 2015.

The funding changes reflect major initiatives in the Aviation Weather programs to support the Next Generation (NextGen) National Air Transportation System. These changes will bring enhancements, including the dissemination of weather products and decision-making information.

United States (U.S.) Code Title 49 Section 44720 (49 U.S.C. 44720) designates the FAA as the Meteorological Authority for domestic and international aviation weather services of the U.S. In this capacity, the FAA provides requirements for the administration of aviation weather services to the National Weather Service (NWS). The FAA is responsible for ensuring compliance with these services and with maintaining International Civil Aviation Organization (ICAO) Standards and Recommended Practices as specified in Annex 3-Meteorological Service for International Air Navigation.

Federal Highway Administration

Due to the extended time for road construction projects, the Department of Transportation does not go through an annual budget process but typically uses a six-year authorization. The current transportation authorization allocates $100 million for the Intelligent Transportation Systems (ITS) Research Program. Of this, for FY 2015, the Road Weather Management Program (RWMP) has budgeted $2.0 million. All of RWMP's funding is for applied research. The majority of RWMP activity involves software development and studies under the research and development category to develop decision-support systems that integrate high-resolution road weather data and products with transportation-oriented applications and management strategies.

ENVIRONMENTAL PROTECTION AGENCY

The anticipated funding level in FY 2015 for directed meteorological research is about $6.4 million and 43 FTE, a 1.1% decrease from FY 2014. All of the Environmental Protection Agency's (EPA) funding of meteorological and air quality programs is for supporting basic and applied research.

Continued attention is being paid to the effects of airborne toxins and fine particulate matter on human health, on the effect of climate change on air quality, and the impact of air pollution on human health and sensitive ecosystems. In addition, to promote excellence in environmental science and engineering, the EPA supports a national research grants program for investigator-initiated research. The funding for grants (with reliance on quality science and peer review) and for graduate fellowships (to support the education and careers of future scientists) will provide for a more balanced, long-term capital investment in improved environmental research and development. The funding for the grants program in FY 2015 is anticipated to be comparable to that in FY 2014.

The EPA's Research Grants Program will fund research in areas, including ecological assessment, air quality, environmental fate and treatment of toxins and hazardous wastes, effects of global climate change on air quality, and exploratory research. The portion of these grants that will be awarded for meteorological research during FY 2015 cannot be foreseen, but it is probable that the grant awards will increase the base amount of $6.4 million listed above for directed meteorological research.

The EPA continues its development and evaluation of air quality models for air pollutants on all temporal and spatial scales as mandated by the Clean Air Act as amended in 1990. Research will focus on urban, mesoscale, regional, and multimedia models, which will be used to develop air pollution control policies, human and ecosystem exposure assessments, and air quality forecasts. There will be increased emphasis placed on meteorological research into global-to-regional-to-urban-local formation and intercontinental transport of air contaminants in support of the revisions to the National Ambient Air Quality Standards and ecosystem protection strategies. Increased efficiency of computation and interpretation of model results are being made possible by means of supercomputing and scientific visualization techniques.

NATIONAL AERONAUTICS AND SPACE ADMINISTRATION

The National Aeronautics and Space Administration's (NASA) estimated meteorological operations and research budget request for FY 2015 is $573.6 million and 4 FTE. This request is an increase of 16.4 percent from the FY 2014 budget of $492.7 million. The budget figures reported are based on relevant missions and programs in the Earth Science Division[1] and the Heliophysics Division[2] within the Science Mission Directorate (SMD), and the Human Exploration and Operations Mission Directorate (HEOMD). Across the directorates, NASA estimates the extent to which each mission and program contributes and relates to meteorological operations and research activities.

NASA Earth Science advances understanding of the Earth system, its components and their interactions, its changes, and the consequences of these changes for life. The program pioneers the use of remote sensing data, primarily space-based, in new and innovative ways, and leverages NASA's unique capabilities in global Earth observation. Earth Science Research sponsors basic disciplinary and interdisciplinary research, Earth system modeling efforts, the Airborne Science project (which provides access to aircraft and unmanned aircraft systems), and supercomputing efforts supporting a variety of programs. At least 90 percent of the funds of the Earth Science Research program are competitively awarded to investigators from academia, the private sector, and NASA Centers. The program uses satellite and airborne measurements, coupled with cutting-edge analyses and numerical models, to turn observations into information and understanding.

NASA takes an organized approach to address complex, interdisciplinary Earth science problems, integrating science across the programmatic elements in pursuit of a comprehensive understanding of the Earth system. The resulting programmatic structure comprises six interdisciplinary and interrelated science focus areas. These areas are:

- Climate Variability and Change: Improve the ability to predict climate changes by better understanding the roles and interactions of oceans, atmosphere, land, and ice in the climate system;
- Atmospheric Composition: Advance the understanding of changes in the Earth's radiation balance, air quality and the ozone layer that result from changes in atmospheric composition;
- Carbon Cycle and Ecosystems: Detect and predict changes in Earth's ecological and chemical cycles, including land cover, biodiversity, and the global carbon;
- Water and Energy Cycle: Enable better assessment and management of water quality and quantity to accurately predict how the global water cycle evolves in response to climate change;
- Weather: Improve the capability to predict weather and extreme weather events; and
- Earth Surface and Interior: Characterize the dynamics of Earth's surface and interior, improving the capability to assess and respond to natural hazards and extreme events.

---

[1] The Earth Science Division (ESD) reported budget includes an estimate of weather observations and research and closely related program activities. Research and satellite mission budgets are estimated proportionally to their overall contributions to activities reported, noting that the objective of ESD's program is to advance Earth System science.

[2] This report includes Heliophysics Division research assets and programs that contribute significantly to the advancement of space weather knowledge and to the transfer of that knowledge into space weather prediction systems.

NASA also supports space weather research through the Heliophysics Division within NASA's SMD. The objective of the Heliophysics Division is to explore and understand the dynamic processes connecting the Sun, the Earth and planetary space environments, and the outer reaches of our solar system and develop the knowledge to predict extreme conditions in space to protect life and society. Research areas in space weather for the coming year include:

- Prediction of the polarity and magnitude of the north-south component of the interplanetary magnetic field as it approaches Earth;
- Physics-based methods to predict connectivity of solar energetic particle (SEP) sources to points in the inner heliosphere, tested by location, timing, and longitudinal separation of SEPs;
- Ion-neutral interactions in the topside ionosphere.

In addition to research, NASA also engages in space weather operations within the HEOMD. The HEOMD objective is weather-related safety of manned spacecraft, satellites, scientific instruments, and launch vehicles. The greatest challenge is to accurately measure and forecast tropospheric and space weather events that strongly impact launch and landing operations.

NUCLEAR REGULATORY COMMISSION

For FY 2015, the Nuclear Regulatory Commission's (NRC) total planned contract expenditures of $1.73 million ($337,000 for operations and $1.39 million for supporting research) and 7 FTE is for meteorological operations to continue:
- technical assistance for the analysis of atmospheric dispersion for routine and postulated accidental releases from nuclear facilities
- conducting meteorological research in support of licensing activities, including consideration of site specific probable maximum precipitation analysis at certain facilities
- preparation of guidance on meteorological issues in licensing actions
- review of proposed sites for possible construction of new nuclear power plants

The meteorological support program in the NRC includes analyzing and utilizing meteorological data in atmospheric transport and dispersion models. These models provide insight on plume pathways in the near- and far-fields for building wake and dispersion characteristics to perform dose calculations on postulated releases into the environment. Meteorological information is used as input to the probabilistic safety assessment, the assessment of the radiological impacts of routine releases from normal operations, the assessment of other (non-radiological) hazards (including rare external events such as extreme storms) that may impact safe operation of the facility, and the assessment of design or operational changes proposed for the facility.

Current and projected research activities include updating the methods used to estimate the effects of extreme precipitation events, and developing an integrated approach for probabilistic flood hazard assessment (PFHA). During FY 2012, research was initiated to address the influence of orographic features on extreme precipitation events. This work was continued in FY 2013 and FY 2014 and is prioritized for those areas of the United States where new nuclear power plants are proposed or where existing nuclear power plants are utilizing these concepts. This work will provide the design basis for flood protection systems. Site specific probable maximum precipitation analysis research was also initiated during FY 2014. This work will compare how these analyses compare to NOAA

Hydrometeorological Reports (HMRs) and what acceptance criteria the NRC should apply to determine the licensing acceptability of these methods.

The nuclear power industry continues to pursue approval for new nuclear power plants. Numerous early site permit, combined license, and design certification applications have been received and are currently under review. These reviews will also consider regional climatology and local meteorology. In addition to its internal review activities, the NRC may seek assistance from other Federal agencies to support its safety and environmental reviews.

The March 2011 accident at the Fukushima Dia-ichi nuclear power plant, which was caused by an earthquake and subsequent tsunami, resulted in the NRC issuing letters to all its operating nuclear power plant licensees requesting that they reevaluate flooding hazards at their sites using updated information and present-day regulatory guidance and methodologies. The NRC licensees were requested to provide their responses on a staggered three-year basis, beginning in March 2013. The NRC staff, along with support from its consultants, has begun reviewing the flood hazard reevaluations that have been submitted to date to determine whether additional regulatory actions are necessary to protect the site against the reevaluated flooding hazards.

BUDGET TABLES

Table 1 Meteorological Operations and Supporting Research Costs* by Agency

This table summarizes, by agency, *both* meteorological operations and supporting research costs. Funding levels for FY 2014 are congressionally appropriated funds. FY 2015 numbers indicate the funding requested in the President's FY 2015 budget. The change percentages between FY 2014 and FY 2015 are also shown.

TABLE 1   Meteorological Operations and Supporting Research Costs* by Agency
(Thousands of Dollars)

AGENCY	Operations FY14	Operations FY15	%CHG	% of FY14 TOTAL	Supporting Research FY14	Supporting Research FY15	%CHG	% of FY14 TOTAL	Total FY14	Total FY15	%CHG	% of FY14 TOTAL	% of FY15 TOTAL
Agriculture	20340	20034	-1.5	0.5	51163	49917	-2.4	5.5	71503	69951	-2 2	1.5	1.4
Commerce/NOAA(Subtot)	3150094	3332023	5.8	83.7	200577	201341	0.4	22.0	3350671	3533364	5 5	72 5	72.1
NWS	1026400	1048354	2.1	26.3	36187	14993	-58.6	1.6	1062587	1063347	0.1	23.0	21.7
NESDIS	2062635	2221926	7.7	55.8	24459	26000	6.3	2.8	2087094	2247926	7.7	45.2	45.9
OAR	0	0	0	0	139931	160348	14.6	17.5	139931	160348	14 6	3.0	3.3
NOS	29907	30143	0.8	0.8	0	0	0	0	29907	30143	0 8	0.6	0.6
OMAO	31152	31600	1.4	0.8	0	0	0	0	31152	31600	1.4	0.7	0.6
Defense(Subtot)	227031	220783	-2.8	5.5	52483	55184	5.1	6.0	279514	275967	-1 3	6.1	5.6
Air Force	112161	110412	-1.6	2.8	24667	25102	1.8	2.7	136828	135514	-1 0	3.0	2.8
Navy	99213	96301	-2.9	2.4	14195	17006	19.8	1.9	113408	113307	-0.1	2.5	2.3
Army	15657	14070	-10.1	0.4	13621	13076	-4.0	1.4	29278	27146	-7 3	0 6	0.6
Homeland Security (Subtot)	29720	29887	0.6	0.8	0	0	0	0	29720	29887	0.6	0 6	0.6
USCG	29720	29887	0.6	0.8	0	0	0	0	29720	29887	0 6	0.6	0.6
Interior/BLM (Subtot)	6602	6708	1.6	0.2	915	900	-1.6	0.1	7517	7608	1.2	0 2	0.2
BLM	4563	4669	2.3	0.1	0	0	0	0	4563	4669	2 3	0.1	0.1
NPS	1973	1973	0.0	0.0	915	900	-1.6	0.1	2888	2873	-0 5	0.1	0.1
USGS	66	66	0.0	0.0	0	0	0	0	66	66	0 0	0 0	0.0
Transportation(Subtot)	351680	370570	5.4	9.3	27030	29013	7.3	3.2	378710	399583	5 5	8.2	8.2
FAA	351680	370570	5.4	9.3	23030	27013	17.3	3.0	374710	397583	6.1	8.1	8.1
FHWA	0	0	0	0	4000	2000	-50.0	0.2	4000	2000	-50 0	0.1	0.0
EPA	0	0	0	0	6500	6430	-1.1	0.7	6500	6430	-1.1	0.1	0.1
NASA	2327	2189	-5.9	0.1	490390	571440	16.5	62.4	492717	573629	16.4	10.7	11.7
NRC	284	337	18.7	0.0	1410	1390	-1.4	0.2	1694	1727	1 9	0 0	0.0
DOE	0	0	0	0	0	0	0	0	0	0	0	0	0
TOTAL	3788078	3982531	5.1	100.0	830468	915615	10.3	100.0	4618546	4898146	6.1	100 0	100.0
% of FY TOTAL	82 0%	81.3%			18 0%	18.7%			100.0%	100.0%			

*The FY 2014 funding reflects Congressionally appropriated funds; the FY 2015 funding reflects the amount requested in the President's FY 2015 budget submission to Congress.

Table 2 **Operational Costs by Budget Category**

This table describes the agency plans to obligate their funds for *meteorological operations* by budget category. The two major categories are Operations Support and Major Systems Acquisition. To a large degree, these categories correspond to non-hardware costs (Operations Support) and hardware costs (Systems Acquisition).

TABLE 2　Operational Costs by Budget Category
(Thousands of Dollars)

AGENCY	Operations Support		Major Systems Acquisition		Total			% of FY15
	FY14	FY15	FY14	FY15	FY14	FY15	%CHG	TOTAL
Agriculture	20340	20034	0	0	20340	20034	-1.5	0.5
Commerce/NOAA(Subtot)	1142421	1169087	2007673	2162936	3150094	3332023	5.8	83.7
NWS	920946	942735	105454	105619	1026400	1048354	2.1	26.3
NESDIS	160416	164609	1902219	2057317	2062635	2221926	7.7	55.8
OAR	0	0	0	0	0	0	0	0
NOS	29907	30143	0	0	29907	30143	0.8	0.8
OMAO	31152	31600	0	0	31152	31600	1.4	0.8
Defense(Subtot)	187040	183154	39991	37629	227031	220783	-2.8	5.5
Air Force	82122	80338	30039	30074	112161	110412	-1.6	2.8
Navy	91061	88746	8152	7555	99213	96301	-2.9	2.4
Army	13857	14070	1800	0	15657	14070	-10.1	0.4
Homeland Security (Subtot)	29720	29887	0	0	29720	29887	0.6	0.8
USCG	29720	29887	0	0	29720	29887	0.6	0.8
Interior/BLM	6602	6708	0	0	6602	6708	1.6	0.2
BLM (Subtot)	4563	4669	0	0	4563	4669	2.3	0.1
SWA	1450	1550	0	0	1450	1550	6.9	0.0
OFA	3113	3119	0	0	3113	3119	0.2	0.1
NPS	1973	1973	0	0	1973	1973	0.0	0.0
USGS	66	66	0	0	66	66	0.0	0.0
Transportation(Subtot)	296255	296350	55425	74220	351680	370570	5.4	9.3
FAA	296255	296350	55425	74220	351680	370570	5.4	9.3
FHWA	0	0	0	0	0	0	0	0
EPA	0	0	0	0	0	0	0	0
NASA	2229	2189	98	0	2327	2189	-5.9	0.1
NRC	284	337	0	0	284	337	18.7	0.0
DOE	0	0	0	0	0	0	0	0
TOTAL	1684891	1707746	2103187	2274785	3788078	3982531	5.1	100.0
% of FY TOTAL	44.5%	42.9%	55.5%	57.1%	100.0%	100.0%		

Table 3 **Supporting Research and Development Costs by Budget Category**

This table describes the agency plans to obligate their funds for *meteorological supporting research and Development* by budget category. Similar to operational funding in table 2, these categories are Research and Development (non-hardware) and Systems Development (hardware).

TABLE 3   **Supporting Research and Development Costs by Budget Category**
(Thousands of Dollars)

AGENCY	Recurring R&D Costs FY14	FY15	Systems Development FY14	FY15	Total FY14	FY15	%CHG	% of FY15 TOTAL
Agriculture	51163	49917	0	0	51163	49917	-2.4	5.5
Commerce/NOAA(Subtot)	189143	198821	11434	2520	200577	201341	0.4	22.0
NWS	26623	14343	9564	650	36187	14993	-58.6	1.6
NESDIS	24459	26000	0	0	24459	26000	6.3	2.8
OAR	138061	158478	1870	1870	139931	160348	14.6	17.5
NOS	0	0	0	0	0	0	0	0
OMAO	0	0	0	0	0	0	0	0
Defense(Subtot)	36114	34108	16369	21076	52483	55184	5.1	6.0
Air Force	10963	7074	13704	18028	24667	25102	1.8	2.7
Navy	14195	17006	0	0	14195	17006	19.8	1.9
Army	10956	10028	2665	3048	13621	13076	-4.0	1.4
Homeland Security (Subtot)	0	0	0	0	0	0	0	0
USCG	0	0	0	0	0	0	0	0
Interior/BLM	915	900	0	0	915	900	-1.6	0.1
BLM (Subtot)	0	0	0	0	0	0	0	0
SWA	0	0	0	0	0	0	0	0
OFA	0	0	0	0	0	0	0	0
NPS	915	900	0	0	915	900	-1.6	0.1
USGS	0	0	0	0	0	0	0	0
Transportation(Subtot)	22200	23848	4830	5165	27030	29013	7.3	3.2
FAA	18200	21848	4830	5165	23030	27013	17.3	3.0
FHWA	4000	2000	0	0	4000	2000	-50.0	0.2
EPA	6500	6430	0	0	6500	6430	-1.1	0.7
NASA	175315	187487	315075	383953	490390	571440	16.5	62.4
NRC	1410	1390	0	0	1410	1390	-1.4	0.2
DOE	0	0	0	0	0	0	0	0
TOTAL	482760	502901	347708	412714	830468	915615	10.3	100.0
% of FY TOTAL	58.1%	54.9%	41.9%	45.1%	100.0%	100.0%		

Tables 4 and 5, Operational Costs and Supporting Research Costs by Service Category

These tables indicate how the funds identified in Tables 2 and 3 are divided among the eleven service categories. Table 4 reflects how the agencies plan to obligate operational funds and Table 5 supporting research funds. The service categories evolve over time, as applications change with technology and with societal needs. The service category definitions are described below:

Service Category Definitions

- *Basic Services.* Basic services include the basic meteorological service system, to include observations, public weather forecasts, severe weather warnings and advisories, and the meteorological satellite activities of NOAA. Basic services also include the operations and supporting research of other Federal agencies that have been identified as contributing to basic meteorological services.

- *Agriculture and Land Management Meteorological Services.* Agricultural and land management meteorological services are those services and facilities established to meet the requirements of the agricultural industries and Federal, state, and local agencies charged with the protection and maintenance of the Nation's land areas. Meteorological services specifically tailored for wildland fire management are reported under the wildland fire weather service category.

- *Aviation Services.* Aviation services are those specialized meteorological services and facilities established to meet the requirements of general, commercial, and military aviation. Civil programs that are directly related to services solely for aviation and military programs in support of land-based aviation and medium- or long-range missile operations are included. Detailed aviation services/products for specific areas include, but are not limited to, ceiling and visibility, convective hazards, en route winds and temperatures, ground de-icing, in-flight icing, terminal winds and temperatures, turbulence, volcanic ash, and other airborne hazardous materials.

- ***Climate Services.*** Climate services are specialized meteorological and hydrological services established to meet the requirements of Federal, state, and local agencies for information on the historical, current, and future state of the earth system. Climate services include observations, monitoring, assessments, predictions, and projections of the atmosphere, hydrosphere, and land surface systems.

- ***Emergency Response and Homeland Security Services.*** Emergency response and homeland security services are those specialized meteorological services and facilities established to meet the requirements of Federal, state, and local agencies responding to natural disasters and security incidents. This category includes the use of atmospheric transport and diffusion (ATD) models for predicting the dispersion of airborne toxic substances; it also includes natural disaster monitoring and prediction services and the transport of water-borne toxic substances not included in basic services.

- ***Hydrometeorology and Water Resources Services.*** Hydrometeorology and water resources services are those specialized meteorological services and facilities that combine atmospheric science, hydrology, and water resources in order to meet the requirements of Federal, state, and local agencies for information on the effects of precipitation events on infrastructure, water supplies, and waterways. These products and services also meet the needs of the general public in the conduct of everyday activities and for the protection of lives and property.

- ***Military Services.*** Military services are those meteorological operations, services, and capabilities established to meet the unique requirements of military user commands and their component elements. Programs and services that are not uniquely military in nature are reported under another service category (e.g., Basic Services, Aviation Services [civilian], Surface transportation Services, or Emergency Response and Homeland Security Services).

- ***Space Weather Services.*** Space Weather Services are those specialized meteorological services and facilities established to meet the needs of users for information on space weather conditions and space weather storms that can affect terrestrial systems, space systems, Earth's atmosphere, and the space environment. Space weather services include monitoring and alerting of space weather storms and their effects on technological infrastructure and human safety. Early warning of an approaching space weather storm, so that timely protective response is possible, is an important part of space weather services.

- ***Surface Transportation.*** Surface transportation services are those specialized meteorological services and facilities established to meet the weather information needs of the following surface transportation sectors: roadways, long-haul railways, the marine transportation system, rural and urban transit, pipeline systems, and airport ground operations. The roadway sector includes state and Federal highways and all state and local roads and streets. The marine transportation system includes coastal and inland waterways, ports and harbors, and the intermodal terminals serving them. Rural and urban transit includes bus and van service on roadways and rail lines for metropolitan subway and surface "light-rail" systems.

- ***Wildland Fire Weather Services.*** Wildland fire weather services are those specialized meteorological services and facilities established to meet the requirements of the wildfire management community at the Federal, state, tribal, and local levels. The primary areas of service are to support the reduction of wildfire initiation potential and the mitigation of both human and environmental impacts once initiation does occur. Services can include support to first responders and land managers and climate services tailored to wildland fire management.

- ***Other Specialized Services.*** Other specialized services include weather and climate information services and facilities established to meet the special needs of user agencies or constituencies not included in basic services or the preceding service categories. This service category includes any efforts to integrate the social sciences into meteorological operations, applications, and services not already described in the preceding sections.

TABLE 4 Operational Costs by Service Category
(Thousands of Dollars)

AGENCY	Basic Services FY14	FY15	Agriculture & Land Management FY14	FY15	Aviation FY14	FY15	Climate FY14	FY15	Emergency Response & Homeland Security FY14	FY15	Hydrometeorology & Water Resources FY14	FY15	Military FY14	FY15	Space Weather FY14	FY15	Surface Transportation FY14	FY15	Wildland Fire Weather FY14	FY15	Other Specialized FY14	FY15	Total FY14	FY15
Agriculture	0	0	2634	2611	0	0	0	0	0	0	9300	8937	0	0	0	0	0	0	8406	8486	0	0	20340	20034
Commerce/NOAA(Subtot)	2957649	3062334	0	0	74553	65293	13198	31442	0	0	53750	82232	0	0	10835	8840	29907	30143	1552	38029	8650	13710	3150094	3332023
NWS	864462	809458	0	0	74553	65293	13198	31442	0	0	53150	81582	0	0	10835	8840	0	0	1552	38029	8650	13710	1026400	1048354
NESDIS	2062635	2221926	0	0	0	0	0	0	0	0	0	0	0	0	0	0	0	0	0	0	0	0	2062635	2221926
OAR	0	0	0	0	0	0	0	0	0	0	0	0	0	0	0	0	0	0	0	0	0	0	0	0
NOS	0	0	0	0	0	0	0	0	0	0	0	0	0	0	0	0	29907	30143	0	0	0	0	29907	30143
OMAO	30552	30960	0	0	0	0	0	0	0	0	600	650	0	0	0	0	0	0	0	0	0	0	31152	31600
Defense(Subtot)	1527	1650	0	0	437	590	102	110	0	0	0	0	213476	201881	11489	16552	0	0	0	0	0	0	227031	220783
Air Force	0	0	0	0	0	0	0	0	0	0	0	0	100672	93860	11489	16552	0	0	0	0	0	0	112161	110412
Navy	0	0	0	0	0	0	0	0	0	0	0	0	99213	96301	0	0	0	0	0	0	0	0	99213	96301
Army	1527	1650	0	0	437	590	102	110	0	0	0	0	13591	11720	0	0	0	0	0	0	0	0	15657	14070
Homeland Security (Subtot)	0	0	0	0	0	0	0	0	29720	29887	0	0	0	0	0	0	0	0	0	0	0	0	29720	29887
USCG	0	0	0	0	0	0	0	0	29720	29887	0	0	0	0	0	0	0	0	0	0	0	0	29720	29887
Interior/BLM	0	0	1450	1550	0	0	0	0	0	0	0	0	0	0	0	0	0	0	3179	3185	1973	1973	6602	6708
BLM (Subtot)	0	0	1450	1550	0	0	0	0	0	0	0	0	0	0	0	0	0	0	3113	3119	0	0	4563	4669
SWA	0	0	1450	1550	0	0	0	0	0	0	0	0	0	0	0	0	0	0	0	0	0	0	1450	1550
OFA	0	0	0	0	0	0	0	0	0	0	0	0	0	0	0	0	0	0	3113	3119	0	0	3113	3119
NPS	0	0	0	0	0	0	0	0	0	0	0	0	0	0	0	0	0	0	0	0	1973	1973	1973	1973
USGS	0	0	0	0	0	0	0	0	0	0	0	0	0	0	0	0	0	0	66	66	0	0	66	66
Transportation(Subtot)	0	0	0	0	351680	370570	0	0	0	0	0	0	0	0	0	0	0	0	0	0	0	0	351680	370570
FAA	0	0	0	0	351680	370570	0	0	0	0	0	0	0	0	0	0	0	0	0	0	0	0	351680	370570
FHWA	0	0	0	0	0	0	0	0	0	0	0	0	0	0	0	0	0	0	0	0	0	0	0	0
EPA	0	0	0	0	0	0	0	0	0	0	0	0	0	0	0	0	0	0	0	0	0	0	0	0
NASA	0	0	0	0	0	0	0	0	0	0	0	0	0	0	1000	1000	0	0	0	0	1326	1189	2326	2189
NRC	0	0	0	0	0	0	0	0	0	0	204	257	0	0	0	0	0	0	0	0	80	80	284	337
DOE	0	0	0	0	0	0	0	0	0	0	0	0	0	0	0	0	0	0	0	0	0	0	0	0
TOTAL	2959176	3063984	2634	2611	426670	436453	13300	31552	29720	29887	63254	91426	213476	201881	23324	26392	29907	30143	13137	49700	12029	16952	3788077	3982531
% of FY TOTAL	78.1%	76.9%	0.1%	0.1%	11.3%	11.0%	0.4%	0.8%	0.8%	0.8%	1.7%	2.3%	5.6%	5.1%	0.6%	0.7%	0.8%	0.8%	0.3%	1.2%	0.3%	0.4%	100.0%	100.0%

TABLE 5 Supporting Research and Development Costs by Service Category
(Thousands of Dollars)

AGENCY	Basic Services FY14	FY15	Agriculture & Land Management FY14	FY15	Aviation FY14	FY15	Climate FY14	FY15	Emergency Response & Homeland Security FY14	FY15	Hydrometeorology & Water Resources FY14	FY15	Military FY14	FY15	Space Weather FY14	FY15	Surface Transportation FY14	FY15	Wildland Fire Weather FY14	FY15	Other Specialized FY14	FY15	Total FY14	FY15
Agriculture	0	0	15876	14982	0	0	27368	27489	0	0	35	35	0	0	0	0	0	0	7864	7411	0	0	51163	49917
Commerce/NOAA(Subtot)	112178	107160	0	0	4298	2250	58858	75454	1025	1025	13904	14052	0	0	750	750	0	0	0	0	9564	650	200577	201341
NWS	23200	12968	0	0	2673	625	0	0	0	0	0	0	0	0	750	750	0	0	0	0	9564	650	36197	14993
NESDIS	24459	26000	0	0	0	0	0	0	0	0	0	0	0	0	0	0	0	0	0	0	0	0	24459	26000
OAR	64519	68192	0	0	1625	1625	58858	75454	1025	1025	13904	14052	0	0	0	0	0	0	0	0	0	0	139931	160348
NOS	0	0	0	0	0	0	0	0	0	0	0	0	0	0	0	0	0	0	0	0	0	0	0	0
OMAO	0	0	0	0	0	0	0	0	0	0	0	0	0	0	0	0	0	0	0	0	0	0	0	0
Defense(Subtot)	10956	10028	0	0	0	0	2665	3048	0	0	0	0	32553	35625	6309	6483	0	0	0	0	0	0	52483	55184
Air Force	0	0	0	0	0	0	0	0	0	0	0	0	18358	18619	6309	6483	0	0	0	0	0	0	24667	25102
Navy	0	0	0	0	0	0	0	0	0	0	0	0	14195	17006	0	0	0	0	0	0	0	0	14195	17006
Army	10956	10028	0	0	0	0	2665	3048	0	0	0	0	0	0	0	0	0	0	0	0	0	0	13621	13076
Homeland Security (Subtot)	0	0	0	0	0	0	0	0	0	0	0	0	0	0	0	0	0	0	0	0	0	0	0	0
USCG	0	0	0	0	0	0	0	0	0	0	0	0	0	0	0	0	0	0	0	0	0	0	0	0
Interior/BLM	0	0	0	0	0	0	0	0	0	0	0	0	0	0	0	0	0	0	0	0	915	900	915	900
BLM (Subtot)	0	0	0	0	0	0	0	0	0	0	0	0	0	0	0	0	0	0	0	0	0	0	0	0
SWA	0	0	0	0	0	0	0	0	0	0	0	0	0	0	0	0	0	0	0	0	0	0	0	0
OFA	0	0	0	0	0	0	0	0	0	0	0	0	0	0	0	0	0	0	0	0	0	0	0	0
NPS	0	0	0	0	0	0	0	0	0	0	0	0	0	0	0	0	0	0	0	0	0	0	0	0
USGS	0	0	0	0	0	0	0	0	0	0	0	0	0	0	0	0	0	0	0	0	915	900	915	900
Transportation(Subtot)	0	0	0	0	23030	27013	0	0	0	0	0	0	0	0	0	0	4000	2000	0	0	0	0	27030	29013
FAA	0	0	0	0	23030	27013	0	0	0	0	0	0	0	0	0	0	0	0	0	0	0	0	23030	27013
FHWA	0	0	0	0	0	0	0	0	0	0	0	0	0	0	0	0	4000	2000	0	0	0	0	4000	2000
EPA	0	0	0	0	0	0	0	0	0	0	0	0	0	0	0	0	0	0	0	0	6500	6430	6500	6430
NASA	0	0	0	0	0	0	258923	286413	670	650	740	740	0	0	229162	288023	0	0	1300	1100	0	0	490390	571440
NRC	0	0	0	0	0	0	0	0	0	0	0	0	0	0	0	0	0	0	0	0	1005	904	1410	1390
DOE	0	0	0	0	0	0	0	0	0	0	0	0	0	0	0	0	0	0	0	0	0	0	0	0
TOTAL	123134	117188	15876	14982	27328	29263	347814	392404	1695	1675	14679	14827	32553	35625	236221	290056	4000	2000	9184	8511	17984	8884	830468	915615
% of FY TOTAL	14.8%	12.8%	1.9%	1.6%	3.3%	3.2%	41.9%	42.9%	0.2%	0.2%	1.8%	1.6%	3.9%	3.9%	28.4%	31.7%	0.5%	0.2%	1.1%	0.9%	2.2%	1.0%	100.0%	100.0%

Table 6 Personnel Authorizations Supporting Meteorological Operations and Research

This table includes both current and FY 2015 requested personnel authorizations (Federal (FTE), Contractor, Military) required to perform or support meteorological operations and supporting research. The breakdown among federal, contractor, and military authorizations is shown for the first time in this Plan. The total, federal agency, staffing requested for FY 2015 is 13,928, a 2.6% decrease from FY 2014.

TABLE 6 Personnel Authorizations Supporting Meteorological Operations and Research
(Full Time Equivalent Staff Years)

Agency		Federal	Contractor	Military*	Current FY 2014	Federal	Contractor	Military*	Requested FY 2015	%CHG	% of FY 15 TOTAL
Agriculture		103	8	0	111	103	8	0	111	0.0	0.8
Commerce/NOAA	Subtotal	6262	1901	0	8163	6172	1901	0	8073	-1.1	58.0
NWS		4721	741	0	5462	4617	741	0	5358	-1.9	38.5
NESDIS		907	897	0	1804	909	897	0	1806	0.1	13.0
OAR**		399	201	0	600	408	201	0	609	1.5	4.4
NOS		114	62	0	176	117	62	0	179	1.7	1.3
OMAO		121	0	0	121	121	0	0	121	0.0	0.9
Defense	Subtotal	573	110	3993	4676	561	103	3733	4397	-6.0	31.6
Air Force		447	71	3208	3726	440	66	2948	3454	-7.3	24.8
Navy		79	20	415	514	75	18	415	508	-1.2	3.6
Marine Corps		0	0	370	370	0	0	370	370	0.0	2.7
Army		47	19	0	66	46	19	0	65	-1.5	0.5
Homeland Security-USCG		1	0	107	108	1	0	107	108	0.0	0.8
Interior/BLM	Subtotal	46	7	0	53	46	7	0	53	0.0	0.4
BLM Soil/Water/Air Program		10	0	0	10	10	0	0	10	0.0	0.1
BLM Fire Weather Program		28	4	0	32	28	4	0	32	0.0	0.2
NPS		8	3	0	11	8	3	0	11	0.0	0.1
Transportation	Subtotal	152	969	0	1121	152	969	0	1121	0.0	8.0
FHWA		3	5	0	8	3	5	0	8	0.0	0.1
FAA		149	964	0	1113	149	964	0	1113	0.0	8.0
EPA		39	11	0	50	43	8	0	51	2.0	0.4
NASA		4	3	0	7	4	3	0	7	0.0	0.1
NRC		7	0	0	7	7	0	0	7	0.0	0.1
DOE		0	0	0	0	0	0	0	0	0	0
Totals					14296				13928	-2.6	100.0

* Active Duty (does not include National Guard or Reserve)

** all research

Table 7 Interagency Fund Transfers

This table summarizes the reimbursement of funds from one agency to another during FY 2014. Agencies routinely enter into reimbursable agreements when they determine an agency can provide a given activity more cost-effectively. While specific activities and amounts may vary from year-to-year, the pattern reflects a significant level of interagency cooperation.

**TABLE 7 Interagency Fund Transfers
for Meteorological Operations and Supporting Research**

FY2014 Funds ($K) Estimated or Planned

Transferred from:	To:	Operations	Supporting Research	
DOC/NOAA				
NESDIS		0	0	
OAR	DOE		4076	
OAR	USDA		231	
DHS				
USFA	DOI/BLM-FA	125		Lightning
USDA				
USDA/USFS	DOI/BLM-OFA	629		RAWS maintenance contracts
USDA/USFS	DOI/BLM	101		Lightning
USDA/USFS	DOI/BLM	0		Smoke Monitoring Cache Management
DOD				
Air Force	DOC/NOAA/NWS	1518	189	
Air Force	DOC/NOAA/NESDIS	100		
Air Force	DOC/NOAA/OAR		125	
Air Force	DOC/OFCM	165		
Air Force	DOI/USGS	450		
Air Force	NASA	396	700	
Air Force	NSF			
Air Force	NSF/UCAR/NCAR		2200	
Navy	DOC/NOAA/OAR		2508	
Army	OFCM			
USACE	DOC/NOAA/NWS	231		
USACE	DOI/USGS	19738		
USACE	USDA			NRCS
USACE	DOC/NOAA/OAR		300	
DOD	DOI/BLM OFA	0		
DOI				
USGS	DOC/NOAA/OAR		100	
DOI/NPS	USDA	397		National Atmospheric Deposition Program
DOI/NPS	EPA	288		CASTNet Filter Pack Analysis
DOI/BOR	USDA			NRCS
DOI	DOC/NOAA/OAR		35	
DOI/BLM SWA	USDA-NIFA	62		
DOI/BLM SWA	USDA	50		NRCS
DOT				
FAA Weather	DOC/NOAA/NWS	13653		IAA-Center Weather Service Unit
FAA Weather	DOC/NOAA/NWS	65		IAA
DOT	DOC/NOAA/OAR		446	
NASA				
	DOD/USAF/45th Space Wing	800	0	
	DOC/NOAA/NDBC	0	100	
DOE	DOC/NOAA/OAR		2481	
State/Local	USDA	103		NRCS
	DOI/BLM-FA	250		RAWS maintenance contracts

Table 8 Facilities/Locations/Systems Taking Meteorological Observations

This table shows the number of facilities/locations/systems at which the Federal agencies carry out or oversee the taking of various types of meteorological observations.

TABLE 8 Facilities/Locations/Systems Taking Meteorological Observations

TYPE OF OBSERVATION by AGENCY	No. of 2014 Locations	
Surface, land		
Commerce (NOAA/NWS all types)	29315	ASOS: 315 COOP: 9000 MESONET: 20000
Commerce (NOAA/OAR manned Atmospheric Baseline Observations)	6	
Commerce (NOAA/OAR Climate Reference Network)	114	
Commerce (NOAA/NWS/INL Mesonet)	35	Idaho National Laboratory (INL) Mesonest used by Pocatello Weather Service Forecast Office. 35 according to the NOSC.
Commerce (NOAA/NWS Weather Monitoring Stations)	16	
Commerce (NOAA/OAR/ARL SORD Mesonet)	22	Special Operations And Research Division (SORD)
Air Force (U.S. & Overseas)	231	
Navy (U.S. & Overseas)	68	
Marine Corps (U.S. & Overseas)	24	
Army (U.S. & Overseas)	18	
Transportation (FAA Contract Wx Obsg Stn)*	136	
Transportation (FAA Auto Wx Obsg Stn - AWOS)	188	
Transportation (FAA Auto Wx Sensor Sys - AWSS)	44	
Transportation (FAA Auto Sfc Obsg Sys - ASOS)**	571	
Transportation (FAA Flight Service Stations in Alaska)***	17	
Transportation (FHWA-Road Wx Obsg Stn)	2437	
Homeland Security (USCG Coastal)	50	
Interior (NPS Air Program)	32	
Interior (BLM Soil/Water/Air Program)	481	
Interior (BLM Office of Fire and Aviation)	971	
Agriculture (Forest Service Smoke Monitors)	30	
Agriculture (Forest Service RAWS)	944	
Agriculture (Forest Service Research)	78	
Agriculture (NRCS active manual snow courses)	1160	
Agriculture (NRCS automated SNOTEL stations)	885	
Agriculture (NRCS automated SCAN stations)	203	
Agriculture (Agricultural Research Service)	25	
NASA (HEOMD)	1	
Total	38102	

*Note: All 136 FAA Contract Wx Obsg Stations are colocated with a FAA or Commerce (NWS) ASOS

**Note: Transportation (FAA oversight Auto Sfc Obsg Sys, non-Fed inspected 1400)
***Note: 17 Flight Service Stations in Alaska since 2007 owned and supported by Harris Corporation

Surface, marine

Commerce (NOAA/NOS/SEAS-equipped ships)	622	
Commerce (NOAA/NOS/PORTS)	67	stations with met sensors
Commerce (NOAA/NOS/NWLON)	180	stations with met sensors
Commerce (NOAA/NWS Coastal-Marine Autom Network)	47	
Commerce (NOAA/NWS/NDBC Buoys--moored)	102	does not include TAO and DART
Commerce (NOAA/OAR/AOML Buoys--drifting)	1238	
Navy (Ships with met personnel)	29	
Navy (Ships without met personnel)	259	
Homeland Security (USCG Cutters)	249	
NASA (Buoys - moored)	0	
Total	2793	

Upper air, balloon

Commerce (NOAA/NWS U.S.)	92	
Commerce (NOAA/NWS Foreign, Cooperative)	10	Cooperative Hurricane Upper-Air Stations (CHUAS)
Commerce (NOAA/NWS met/ozone/ water vapor)	25	
Air Force, Mobile	1	
Army, Fixed (U.S. & Overseas)	10	
Army, Mobile (U.S. and Overseas)	8	
Navy, Fixed (U.S. & Overseas)		
Navy, Mobile(U.S. & Overseas)		
Navy, Ships		
Marine Corps, Mobile	13	
NASA (U.S. and Overseas)	0	
Total	159	

Atmospheric Profilers

Commerce (NOAA/NWS) (404 and 449 MHz)	32
Air Force (Eastern Range) (915 MHz)	
Air Force (Eastern Range) (SODARS)	
Air Force (Western Range) (915 MHz)	
Air Force (Western Range) (50 MHz)	
Air Force (Western Range SODARS)	
Army	4
NASA (50 MHz) HEOMD	1
Total	37

Doppler weather radar (WSR-88D) sites

Commerce (NOAA/NWS)	122
Air Force (U.S. & Overseas)	26
Transportation-FAA (Off CONUS)	13
Total	161

Doppler weather radar (Not WSR-88D) sites

Air Force (Fixed, Deployed))	20

Army		3	
Navy (Fixed)			
Commerce (NOAA/NWS Research Phased Array Radar- NWRT)		1	
	Total	24	
Airport Terminal Doppler weather radars			
Transportation-FAA (Commissioned)		45	
	Total	45	
Conventional radar (non-Doppler) sites			
Commerce (NOAA/NWS)		0	legacy radars removed in 2013
Air Force, Mobile Units			
Army (U.S. and Overseas)			
Transportation (FAA (WSP))		34	
	Total	34	
Off-site WSR-88D Principle User Processors (PUPs)			
Air Force (OPUPs only)		7	
Marine Corps (U.S. & Overseas)			
Army			
	Total	7	
Weather reconnaissance Aircraft			
Commerce (NOAA/OMAO)		3	
Air Force Reserve Command (AFRC) – WC-130J			
	Total	3	
Geostationary meteorological satellites (No. operating)			
GOES 12 - South America	Decommissioned		
GOES 13 - Operational East		1	
GOES 14 - On-Orbit Storage		1	
GOES 15 - Operational West		1	
NESDIS	Total	3	http://www.oso.noaa.gov/goesstatus/
Polar meteorological satellites (No. operating)			
METOP-B - AM Primary		1	
METOP-A - AM Backup		1	
NOAA 15 - AM Secondary		1	
NOAA 16 - PM Secondary	Decommissioned		
NOAA 17 - AM Backup	Decommissioned		
NOAA 18 - PM Secondary		1	
NOAA 19 - PM Prime Service Mission		1	
SNPP - PM Primary		1	
NESDIS	Sub-Total	6	http://www.oso.noaa.gov/poesstatus/
Air Force (2 primary, 2 secondary, 2 tactical)		6	
Navy (WINDSAT and GFO)		1	
	Total	13	
Electric Field Mills (Surface)			

Air Force (Eastern Range - (KSC)		1
Army		5
	Total	1
Lightning Detection Systems		
Air Force (Eastern and Western Range Cloud-Ground)		2
Air Force (Eastern and Western Range- NLDN & Total Ltng Sys)		2
Army		7
NASA (KSC Lightning Mapping Array)		1
	Total	11
Rocketsondes		
Army	Total	0

SECTION 2

FEDERAL METEOROLOGICAL SERVICES AND SUPPORTING RESEARCH PROGRAMS

FEDERAL COORDINATION AND PLANNING
FOR METEOROLOGICAL SERVICES AND SUPPORTING RESEARCH

The mission of the Office of the Federal Coordinator for Meteorological Services and Supporting Research (OFCM) is to ensure the effective use of Federal meteorological resources by leading the systematic coordination of operational weather requirements, services, and supporting research among the Federal agencies. Its high-level focus includes cross-agency needs and requirements, issues and problems, studies, reports, plans, handbooks, and crosscut reviews, assessments, and analyses.

The OFCM operates with policy guidance from the Federal Committee for Meteorological Services and Supporting Research (FCMSSR). The principal work in coordinating meteorological activities and in the preparation and maintenance of OFCM reports, plans, and other documents is accomplished by the OFCM staff with the advice and assistance of the Interdepartmental Committee for Meteorological Services and Supporting Research (ICMSSR) and more than 30 program councils, committees, working groups, and joint action groups. The members who serve on these entities are Federal agency representatives.

STATUTORY BASIS FOR THE FEDERAL COORDINATION PROCESS

In 1963, Congress and the Executive Office of the President expressed concern about the adequacy of the coordination of Federal meteorological activities. In response, Congress directed in Section 304 of Public Law 87-843 (the Appropriations Act for State, Justice, Commerce, and Related Agencies) that the Bureau of the Budget prepare an annual horizontal budget for all meteorological programs in the Federal agencies. The Bureau of the Budget (now the Office of Management and Budget, OMB) issued a report in 1963 entitled *Survey of Federal Meteorological Activities*. That report described each agency's program for meteorological services and products and detailed the relationships among the programs of the various agencies. The report revealed close cooperation but little evidence of systematic coordination. Based on its survey, the Bureau of the Budget issued a set of ground rules to be followed in the coordination process. It established a permanent general philosophy for assignment and assessment of agency roles in the field of meteorology and set certain goals to be achieved by the coordination process. The Bureau of the Budget tasked the Department of Commerce (DOC) to establish the coordinating mechanism in concert with the other Federal agencies. It also reaffirmed the concept of having a central agency—the DOC—responsible for providing common meteorological facilities and services and clarified the responsibilities of other agencies for providing meteorological services specific to their mandated missions.

The implementation of these directives by DOC led to the creation of OFCM and the appointment of the first Federal Coordinator for Meteorological Services and Supporting Research (the Federal Coordinator). The Federal Committee for Meteorological Services and Supporting Research (FCMSSR) was established in 1964 to provide policy-level agency representation and guidance to the Federal Coordinator in addressing agency priorities, requirements, and issues related to services, operations, and supporting research. The FCMSSR also resolves agency differences that arise during the coordination of meteorological activities and the preparation of Federal plans.

FCMSSR comprises representatives of the 15 Federal agencies that engage in meteorological activities or supporting research, have a major need for meteorological services, or set policy and direction for such services and research. These 15 agencies are the Departments of Agriculture (USDA), Commerce (DOC), Defense (DOD), Energy (DOE), Homeland Security (DHS), the Interior (DOI), State (DOS), and Transportation (DOT); the Environmental Protection Agency (EPA), National Aeronautics and Space Administration (NASA), National Science Foundation (NSF), National Transportation Safety Board (NTSB), and Nuclear Regulatory Commission (NRC); and OMB and the Office of Science and Technology Policy (OSTP). The Under Secretary of Commerce for Oceans and Atmosphere, who is also the Administrator of the National Oceanic and Atmospheric Administration (NOAA), serves as the FCMSSR Chairperson.

OFCM COORDINATING INFRASTRUCTURE

The OFCM coordinating infrastructure diagram on page 2-5 shows the current program councils, committees, working groups (WG), and joint action groups (JAG) through which OFCM carries out its mission of ensuring the effective use of Federal meteorological resources by coordinating operational weather requirements, services, and supporting research among the Federal agencies. FCMSSR is shown at the top of the diagram, as the policy guidance advisor to the Federal Coordinator.

- The Interdepartmental Committee for Meteorological Services and Supporting Research (ICMSSR), which is chaired by the Federal Coordinator, is the primary program management body of the Federal coordinating structure. ICMSSR provides advice to OFCM, implements FCMSSR policies, and oversees the committees and working groups that address observing systems, weather operations and services, operational processing centers, and automated weather information systems. The full membership of ICMSSR is shown on the inside cover of this Federal Plan.

- The Program Councils, which are directly under FCMSSR and are each chaired by the Federal Coordinator, coordinate key programs at the highest interagency policy decision-making level, and ensure that the programs meet joint requirements. In addition to establishing policy, the program councils coordinate development and oversee the preparation and implementation of national program plans, which include research and development (R&D), systems development, validation and integration, acquisition strategy, operational concepts, agency roles, and management.

- The committees and their working groups and joint action groups operate at the program and working levels to provide: (1) a forum for each agency to report activities, challenges, and achievements; (2) a mechanism to coordinate change and problem solving; (3) a way to collect, document, and consolidate agency requirements and inventories; (4) oversight for coordinated system development; (5) a vehicle for coordinating with other groups internal and external to the coordinating infrastructure;

FEDERAL METEOROLOGICAL COORDINATING INFRASTRUCTURE

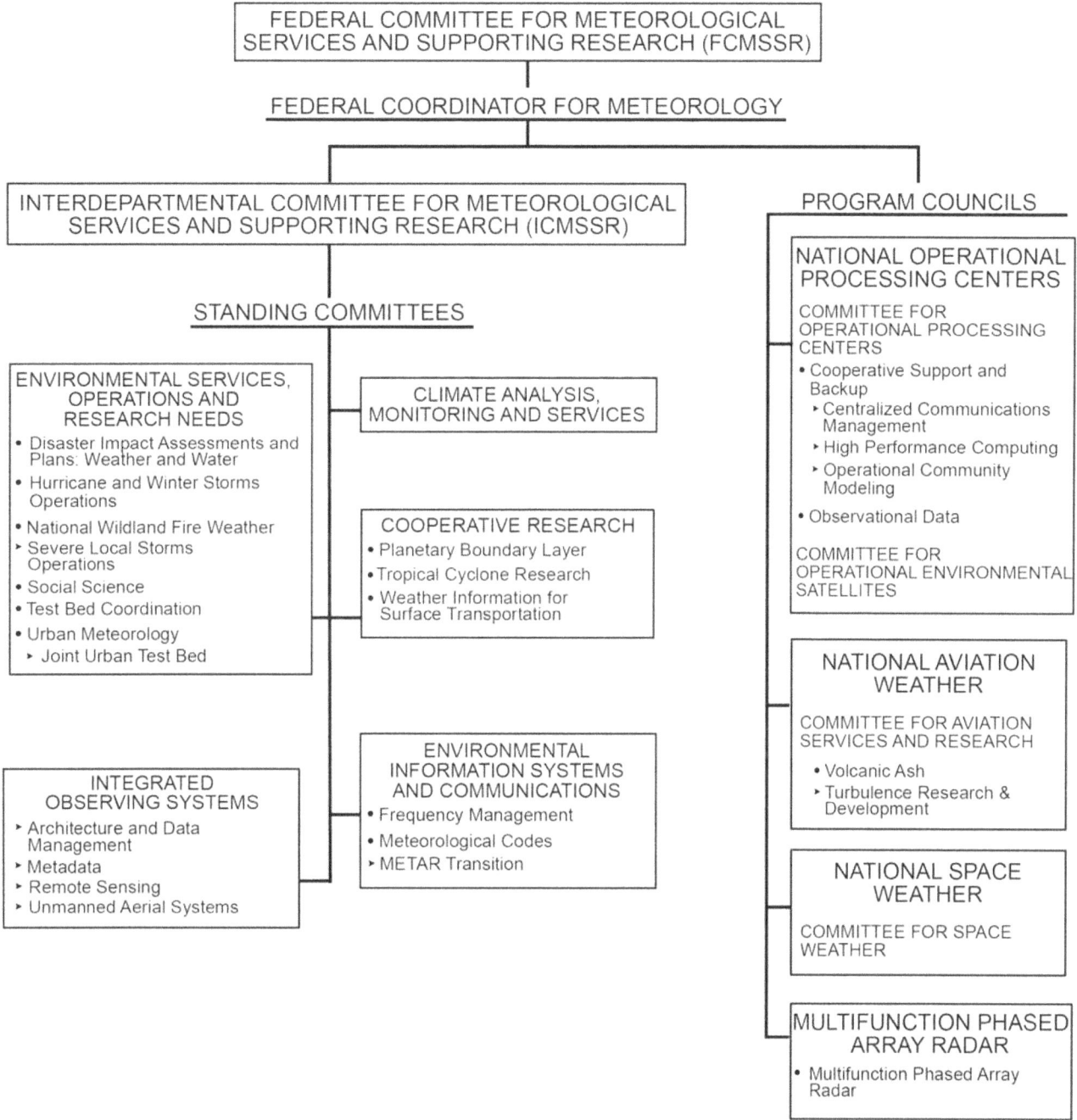

FEDERAL COMMITTEE FOR METEOROLOGICAL SERVICES AND SUPPORTING RESEARCH (FCMSSR)

FEDERAL COORDINATOR FOR METEOROLOGY

INTERDEPARTMENTAL COMMITTEE FOR METEOROLOGICAL SERVICES AND SUPPORTING RESEARCH (ICMSSR)

STANDING COMMITTEES

ENVIRONMENTAL SERVICES, OPERATIONS AND RESEARCH NEEDS
- Disaster Impact Assessments and Plans: Weather and Water
- Hurricane and Winter Storms Operations
- National Wildland Fire Weather
- ▸ Severe Local Storms Operations
- Social Science
- Test Bed Coordination
- Urban Meteorology
  - ▸ Joint Urban Test Bed

CLIMATE ANALYSIS, MONITORING AND SERVICES

COOPERATIVE RESEARCH
- Planetary Boundary Layer
- Tropical Cyclone Research
- Weather Information for Surface Transportation

INTEGRATED OBSERVING SYSTEMS
- ▸ Architecture and Data Management
- ▸ Metadata
- ▸ Remote Sensing
- ▸ Unmanned Aerial Systems

ENVIRONMENTAL INFORMATION SYSTEMS AND COMMUNICATIONS
- Frequency Management
- Meteorological Codes
- METAR Transition

PROGRAM COUNCILS

NATIONAL OPERATIONAL PROCESSING CENTERS

COMMITTEE FOR OPERATIONAL PROCESSING CENTERS
- Cooperative Support and Backup
  - ▸ Centralized Communications Management
  - ▸ High Performance Computing
  - ▸ Operational Community Modeling
- Observational Data

COMMITTEE FOR OPERATIONAL ENVIRONMENTAL SATELLITES

NATIONAL AVIATION WEATHER

COMMITTEE FOR AVIATION SERVICES AND RESEARCH
- Volcanic Ash
- ▸ Turbulence Research & Development

NATIONAL SPACE WEATHER

COMMITTEE FOR SPACE WEATHER

MULTIFUNCTION PHASED ARRAY RADAR
- Multifunction Phased Array Radar

September 2014

LEGEND: • Designates a Working Group
▸ Designates a Joint Action Group

and (6) a mechanism for the preparation of studies, agreements, standards, protocols, reports, and national plans.

Using these multiagency groups, OFCM pursues the following objectives as the means to achieve its mission:

- Document agency programs and activities in a series of national plans and reports that enable agencies to adjust their individual ongoing programs, and provide a means for communicating new ideas and approaches to fulfill requirements.

- Provide structure and programs to promote continuity in development and coordination of interagency plans and procedures for meteorological services and supporting research.

- Prepare analyses, summaries, or evaluations of agency meteorological programs and plans that provide a factual basis for the executive and legislative branches to make appropriate decisions on the allocation of funds.

- Review federal weather programs and federal requirements for meteorological services and supporting research. This review may suggest additions or revisions to current or proposed programs or identify opportunities for improved efficiency, reliability, or cost avoidance through coordinated actions and integrated programs.

OFCM HIGHLIGHTS FOR FISCAL YEAR 2014 AND PLANS FOR FISCAL YEAR 2015

Federal coordination activities during FY 2014 and plans for FY 2015 are described here, organized by program council and ICMSSR standing committee. The highlights begin with activities reporting directly to the Federal Coordinator, followed by the program councils and the ICMSSR standing committees.

Activities Reporting to the Federal Coordinator

Support for Plans and Analyses Mandated by the COASTAL Act of 2012

Congress passed the Consumer Option for an Alternative System to Allocate Losses (COASTAL) Act in June 2012 to lower the cost of the National Flood Insurance Program by better discerning wind versus storm surge damages in cases where little tangible evidence beyond a building's foundation remains for the proper adjustment of insurance claims after a hurricane or tropical storm. This law requires NOAA to produce detailed post-storm analyses following named storms that affect the coastal zone of the United States. These analyses are to be submitted to the Federal Emergency Management Agency (FEMA) within 90 days. NOAA is required to make all data and post-storm assessments available to the public and to maintain an online database. The law mandates that the database and post-storm model be operational by December 28, 2013 (540 days after enactment).

The COASTAL Act also requires that NOAA, in consultation with OFCM, provide several interim deliverables to Congress, including an assessment of current capabilities and needs to provide the required post-storm analyses. The Federal Coordinator established the Joint Action Group for the COASTAL Act Post-Storm Analysis (JAG/CAPSA) to develop a COASTAL Act Capabilities Development Plan (CACDP) in collaboration with NOAA. The plan, which was submitted to the NOAA Administrator on April 15, 2013 and released to the Department of

Commerce on June 7, 2013, provides for the collection of the required observational (covered) data to support post-storm assessments.

The JAG/CAPSA, working with OFCM's Working Group for Disaster Impact Assessments and Plans: Weather and Water (WG/DIAP) also developed an observing protocol to support collection of data for the post-storm analyses required by the COASTAL Act. This protocol was published in December 2013. The protocol will be included in subsequent editions of the *National Plan for Disaster Impact Assessments: Weather and Water Data* (Publication FCM-P33).

The JAG/CAPSA also facilitated population of the Coastal Wind and Water Event Database for the Named Storm Event Model.

National Plan for Civil Earth Observations

In July 2014 the Director of the Office of Science and Technology Policy (OSTP) in the Executive Office of the President issued the *National Plan for Civil Earth Observations* on behalf of the National Science and Technology Council (NSTC). The plan was the culmination of interagency coordination in support of the 2013 *National Strategy for Civil Earth Observations* that began with an assessment of the federal Earth observation enterprise organized around 13 societal benefit areas (SBAs). OFCM led teams of subject matter experts through the assessment process for the weather and space weather SBAs and participated as a member of the Assessment Working Group (AWG). The original assessment was conducted under the auspices of the temporary National Earth Observations Task Force and subsequent triennial assessments will be conducted by a new AWG aligned under the U.S. Group on Earth Observations (USGEO). USGEO is a subcommittee of the NSTC's Committee on Environment, Natural Resources, and Sustainability (CENRS). The Federal Coordinator for Meteorology is a member of CENRS.

Planning for the second national Earth observation assessment began in FY 2014 with the assessment to follow in FY 2015, again organized along the same 13 SBAs used in the first assessment. OFCM anticipates playing a significant role again in the space weather SBA analysis in FY 2015.

National Aviation Weather Program Council

OFCM supports implementation of the National Aviation Weather Program, which is a broad interagency effort to advance meteorological standards, improve products, enhance services, and conduct research that contributes to the overall goal of providing state-of-the-art information to aviation end users where and when they need it. OFCM also participates in the Next Generation Air Transportation System (NextGen) Weather Working Group (NWWG) and Friends/Partners in Aviation Weather (FPAW).

In FY 2014, OFCM supported the Research, Engineering, and Development (RE&D) initiative managed by FAA's Interagency Planning Office (IPO), the successor organization to the NextGen Joint Planning and Development Office. The NextGen Senior Policy Committee (SPC) directed the RE&D initiative to identify mid-term programs that are in an R&D phase and that may be key to the implementation of NextGen capabilities. OFCM staff provided subject matter expertise for the identification of potential projects, served on the RE&D team charged with selecting the final projects from among dozens of nominations, and helped prepare the RE&D report to the NextGen Executive Board and SPC. OFCM staff also successfully advocated selection of the Multifunction Phased Array Radar project and developed the MPAR-related sections of the report.

OFCM staff also supported aviation-related initiatives related to meteorological codes and observing practices under the auspices of other OFCM coordinating groups.

National Operational Processing Centers Program Council

In FY 2014, OFCM continued to support and lead the National Operational Processing Centers Program Council (NOPC). The NOPC provides senior management oversight for the Committee for Operational Processing Centers (COPC) and facilitates improved collaboration among the NOPC member organizations that have direct oversight of COPC members. The NOPC principal members are NOAA National Weather Service (NWS); NOAA National Environmental Satellite, Data, and Information Service (NESDIS); the U.S. Navy Meteorology and Oceanography Command; and the U.S. Air Force Directorate of Weather. The NOPC sets policy; provides strategic vision, planning, program guidance, and interagency funding authority; identifies future roadmap capabilities for their own agencies; determines how to effectively position themselves for future requirements and collaboration; and identifies coordinated approaches to solving the Nation's highest priority environmental information needs.

NOPC held two meetings in FY 2014 focused on three broad areas: (1) space-based environmental monitoring, (2) operational processing center (OPC) data exchange, and (3) numerical weather prediction. The council received updates on the Joint Polar Satellite System (JPSS), Geostationary Operational Environmental Satellite R Series (GOES-R), and Committee for Operational Environmental Satellites (COES) activities. The OPC data exchange topic focused on the status of the Non-secure Internet Protocol Router Network (NIPRNet) Federated Gateway implementation and the continued need to establish and fund a second point of presence to exchange data between DOD and NOAA. Lastly, the numerical weather prediction updates focused on review of the Earth System Prediction Capability (ESPC) program, NWS activities funded by Hurricane Sandy disaster relief, and a review of NSF's visiting scientist program and the university-based ensemble project.

Committee for Operational Processing Centers

In FY 2014, COPC continued to facilitate improved processing and backup capabilities for NOAA's National Centers for Environmental Prediction and Office of Satellite and Product Operations, the Air Force Weather Agency, and the U.S. Navy's Fleet Numerical Meteorology and Oceanography Center and Naval Oceanographic Office. Semiannual meetings of COPC continued in FY 2014 and are expected to continue in FY 2015.

Activities under COPC and its Working Group for Cooperative Support and Backup included the work of three subgroups. The Joint Action Group for Operational Community Modeling has and will continue to enhance the implementation of ensembles among the centers. The Joint Action Group for Centralized Communications Management worked to create a second data exchange point between NOAA and DOD in Boulder, Colorado. And the Working Group for Observational Data (WG/OD), formerly the Joint Action Group for Operational Data Acquisition for Assimilation (JAG/ODAA), updated their terms of reference to reflect their long-term, ongoing activities and the range of areas they are addressing for data exchange among the OPCs. These groups will continue their respective activities in FY 2015.

Observational Data

The Working Group for Observational Data (WG/OD) facilitates the acquisition, processing, and exchange of observational data among the national operational processing centers (OPCs) and other related data centers. The primary focus areas are the following:

- Acquisition, processing, and exchange of meteorological, oceanographic, and space environmental data.

- Interface among the OPCs, their research and development partners in the Joint Center for Satellite Data Assimilation (JCSDA), and the other national data and prediction centers for the purpose of coordinating and satisfying national requirements for observational data.

- Coordination of data formatting standards where practical and the implementation of approved data product enhancements and new data products.

- Coordination of observational data issues that overlap related responsibilities of other OFCM committees and COPC groups.

- Assurance that the OPCs and related data centers are provided the maximum quality and optimum quantity of observational environmental data streams required for assimilation into their respective processes.

During FY 2014, the working group completed a number of satellite data-related activities. It satisfied a number of interagency satellite data flow requests, most notably the acquisition of GOES hourly winds. It also initiated action to acquire additional satellite data sources, including Indian National Satellite (INSAT) sea surface temperature data for the Navy, Global Precipitation Measurement (GPM) data for the Air Force and Navy, and Oceansat-2 ocean color data for the Navy. The Working Group also initiated an update to the Environmental Satellite Data Annex under the Data Acquisition, Processing, and Exchange (DAPE) Memorandum of Agreement (MOA).

In FY 2015, the group will pursue a number of non-satellite (conventional) data and satellite data-related activities. The group will develop a Conventional Data Annex for the DAPE MOA, outlining the principal guidelines for interagency conventional data acquisition, processing, and exchange. They will also seek improvements in the conventional data bulletin subscription process, continue to monitor issues arising from the World Meteorological Organization (WMO)-mandated conventional data product transition to binary format (BUFR), and assist in the development of data flow through the DOD-mandated Nonsecure Internet Protocol Router Network (NIPRNET) Federated Gateway and NESDIS's Product Distribution and Access (PDA) data communication and dissemination system. In the satellite data area, the working group will pursue access to pre-operational meteorological satellite data for testing, as well as development and implementation of a data flow pathway for imagery and data from Japan's new Himawari-8 satellite scheduled for launch in October 2014. Lastly, spanning both conventional and satellite data, the group will continue development of OPC data inventories to identify significant data gaps.

Committee for Operational Environmental Satellites

Responding to stakeholder desire to improve interagency communication and coordination of matters regarding the use environmental satellites, OFCM reactivated the Committee for Operational Environmental Satellites (COES). In FY 2014, the NOPC approved establishment of COES under NOPC as a sister committee to COPC. COES will help advance the goals of the NOPC to achieve interagency coordination of operational environmental satellite systems planning. The objectives of the COES are the following:

- Foster interagency review and coordination of approved requirements for operational environmental satellite programs.

- Promote free and open exchange of information on environmental satellite systems development, satellite data systems architecture, continuity plans, and data exploitation readiness plans.

- Consider potential use of research satellite capabilities to augment operational systems in meeting user needs, and develop plans to transition research data into operational products and new applications.

- Facilitate working-level relationships between federal members and other stakeholders to effectively resolve interagency issues on the availability of environmental satellite data and products from future systems.

- Establish dialog with other groups currently engaged in various aspects of environmental satellites, data readiness, and data exploitation.

National Space Weather Program Council

Over the past year, the National Space Weather Program Council (NSWPC) has sought closer coordination of space weather science, research, and services to the Nation. The NSWPC directed the Committee for Space Weather (CSW) to develop an implementation plan to achieve the goals described in the 2010 *National Space Weather Program Strategic Plan* and, in response, CSW established the Joint Action Group for Space Weather Implementation Planning. The action group comprises six working groups to address the five goals of the *Strategic Plan* as well as disaster preparedness, readiness, and response. Development of the implementation plan began in April 2014, supported by more than 50 subject matter experts from all nine NSWPC agencies, with projected completion by the end of calendar year 2014.

The NSWPC also supported the review and publication of the 2014 *National Plan for Civil Earth Observations* by OSTP (see pg. 2-7) following significant support and leadership of the first national Earth observation assessment's space weather SBA team in FY 2012. The NSWPC agencies expect to participate in the space weather SBA team for the next triennial Earth observation assessment beginning in FY 2015.

Executive Council for Multifunction Phased Array Radar

The Multifunction Phased Array Radar (MPAR) initiative seeks to consolidate the radar surveillance missions of four agencies: DOD, DHS, DOC/NOAA, and DOT/Federal Aviation Administration (FAA), reducing the number of radars required and consolidating operations and logistics. The MPAR Working Group, comprising various elements of the four stakeholder agencies as well as representatives from several other interested agencies, meets several times a year to review program progress and plan interagency activities. In April 2014 the Executive Council directed preparation of a plan for development and testing of an MPAR Advanced Technology Demonstrator (ATD) with a full-scale, dual-polarization, active electronically-scanned array (AESA) antenna and processing system capable of demonstrating dual polarization and multifunctionality. NOAA and FAA, who have been investing in MPAR risk reduction, participate in a Government Engineering Team (GET), with OFCM sharing leadership responsibilities and hosting face-to-face meetings. The GET supports interagency collaboration to determine priorities, review study results, and make decisions on resources. In advance of the ATD, NOAA and FAA paid for the development of a proof-of-concept system with a small phased array being built from ten 64-element dual-polarization panels. MIT Lincoln Laboratory has been developing the dual polarization panel technology for several years and is building the trailer-mounted array for delivery in winter 2015. The panel will be used to confirm the dual-polarization capability of the MIT panels and to test other fundamental aspects of the technology. If proven, the technology will be applied to build the ATD, which will replace the legacy SPY-1A phased array radar at the National Severe Storms Laboratory. Planning for the ATD is underway while proof of concept system assembly is monitored.

Meanwhile, MPAR outreach continues with the following groups: (1) FAA's Interagency Planning Office on the Research, Engineering, and Development initiative (see the earlier section on the National Aviation Weather Program Council); (2) DOD spectrum management leadership; (3) the interagency Wind Turbine Radar Interference Mitigation Working Group; (4) the DOD/FAA terminal radar systems modernization group; and (5) the Air Force NextGen lead service office.

Committee for Environmental Services, Operations and Research Needs

The Committee for Environmental Services, Operations and Research Needs (CESORN) covers a wide range of basic meteorological services and supporting research. Areas for coordination vary from year to year based on agency needs.

Hurricane and Winter Storms Operations and Research and the Interdepartmental Hurricane Conference

In FY 2014, the Working Group for Hurricane and Winter Storms Operations and Research (WG/HWSOR) continued to manage the interagency operations plans for hurricanes and winter storms and address related issues relevant to the federal meteorological community.

The *National Hurricane Operations Plan (NHOP)* describes individual agency responsibilities and prescribes operational procedures, common reference points, formats for data exchange, tropical cyclone names and pronunciations, and other information to achieve economy and efficiency in the provision of tropical cyclone forecasting and warning services to the Nation.

The plan is reviewed and updated annually in advance of the hurricane season to address action items and implement agreements reached at the Interdepartmental Hurricane Conference (IHC). The working group published the 52nd edition of the *NHOP* in FY 2014, including updated guidance on notification procedures among the tropical cyclone forecasting centers, DOD weather reconnaissance support to NOAA, tropical cyclone names and pronunciations, data codes, and air traffic control-related procedures.

WG/HWSOR also manages the *National Winter Storms Operations Plan (NWSOP)* to coordinate the activities of the federal agencies providing enhanced observations of severe winter storms that affect the coastal regions of the United States. The plan focuses on the coordination of requirements for DOD and NOAA aircraft reconnaissance observations with the goal of improving the accuracy and timeliness of winter storm forecasts and warning services. The current *NWSOP* was published prior to the 2012-2013 winter season and minor changes for the 2013-2014 winter season were not deemed significant enough to warrant an early update to the plan. A new edition of the plan will be published in FY 2015.

The WG/HWSOR depends on the annual IHC to meet its hurricane-related responsibilities. The following paragraphs summarize the FY 2014 IHC and Tropical Cyclone Research Forum.

OFCM hosts the IHC each year to provide a forum for the responsible federal agencies, together with representatives from the academic community, industry, and other user communities such as emergency management, to prepare for the upcoming hurricane season and make improvements to the Nation's hurricane forecasting and warning program. OFCM develops the agenda with input from working groups under CESORN and the Working Group for Tropical Cyclone Research (WG/TCR) under the Committee for Cooperative Research (CCR).

OFCM hosted the 68th Interdepartmental Hurricane Conference and Tropical Cyclone Research Forum March 3-6, 2014, in a combination face-to-face and online, virtual conference. The primary face-to-face locations were the NOAA Center for Weather and Climate Prediction in College Park, Maryland, and the National Hurricane Center in Miami, Florida. The distribution of attendees included 73 in-person and 220 virtual attendees representing eight federal agencies as well as academia, private industry, and the media. Notably, a winter storm closed federal offices (including the conference location) in the Washington, DC, area on March 3, but the WG/HWSOR meeting and other conference activities continued via virtual conference. Despite the virtual conference mode enabling the conference to continue on schedule, several participants noted the significant loss of interaction overall when the IHC has a large virtual attendance.

The theme was "Tropical Cyclone Research: Assessing the Past – Planning for the Future" and the IHC and forum sessions included the following topics, examining capabilities improvements and identified gaps for each area with a common thread of assessing the transition of research to operations:

- Research priorities of the operational centers.
  - Intensity and rapid intensification.
  - Structure.
  - Track.

  o Sea state and sea heights.

  o Storm surge.

  o Precipitation and inland flooding.

  o Genesis

- Observations and observing strategies.

- Federal numerical modeling and data assimilation initiatives.

- Social science research results and demonstration projects.

More information on capability achievements and identified gaps may be found in the conference summary available in the special projects section of the ofcm.gov web site.

The forum also included a special session to summarize the assessment results and propose a plan of action for a mid-course assessment of the *Interagency Strategic Research Plan for Tropical Cyclones: The Way Ahead* (FCM-P36-2007).

The action items developed at the conference are listed below:

- OFCM will post conference presentations and the conference summary on the conference web page. (Complete)

- OFCM will publish the annual NHOP, to include changes recommended to and accepted by the WG/HWSOR, no later than May 15, 2014. (Complete)

- OFCM will initiate a mid-course assessment on the *Interagency Strategic Research Plan for Tropical Cyclones: The Way Ahead* (FCM-P36-2007) to be completed by the WG/TCR. (In progress)

Disaster Impact Assessments and Plans: Weather and Water Data

Disaster-relevant data acquired from many routine observing systems often do not provide data coverage and density sufficient to adequately document the effects of a major storm or flood or enable understanding of small-scale, localized weather and water processes. As a result, routine observations are supplemented by mobile observing system data and post-storm analysis of impact features, such as high water marks and wind damage, to fill information gaps and obtain more complete spatial coverage. Collecting this information contributes to determination of the intensity and magnitude of storms and may support Presidential disaster declarations. The information may also be used in a number of other ways: to assess threat potential; validate emergency management and hurricane storm surge models; update FEMA flood insurance rate maps; assist in evaluating indeterminate loss causes; and improve forecasting models. Furthermore, the National Institute of Standards and Technology and various state agencies use the data to improve building codes and construction practices.

To improve the efficiency of data collection and promote sharing of data within an organized, interagency disaster impact assessment process, the Federal Coordinator established the Working

Group for Disaster Impact Assessments and Plans: Weather and Water Data (WG/DIAP) in 2010. The WG/DIAP published the *National Plan for Disaster Impact Assessments: Weather and Water Data* (NPDIA) in October 2010. The plan documents the types of data required, the acquisition processes, and the coordinating procedures to be used leading up to, during, and following a significant storm event.

In response to the COASTAL Act of 2012 requirement to establish a protocol for data collection for post-storm assessments (see pg. 2-6), the WG/DIAP developed a *COASTAL Act Data Protocol Annex* to the NPDIA. This annex outlines the interagency protocol for responding to storms that could come under provisions of the COASTAL Act of 2012. It describes the plan for capturing the necessary data and assembling it in NOAA's Coastal Weather and Water Event Database (CWWED) for assimilation into NOAA's Named Storm Event Model (NSEM). It also specifies how these data are made available to FEMA, other federal, state, and local government agencies, private insurance adjusters, and the general public to help evaluate indeterminate loss causes and insurance coverage responsibilities. The next revision of the NPDIA will include the new COASTAL Act data protocol.

Other upcoming changes to the NPDIA include the latest data collection protocol for response to tsunamis and the revised Enhanced Fujita (tornado) Scale now under development.

Aerial support from the Civil Air Patrol provides one of the sources of pre-event and post-event data to support impact assessments. Under the OFCM-coordinated Air Force-NOAA Memorandum of Agreement for Civil Air Patrol support, in FY 2015 OFCM will continue to coordinate post-storm data acquisition surveys in response to natural disasters and agency requirements. The annual agreement between OFCM and the U.S. Air Force for reimbursable support funds Civil Air Patrol missions such as Alaska glacial lake damming assessments, tornado damage photos and video, and severe flooding imagery.

Atmospheric Transport and Diffusion (ATD)

On June 25, 2014, OFCM cosponsored a special session of the 18th annual George Mason University Atmospheric Transport and Dispersion Modeling Conference. The theme of the session was *Progress in Governmental Atmospheric Transport and Dispersion Modeling and Response*. Subject matter experts from DOC/NOAA (National Weather Service/National Centers for Environmental Prediction, Office of Oceanic and Atmospheric Research/Air Resources Laboratory (ARL), and National Ocean Service/Office of Response and Restoration), DOE (National Atmospheric Release Advisory Center), DOD (U.S. Air Force, U.S. Army, and Defense Threat Reduction Agency (DTRA)), DHS (Interagency Modeling and Atmospheric Assessment Center (IMAAC)), and Nuclear Regulatory Commission (Pacific Northwest National Laboratory) participated in the session's panel discussion.

The session addressed the following topics:

- **Current status**. Agency leaders discussed roles and plans in federal ATD field experiments, modeling, responsibilities, and governance; the status of existing modeling capabilities; and how model output is provided to decision makers.

- **Advances.** Scientists and experts presented and discussed improvements made to observations, models, and processes in the past year.

- **Gaps.** Participants discussed known areas of basic and applied research needs, model development, and tool development that comprise the highest priorities to address.

- **Future plans.** Participants discussed areas of concern the community should pursue or continue to pursue.

This year's session included a sharper focus on experimentation, with presentations on the following topics:

- Project Sagebrush, an effort by ARL's Field Research Division to advance the fundamental observation-based understanding of atmospheric transport and diffusion beyond the still-used results of the 1956 Prairie Grass project.

- The Army's Granite Mountain Atmospheric Sciences Testbed (GMAST).

- The Mountain Terrain Atmospheric Modeling and Observations (MATTERHORN) field campaign that was conducted at GMAST.

Discussion following the presentation centered on the possible application of multiple model outputs for several related reasons: to compare model outputs to select the best output for the particular application; to compare model outputs to help estimate and express the uncertainty of the results in particular situations; or to create an ensemble to perhaps select a composite or hybrid solution. This approach would help address concerns expressed during the session about the lack of error bars on the ATD model output.

In FY 2015, OFCM plans to follow up with the federal agencies on the issues that arose during the GMU conference. In addition, OFCM will continue to participate in FEMA's Scientific Support Working Group, through which an opportunity exists for interagency resources to help support the DHS Radiological and Nuclear Response and Recovery R&D Investment Plan.

Committee for Integrated Observing Systems

In FY 2014, the Committee for Integrated Observing Systems (CIOS) continued its work in support of the Network of Weather and Climate Observing Networks (NOWCON) initiative described in this section in previous-year *Federal Plan for Meteorological Services and Supporting Research* editions. NOWCON grew out of federal meteorological community interest in weather and climate observing networks and the 2009 publication of a National Research Council (NRC) report, cosponsored by OFCM, titled *Observing Weather and Climate from the Ground Up – A Nationwide Network of Networks*. In brief, the theme of the report is that the United States enjoys an effective synoptic-scale weather observing network, but society demands increasingly finer-scale weather and climate information. State and local governments, corporations, academic institutions, and individuals have deployed a rapidly growing array of individual sensors and sensor networks in patchwork fashion across the country and much of the data from these systems remain unknown or inaccessible to a wider audience of potential users. CIOS continues to provide a venue for federal agencies to respond to this report and collaborate with the American Meteorological Society on its national network of networks recommendations. Early attention has centered on metadata, providing a catalyst encouraging the implementation of metadata for surface observations across the federal agencies and the broader non-federal weather and climate observing network community. In FY 2015, CIOS will investigate ways to leverage available resources and work already underway to streamline and standardize the implementation of metadata in compliant formats.

In other activities during FY 2014, CIOS continued to examine potential updates to *Federal Meteorological Handbook No. 1, Surface Weather Observations and Reports* to take advantage of enhanced capabilities of new instrumentation, incorporate other changes that have been awaiting implementation, and address additional emerging issues. This work will continue into FY 2015.

Also in FY 2015, OFCM will begin collaboration with the NWS Radar Operations Center to update the *Federal Meteorological Handbook No. 11, Doppler Radar Meteorological Observations*.

Unmanned Aerial Systems

The Joint Action Group for Unmanned Aerial Systems (UAS) for Environmental Monitoring (JAG/UAS-EM) supported several interagency meetings on UAS utilization. In May 2014, OFCM hosted a NOAA Sensing Hazards with Operational Unmanned Technology (SHOUT) interagency planning meeting. SHOUT is a multi-year program to deploy the NASA Global Hawk aircraft for high impact weather observations. Throughout the year, OFCM also provided a number of venues for the NOAA UAS program to share information and coordinate across the federal agencies.

Additionally, OFCM aligned JAG/UAS-EM activities with other groups working UAS development and utilization issues including the Interagency Coordinating Committee for Airborne Geosciences Research and Applications (ICCAGRA) and the NSTC Subcommittee for Unmanned Systems.

The ICCAGRA meets semiannually to improve cooperation, foster awareness, facilitate communication among airborne instrument-sponsoring agencies, and serve as a resource to senior-level management on airborne geosciences issues. The Committee facilitates cooperation on the use of airborne platforms and payloads for individual investigators as well as national and international field campaigns. OFCM supports ICCAGRA through meeting participation and access to the OFCM federal meteorological coordinating infrastructure.

OFCM continues to support the activities of the CENRS Ocean Science and Technology Subcommittee's work on unmanned systems. OFCM's participation provides a connection to meteorological research and service providers and supports effective coordination of unmanned system applications across the federal enterprise.

Committee for Cooperative Research

Tropical Cyclone Research

For additional information on tropical cyclone research, see the Working Group for Hurricane and Winter Storms Operations and Research and Interdepartmental Hurricane Conference section above.

In 2007, the Joint Action Group for Tropical Cyclone Research completed the *Interagency Strategic Research Plan for Tropical Cyclones: The Way Ahead*. This plan presented a shared set of R&D priorities matched to operational requirements from the tropical cyclone forecast and warning centers and served as a baseline for NOAA's Hurricane Forecast Improvement Program (HFIP). In 2008, 2010, and 2012 the Working Group for Tropical Cyclone Research (WG/TCR) assessed agency research activities mapped against tropical cyclone research needs and operational priorities. The assessments have enabled WG/TCR to establish a successful process to accomplish the following tasks:

- Update the operational priorities.

- Evaluate how research is contributing to meeting operational priorities.

- Support informed research manager decisions on future investments.

- Facilitate interagency collaboration and coordination.

Following the biennial assessment schedule described above and in response to a 68th Interdepartmental Hurricane Conference (IHC) action item, in FY 2014 the working group identified preliminary capability gaps and challenges and in FY 2015 will complete a mid-course assessment of the 2007 plan.

Weather Information for Surface Transportation

In FY 2014, OFCM continued to explore weather services and R&D activities supporting the surface transportation community, building on its earlier publication *Weather Information for Surface Transportation--National Needs Assessment Report*. Part of this activity included participation in AMS Intelligent Transportation Systems/Surface Transportation Committee meetings and activities.

In FY 2015, OFCM will explore development of a coordinated research plan for surface transportation weather information in conjunction with related activities in other parts of the coordinating infrastructure (e.g., CIOS).

Committee for Environmental Information Systems and Communications

In FY 2015, the Committee for Environmental Information Systems and Communications (CEISC) will explore opportunities for interagency collaboration on definition and implementation of geospatial standards for information exchange.

Meteorological Codes

In FY 2014, OFCM formed the Joint Action Group for IWXXM Implementation (METAR Transition) in response to the International Civil Aviation Organization (ICAO) initiative to transition the METAR family of codes (METAR, SPECI, TAF, and SIGMET) from the current alphanumeric format to an Extensible Markup Language (XML)-based code. The JAG is charged with developing and implementing an interagency plan to prepare federal agencies to meet the ICAO code transition phase points over the next 5 years.

In addition, some of the expected revisions of *Federal Meteorological Handbook No. 1, Surface Weather Observations and Reports* described above (see the CIOS section, pg. 2-16) will involve coding practices, which fall under the purview of the Working Group for Meteorological Codes.

OFCM External Collaborations

NAS/NRC Board on Atmospheric Sciences and Climate

OFCM continues its mutually beneficial interactions with the National Research Council (NRC). The Federal Coordinator participates in the NRC Board on Atmospheric Sciences and Climate (BASC) strategic planning workshops and attends regularly scheduled BASC meetings.

NAS/NRC Space Science Board, Committee on Space and Solar Physics

The executive secretary of the NSWPC provided program updates to the NRC Space Science Board (SSB) and its Committee on Space and Solar Physics (CSSP). The OFCM anticipates that the CSSP will continue to invite the NSWPC members and the executive secretary to participate in its semiannual meetings. The OFCM and SSB are seeking ways to leverage each other's strengths in bringing together the federal and nongovernmental space weather communities.

American Meteorological Society

OFCM supports AMS activities by participating in AMS conferences and workshops and other environmental science education and outreach programs. In FY 2014, OFCM co-authored and/or presented two papers at the 94th AMS Annual Meeting. OFCM will present two papers at the 95th Annual meeting in 2015.

FY 2014 OFCM PUBLICATIONS

The publications listed in the table below were prepared in hard copy and/or were added to OFCM's web site (www.ofcm.gov) during FY 2014.

OFCM PUBLICATION	DATE	NUMBER
PLANS (P)		
Federal Plan for Meteorological Services and Supporting Research, Fiscal Year 2014	October 2013	FCM-P1-2013

OFCM PUBLICATION	DATE	NUMBER
National Hurricane Operations Plan • 2014 Build 14.1 Dual Pol WSR-88D Tropical Cyclone Operations Plan (05/29/2014) ◦"QUICK CHECK" List • 2013 Build 13.1-13.2 Dual Pol WSR-88D Tropical Cyclone Operations Plan (05/15/2013) ◦"QUICK CHECK" List	May 2014	FCM-P12-2014
PROCEEDINGS		
2014 Tropical Cyclone Research Forum/ 68th Interdepartmental Hurricane Conference Summary Report	March 2014	
Proceedings of the Special Session, 18th Annual George Mason University (GMU) Atmospheric Transport and Dispersion Modeling Conference	June 2014	

BASIC SERVICES

For purposes of this *Federal Plan*, Basic Services include the basic meteorological service system, to include observations, public weather forecasts, severe weather warnings and advisories, and the meteorological satellite activities of NOAA. Basic Services also include the operations and supporting research of other Federal agencies that have been identified as contributing to basic meteorological services.

OPERATIONAL PROGRAMS INCLUDING PRODUCTS AND SERVICES

NOAA/National Weather Service

The National Oceanic and Atmospheric Administration's (NOAA) National Weather Service (NWS) provides climate, water, weather, ocean and space weather warnings and forecasts for the United States, its territories, adjacent waters, and ocean areas to help protect life and property and enhance the national economy. These services are provided through 122 Weather Forecast Offices (WFO), 13 River Forecast Centers (RFC), and the National Centers for Environmental Prediction (NCEP). These offices collect data, prepare local warnings and forecasts, and disseminate information to the public, both nationally and internationally, through NOAA Weather Radio (NWR), satellite-based telecommunication systems, radiofacsimile, the media, and the internet. NWS forecasters issue short-duration watches and warnings for severe weather, such as tornadoes and severe thunderstorms, as well as long-duration watches, warnings, and advisories for hazardous winter weather conditions, high wind events, dense fog, flooding, and extreme temperatures.

The NWS uses surveillance technologies such as a national network of Doppler weather radars, satellites operated by NOAA's National Environmental Satellite, Data, and Information Service (NESDIS), aircraft observations, data buoys for marine observations and tsunami detection, surface weather observing systems at airports, and weather balloons to obtain vertical measurements of the atmosphere. Some observations are obtained through the Cooperative Observer Program, a nationwide network of volunteer-operated weather observing sites. Many other observations are contributed through arrangements with publicly and privately operated networks. Observations feed sophisticated environmental prediction models running on high performance supercomputers, which provide weather, water, climate, ocean and space weather forecast guidance that is available to users. The NWS' highly trained and skilled workforce uses powerful workstations to analyze these data to issue forecasts and warnings around the clock. A high-speed communications hub allows for the efficient exchange of these data and products among NWS components, partners, and other users. NWS integrated dissemination infrastructure including NOAA Weather Radio, satellite broadcast, and the internet rapidly distributes this information.

The NWS creates forecasts in digital formats and makes them readily available. Forecasters use their expertise to maintain an up-to-date digital forecast database of weather elements. This information is stored in the National Digital Forecast Database (NDFD). Output from the NDFD is publicly available in the form of web graphics on the Internet and in several other digital formats. Outreach, education, and feedback are also critical elements in effective public response and improvements to NWS services.

RFCs routinely generate short range (deterministic) through extended range (probabilistic) river forecasts and (deterministic) flash flood guidance. Information from the RFCs serves as the basis for local flood and flash flood warnings, watches, and advisories issued by the WFOs. Some RFCs, especially those in mountainous regions, also provide water-supply volume and peak-flow forecasts based on snow pack in high elevations. These water supply forecasts are used by a wide range of decision makers, including those in agriculture, hydroelectric dam operation and electricity generation, and water resources management. National Operational Hydrologic Remote Sensing Center (NOHRSC) provides comprehensive snow observations, analyses, data sets and map products for the Nation. NOHRSC products and services are used by RFCs and WFOs to develop a variety of hydrologic products

The Advanced Weather Interactive Processing System (AWIPS) is a technologically advanced information processing, display, and telecommunications system. AWIPS is an interactive computer system that integrates all meteorological and hydrological data, and all satellite and radar data, and enables the forecaster to prepare and issue more accurate and timely forecasts and warnings. (NOAA Image)

such as spring flood outlooks, water supply outlooks, river and flood forecasts, and reservoir inflow forecasts. The capabilities of NWS Hydrologic Services were recently expanded through the implementation of the Community Hydrologic Prediction System (CHPS) at the 13 NWS RFCs. CHPS is the RFC operational framework, allowing for broad systems interoperability to support new water resources-related forecasts.

NCEP consists of nine national centers that provide a backbone of national expertise for both forecast and numerical guidance. The NCEP Storm Prediction Center (SPC) provides forecasts and watches for severe thunderstorms and tornadoes over the contiguous United States. The SPC also monitors heavy rain, heavy snow, and provides national outlooks on fire weather potential. The National Hurricane Center (NHC) provides forecasts of tropical weather systems and issues watches and warnings for the U.S. and surrounding areas for systems in the tropical North Atlantic, Caribbean, Gulf of Mexico and Eastern Pacific and provides educational outreach and guidance for the international community in the region. The Central Pacific Hurricane Center (Honolulu) and the Joint Typhoon Warning Center (Guam) provide additional tropical cyclone coverage for the central and western Pacific, though not part of NCEP. The Weather Prediction Center (WPC) provides analyses and forecast products with a focus on precipitation amount and type, winter precipitation, model diagnostics, surface pressure, frontal analysis and forecast products for the medium range (days 3-7). The Ocean Prediction Center (OPC), along with NHC, issues marine forecasts for the Atlantic and Pacific oceans from the equator to the northern polar regions. The Aviation Weather Center (AWC) provides aviation warnings and forecasts of hazardous flight conditions at all levels within domestic and international air space. The Climate Prediction Center (CPC) provides assessments and forecasts of the impact of short-term climate variability, emphasizing enhanced risks of weather-related extreme events for use in mitigating losses and maximizing economic gain. The Space Weather Prediction Center (SWPC) provides real-time monitoring and forecasting of solar and geophysical events which impact satellites, power grids, communications, navigation, and many other technological systems. The Environmental Modeling Center (EMC) develops and improves numerical weather, climate, hydrological and ocean prediction through applied research in data analysis, modeling and product development. NCEP Central Operations (NCO) sustains and executes the operational suite of numerical environmental analysis and forecast models and prepares NCEP products for dissemination. They also process and manage the flow of data and products to and from the NCEP centers, partners, and customers.

There are other specialized service centers within the NWS, such as the Pacific Tsunami Warning Center (PTWC) and the National Tsunami Warning Center (NTWC), which use data from a worldwide seismic network and Deep-ocean Assessment and Reporting of Tsunamis (DART) buoys located throughout the Pacific Ocean, Atlantic Ocean, and Caribbean Sea to issue tsunami watches and warnings for all U.S. and many international coastal communities. The National Data Buoy Center (NDBC) is responsible for the deployment and maintenance of coastal and ocean buoys and sensors that are used for marine forecasts and analysis of ocean-based storms. The NWS Volcanic Ash Advisory Center (VAAC) located in Anchorage, Alaska, provides worldwide warnings and advisories to aviation interests regarding airborne volcanic ash hazards (see further description of the global system of VAACs in the section on volcanic ash in Aviation Services.)

NWS forecasters support several health-related programs such as Heat Health, and the Ultraviolet Index. Heat Health Watch Warning Systems (HHWWS) have been developed for select cities to provide advance notice of excessive heat events that produce the greatest number of weather-related deaths.

NWS' suite of products and services enable core partners' decisions when weather, water, or climate has a direct impact on the protection of lives and livelihoods. As NWS works to fulfill the vision of a Weather-Ready Nation (WRN) its diverse portfolio of service capabilities will become more focused on a framework approach that will enable the agency to provide the foundational information and Impact-Based Decision Support Services (IDSS) needed. It is at this juncture where NWS' highly skilled workforce provides significant value to achieve its mission. IDSS is the foundational concept of NWS' WRN. Rather than developing and transmitting a suite of products at fixed times during the day and expecting stakeholders to fully understand and take appropriate action; IDSS changes the paradigm so information users drive the update frequency and value-added meaning of the product. IDSS also provides greater flexibility for forecasters to work with key governmental partners and even embed within their emergency operations centers to give first hand support to enhance decision making and public safety.

In 2015, NOAA proposes to restructure the NWS budget to function. This restructure will restore budget credibility and transparency by strengthening internal controls and customer service while improving coordination and collaboration among activities that serve NWS' overall mission. Guided by the recent National Academy of Sciences, "Becoming Second to None," and the National Academy of Public Administration (NAPA), "Forecast for the Future: Assuring the Capacity of the NWS" reports, NWS will continue to evolve and improve its weather, water, and climate products and services to enhance performance. In FY 2015 NWS continues making the bold vision for the WRN, a reality.

NOAA/National Environmental Satellite, Data, and Information Service

NOAA's National Environmental Satellite, Data, and Information Service (NESDIS) operates the Nation's civil operational environmental satellite system, making constant observations of the Earth and its oceans and atmosphere. Satellite observations are collected, processed, and used to develop weather, climate, ocean, and other environmental products, services, and long-term data records that benefit the American public.

NOAA's main satellite constellations are the Geostationary Operational Environmental Satellites (GOES) and Polar-orbiting Operational Environmental Satellites (POES). These two systems provide the U.S. component of a joint environmental monitoring system in partnership with the European Organisation for the Exploitation of Meteorological Satellites (EUMETSAT). NOAA also operates the Suomi National Polar Partnership (S-NPP) mission, a NOAA-NASA mission which is the precursor to the future NOAA Joint Polar Satellite System (JPSS). S-NPP became NOAA's primary afternoon orbit satellite on May 1, 2014. On behalf of the Department of Defense (DoD), NESDIS also operates the Defense Meteorological Satellite Program (DMSP) spacecraft, part of the military's sixth generation of weather satellites. NOAA also operates Jason-2, a joint U.S.-European specialized polar orbiting satellite that measures ocean surface topography. In addition, on behalf of the Department of Commerce, NESDIS licenses the operation of commercial remote-sensing land-imaging satellites. NESDIS also provides long-term stewardship of environmental data, managing the world's largest collection of climatic, geophysical, and oceanographic data derived from both in situ and space-based systems.

Polar-orbiting Operational Environmental Satellites (POES)

POES circle the Earth in a nearly north-south orbit at an altitude of 517 miles, passing close to both poles. These satellites ensure observational data for any region of Earth are no more than six hours old. Data from POES support global weather forecasting models, long-term global climate change research, and hazard detection and mitigation. NESDIS operates polar orbiters. In addition to S-NPP, NOAA-15, NOAA-18, and NOAA-19 satellites continue to transmit data as back-up and secondary satellites. NESDIS also manages the command, control, and communications functions of DoD's DMSP. MetOp-B, operated by EUMETSAT and launched in 2012, provides operational environmental monitoring in the mid-morning orbit. MetOP-A provides secondary coverage. The MetOp satellites carry three instruments provided by NOAA.

In addition, NOAA operates Jason-2, a joint U.S.-European specialized polar-orbiting satellite. This spacecraft's mission is to provide physical data of the ocean surface, including ocean surface altimetry, sea wave height, sea wave period, surface roughness, and other measurements.

This mission, also called the Ocean Surface Topography Mission (OSTM) is a follow-on to the successful Jason-1 mission developed by the French space agency—Centre National d'Etudes Spatiales (CNES)—and the National Aeronautics and Space Administration (NASA).

Joint Polar Satellite System (JPSS)

JPSS is the next generation of U.S. civil polar-orbiting satellites, with the first launch scheduled for FY 2017 and S-NPP providing coverage as a bridge mission between POES and JPSS.

Because weather forecasters rely on data from NOAA's current on-orbit assets, efforts to develop the first of the JPSS platforms will focus on ensuring both short and long-term continuity in crucial weather forecasting, environmental monitoring, and climate monitoring data. S-NPP was successfully launched from Vandenberg Air Force Base, CA, on October 28, 2011. It is intended to provide operational support, serve as a prototype for the next-generation of NOAA's polar-orbiting satellites, and provide continuity for NASA's Earth Observing System mission. In 2013, operation was transferred to NOAA's Office of Satellite Product Operations (OSPO). In FY 2014, instruments for the next satellite to be launched, JPSS-1, will start to be delivered, and the JPSS-1 spacecraft will begin integration. The satellite is scheduled to be launched no later than the second quarter of FY 2017. JPSS continues to enhance the ground segment of the program to support S-NPP operations as well as to prepare for JPSS-1. NOAA-NESDIS maintains a strong partnership with EUMETSAT, which will continue to be a cornerstone of a joint polar-orbiting constellation and will ensure NOAA-NESDIS' ability to provide continuous measurements. JPSS-1 will host the same instrument complement as S-NPP, with the exception of the OMPS-Limb. In FY 2015 JPSS expects to have the JPSS-2 instruments on contract and the JPSS-2 spacecraft contract awarded. JPSS-2 is planned for launch in FY 2022.

Geostationary Operational Environmental Satellites (GOES)

GOES spacecraft orbit Earth in a geosynchronous orbit at an altitude of 22,300 miles, which means they orbit the equatorial plane of the Earth at a speed matching the Earth's rotation. The GOES system provides continuous observations of environmental conditions of North, Central, and South America and the surrounding oceans. These spacecraft provide data critical for fast, accurate weather forecasts and warnings, detection of solar storm activity, and relay of distress signals from emergency beacons. They provide nearly continuous monitoring necessary for effective, detailed, and extensive weather forecasting, prediction, and environmental monitoring.

There are two operational geostationary satellites for the North American region, GOES-East at 75°W and GOES-West, at 135°W, plus an on-orbit spare satellite at 105°W. Each operational satellite continuously views nearly one-third of the Earth's surface. GOES-P was launched on March 4, 2010, and was renamed GOES-15 once it was successfully on orbit. GOES-15 is the third and last in the current series of NOAA geostationary satellites. The current constellation consists of GOES-15 (West), GOES-13 (East), and GOES-14 (on-orbit spare). GOES-12, an older satellite operational as GOES-East starting in 2003 and moved to provide coverage for South America in 2010, was decommissioned in August 2013.

GOES Series R

Geostationary satellites remain the weather sentinels for NOAA—tracking hurricanes, severe storms, clouds, and ocean features. The next-generation geostationary satellite series is called the Geostationary Operational Environment Satellite–R Series (GOES-R). The advanced spacecraft and instrument technology used on the GOES-R Series will result in more timely and accurate weather forecasts. It will improve support for the detection and observations of meteorological phenomena and directly benefit public safety, protection of property, and ultimately, economic health and development. The GOES-R Advanced Baseline Imager (ABI) will scan its field of view on the Earth nearly five times faster, with more than three times the spectral coverage and four times the spatial resolution of the current GOES. GOES-R will provide users such as meteorologists and government agencies

Geostationary Operational Environment Satellite–R Series (GOES-R). Image courtesy GOES-R Series Program and Lockheed Martin Space Systems Company

with approximately 60 times the amount of data currently provided. A new instrument, the Geostationary Lightning Mapper (GLM), will be capable of measuring total lightning activity continuously day and night with a horizontal resolution on the order of 10 km and detection efficiency ranging between 70-90%.

To ensure user readiness, forecasters and other users must have access to prototype advanced products within their operational environment well before launch. The GOES-R Proving Ground (http://www.goes-r.gov/users/proving-ground.html) engages the National Weather Service (NWS) forecast, watch, and warning community and other agency users in pre-operational demonstrations of the new and advanced capabilities that will be available from GOES-R, as compared to the current GOES constellation. Examples of the advanced products include improvements on: volcanic ash detection, lightning detection, dust and aerosol detection, synthetic cloud and moisture imagery, as well as one-minute interval rapid scan imagery. A key component of the GOES-R Proving Ground is the two-way interaction between the researchers who introduce new products and techniques and the forecasters who then provide feedback and ideas for improvements that can best be incorporated into NOAA's integrated observing and analysis operations.

GOES-R is a collaborative development and acquisition effort between NOAA and NASA. In FY 2014, the GOES-R Series Program plans to continue development of the GOES-R spacecraft to meet phased instrument delivery milestones which began in 2013. Development of the Core

Ground and Antenna Systems will also continue. The GOES-R satellite has planned launch readiness for early FY 2016.

NOAA/NESDIS Data Centers

National Climatic Data Center

The National Climatic Data Center is the largest climate data center in the world. See Climate Services for additional details.

National Geophysical Data Center (NGDC)

NOAA's National Geophysical Data Center (www.ngdc.noaa.gov), located in Boulder CO, provides long-term scientific data stewardship for the Nation's geophysical data, ensuring quality, integrity, and accessibility. NGDC builds and maintains long-term archives of scientific data, with a special emphasis on scientific stewardship of data acquired by NOAA observing systems. NGDC provides stewardship, products, and services for data from the sun to Earth's sea floor, including Earth observations from space. NGDC is one of two operational sites for NOAA's Comprehensive Large Array-data Stewardship System (CLASS), which provides a permanent archive for all NOAA satellite observations.

Key NGDC data sets include:

- Global bathymetry
- Long-term solar observations
- Marine geophysical data sets
- Geological hazards
- In situ and remotely-sensed space environment
- Global Nighttime Lights imagery

NGDC plays an integral role in the Nation's research enterprise, at the same time providing public domain data to a wide group of users. NGDC works with contributors of scientific data to prepare documented, reliable data sets and currently maintains more than 850 digital and analog data sets. NGDC also continually develops data management programs that reflect the changing world of geophysics in an era of electronic data access.

NGDC's unique capabilities have attracted other mission-related functions to the Center. NGDC is responsible for the development and maintenance of the World Magnetic Model for DOD. NGDC is the lead for data management and integration to the U.S. Extended Continental Shelf Project under the United Nations Convention on the Law of the Sea. NGDC operates the World Data Service for Geophysics for the International Council of Science under the auspices of the U.S. National Academy of Sciences. NGDC is active in many international programs, offering scientists around the world access to global databases through international exchange.

Natural Hazards Coastal Inundation Modeling and Mapping. NGDC plays an integral role in the Nation's research enterprise, at the same time providing public domain data to a wide group of users. NGDC works with contributors of scientific data to prepare documented, reliable data sets and currently maintains more than 850 digital and analog data sets. NGDC also continually develops data management programs that reflect the changing world of geophysics in an era of electronic data access.

NGDC's unique capabilities have attracted other mission-related functions to the Center. NGDC is responsible for the development and maintenance of the World Magnetic Model for DOD. NGDC is the lead for data management and integration to the U.S. Extended Continental Shelf Project under the United Nations Convention on the Law of the Sea. NGDC operates the World Data Service for Geophysics for the International Council of Science under the auspices of the U.S. National Academy of Sciences. NGDC is active in many international programs, offering scientists around the world access to global databases through international exchange.

Geomagnetic Field Modeling For Improved Navigation. The NGDC geomagnetism group develops and produces magnetic field models for navigation and pointing, which are used in a multitude of defense and civilian applications. Production of the World Magnetic Model, the standard magnetic model for DOD and the North Atlantic Treaty Organization, is sponsored by the National Geospatial-Intelligence Agency. The geomagnetism group also leads the production and distribution of the International Geomagnetic Reference Field. These main magnetic field models represent approximately 90 percent of the magnetic field that influences a compass on or near the surface of Earth. NGDC continues to develop improved magnetic models, addressing additional magnetic influences affecting navigation by land, sea, and air. Making use of its extensive holdings of satellite, airborne, and marine magnetic data, NGDC is developing new high-resolution magnetic field models. Recent products include animations of the model results for the change in the magnetic field from 1590 to 2010, a three-arc-minute World Digital Magnetic Anomaly Map, and the extended magnetic reference model to spherical harmonic degree 720 (NGDC-720). The NGDC-720 model corresponds to a 15-arc-minute model resolution. See http://www.ngdc.noaa.gov/geomag/EMM/index.html for more information

National Oceanographic Data Center (NODC)

For over 50 years, NODC (www.nodc.noaa.gov) has served the Nation with unmatched expertise in the scientific stewardship of marine data and information. NODC manages the world's largest collection of freely available oceanographic data, provides a record of Earth's changing environment, and supports numerous research and operational applications. NODC maintains and updates a national ocean information archive with environmental data acquired from national and international activities. NODC's ocean archive is capable of archiving any type of oceanographic data in the world, and their mission statement is: "To provide scientific stewardship of marine data and information." This information includes physical, biological, and chemical measurements derived from *in situ* oceanographic observations, satellite remote sensing of the oceans, and ocean model simulations, including water temperature data dating back to the late 1700s and measuring thousands of meters deep. The U.S. NODC is also a regular member of the International Council for Science (ICSU) World Data System.

Key NODC data sets include:

- Global Ocean Heat and Salt Content
- Jason-2 Ocean Surface Topography
- Group for High Resolution Sea Surface Temperatures
- Regional Climatologies for the Gulf of Mexico; and Arctic Ocean and surrounding seas
- World Ocean Atlas and World Ocean Database
- Argo Free Drifting Profiling Floats
- NOAA Marine Environmental Buoy Database
- Pathfinder Climate Data Records for sea surface temperature

In addition, NODC has implemented numerous interoperable data technologies to enhance the discovery, understanding, and use of the vast quantities of oceanographic data in the NODC archives. Combined, these technologies enable NODC to provide access to its data holdings and products through some of the commonly-used standardized Web services.

NODC also manages the NOAA Central Library, in Silver Spring, MD and the National Coastal Data Development Center (NCDDC), located in Stennis Space Center, MS. NCDDC provides coastal data management expertise including metadata training, while the NOAA Central Library provides scientific journals, rare books, historical photo collections, and maps. Working cooperatively, the data center divisions provide products and services to scientists, engineers, policy makers, and other users in the United States and around the world.

Library and Information Services Division The NOAA Central Library, under the auspices of NODC, is located in Silver Spring, Maryland with satellite libraries in Miami, Florida, Seattle, Washington, and College Park, Maryland. They provide information and research support to NOAA staff and the public. The library also networks with over 30 NOAA libraries across the nation. Disciplines covered include weather and atmospheric sciences, oceanography, ocean engineering, nautical charting, marine ecology, marine resources, ecosystems, coastal studies, aeronomy, geodesy, cartography, mathematics and statistics. See www.lib.noaa.gov for more information.

National Coastal Data Development Center (NCDDC) In 2000, NOAA established the National Coastal Data Development Center (NCDDC) as the coastal division of the NODC at Stennis Space Center, Mississippi to provide a national capability for the "archive of, and access to the long-term coastal data record." Since its inception, NCDDC has been tasked to work closely with coastal data providers to ensure applicable data standards are employed on all data and associated metadata delivered through NCDDC as well to maintain a catalog of available coastal data and provide access to that data. NCDDC was chartered to broadly work with coastal data providers including Federal, State, local agencies, academic institutions, non-profit organizations and the private sector to create a "unified", long-term collection of coastal data. Most significantly, NCDDC is tasked to have strong science-based capacity and to produce retrospective analyses and trend information to help form the basis of environmental predictions and public policy. See www.ncddc.noaa.gov for more information.

Marine Data Stewardship and Ocean Climate Laboratory Divisions The NODC Marine Data Stewardship Division (MDSD) and Ocean Climate Laboratory archive and distribute oceanographic data and develop products to better understand marine ecosystem dynamics for improved near real-time decision support and understanding of the processes affecting climate change. Division personnel ingest a wide variety of data from *in situ* ocean observation programs and satellite platforms; these data are managed, preserved, disseminated, and utilized in products that improve the understanding of the world oceans.

The goal of the Marine Data Stewardship Division and the Ocean Climate Laboratory is to provide open access to archived oceanographic data and products to better understand oceanic processes for improved decision support. The Division focuses on end-to-end data stewardship, preserving and archiving oceanographic data and developing value-added data products. This includes: (1) Ensuring the data are properly archived and easily accessed by a wide range of users; (2) Generating authoritative long-term records; and (3) Using those climate data records to characterize and place the current state of the environment in its proper historical perspective. The Division improves the quality and utility of NODC's oceanographic data archives by using the data to perform scientific analyses, develop improved ocean climatologies for annual, seasonal, and monthly compositing periods, investigate interannual-to-decadal ocean climate variability using historical oceanographic data, build scientifically, quality-controlled global oceanographic databases, and facilitate international exchange of oceanographic data.

NOAA/Office of Marine and Aviation Operations

The NOAA Office of Marine and Aviation Operations (OMAO) operates a fleet of survey ships and aircraft to support NOAA's mission goals. NOAA's ship fleet includes oceanographic and atmospheric research vessels. The NOAA aircraft fleet includes aircraft that collect environmental and geographic data essential to NOAA hurricane and other severe weather and atmospheric research and aircraft that conduct aerial surveys to forecast water supply and flooding potential from snow melt.

NOAA vessels make weather and ocean observations in the marine environment. Over 50,000 automated observations are submitted per year through the World Meteorological Organization's (WMO) Voluntary Observing Ships scheme. NOAA vessels also support NOAA's NDBC in recovery of buoys that have been disabled or gone adrift.

NOAA supports a broad range of meteorological activities and projects with its fleet of aircraft, based at MacDill Air Force Base in Tampa, Florida. Three of its nine aircraft are dedicated to this purpose throughout the year, providing valuable information to NOAA and the Nation. The NOAA Gulfstream, G-IV (SP) (N49RF), provides scientists with a platform for the investigation of processes in the upper troposphere and lower stratosphere. With an operating ceiling of 45,000 ft, the G-IV is a critical tool for obtaining the data necessary to improve hurricane track forecasts and for research leading to improvements in hurricane intensity forecasts.

The NOAA G-IV annually supports hurricane synoptic surveillance missions; the aircraft flies in the environment surrounding the storm at a high altitude, releasing Global Positioning System (GPS)–equipped dropsondes at preselected locations. The data from these vertical atmospheric soundings are received, processed, and transmitted from the aircraft to a NOAA NCEP computer site, where they are incorporated into computer models of hurricanes to improve hurricane track forecasts. Each dropsonde directly measure temperature, pressure, and humidity at the rate of two samples per second as it falls through the atmosphere to the surface and computes wind speed and wind direction at a rate of four samples per second, using a full-up GPS receiver. Recent estimates of the improvement in hurricane track predictions utilizing this technology show an improvement of between 20 and 30 percent, which represents a savings of $10 million or more per hurricane in warning and preparedness costs.

Two NOAA WP-3D Lockheed Orion aircraft (N42RF and N43RF) support NOAA's atmospheric and oceanographic research, as well as its tropical storm and hurricane reconnaissance operations. The aircraft's research and navigation systems provide detailed spatial and temporal observations of a wide range of atmospheric and oceanic parameters. NOAA's Aircraft Operations Center (AOC) develops and calibrates specialized instruments, installs and integrates user-supplied instrumentation into the aircraft and data network, and processes data for immediate satellite transmission or future analysis. The NOAA WP-3D aircraft, while executing the complex flight patterns required for hurricane research, also provide storm data to the NHC in real time, transmitting flight level data and GPS dropsonde messages, as well as radar images transmitted via their multiple aircraft-satellite data links. The stepped frequency microwave radiometers (SFMR) on the NOAA WP-3D are used to map the surface wind fields in and around hurricanes and tropical storms. Real-time surface wind speed maps are critical to providing more accurate forecasts of the extent of hurricane and tropical storm force winds.

During each hurricane season, the two NOAA WP-3Ds support several major research experiments in support of NOAA's Hurricane Research Division (HRD) of NOAA's Atlantic Oceanographic and Meteorological Laboratories in its Intensity Forecast Experiment (IFEX), an ongoing program studying hurricane genesis, rapid intensification, and other related experiments. A promising part of this research effort is the collaboration with NOAA's EMC in a program to obtain three-dimensional horizontal wind fields in developing tropical systems and hurricanes, utilizing the Tail Doppler Radars of the WP-3D aircraft. The objective of this effort is to obtain data that can be assimilated into the Hurricane Weather Research and Forecasting (HWRF) model for the purpose of improving hurricane intensity forecasts. Extensive descriptions of the various hurricane research experiments may be found in the HRD Field Program Plan for 2012.

The NOAA WP-3D aircraft annually participate in both summer and winter operations supporting the NESDIS satellite validation program. Operating in regions of high winds and heavy precipitation, one of the WP-3Ds, equipped with microwave scatterometers and radiometers, provides under-flight validation of the ocean surface wind vectors sensed by the European ASCAT and Indian OceanWind2 satellites. Traditional venues for these satellite validation operations are Alaska or Newfoundland in the winter and the Atlantic and Caribbean regions during the summer hurricane season.

These versatile aircraft also support international and interagency projects. Most recently the N43RF aircraft supported the FY 2012 Dynamics of the Madden-Julian Oscillation (DYNAMO) project from Diego Garcia. This project collected coordinated measurements in the atmosphere and ocean to capture process involved with the intra-seasonal variability such as the Madden-

Julian Oscillation. The aircraft data will be used for evaluation and improvement of coupled atmosphere-ocean models.

National Aeronautics and Space Administration Basic Meteorological Services

The National Aeronautics and Space Administration (NASA) is a long-term partner with the National Oceanic and Atmospheric Administration (NOAA) for building and launching the U.S. Polar Operational Environmental Satellite (POES) and Geostationary Operational Environmental Satellite (GOES), civilian weather satellites, under reimbursable arrangements. Additionally, after the termination of the National Polar-orbiting Operational Environmental Satellite System (NPOESS) Program, NASA took on the role of the acquisition agency for the NOAA-funded *Joint Polar Satellite System* (JPSS) Program. Several missions, including the Ocean Surface Topography Mission (OSTM) and Suomi National Polar-orbiting Partnership (S-NPP), are NASA research missions that were developed in partnership with NOAA and are now being operated by NOAA as part of the operational weather system. Jason 3, which is under development by NOAA, is a follow-on to the TOPEX/Poseidon, Jason-1, OSTM series. In the President's FY 2015 budget request, NASA assumes responsibility for the sustained climate measurements such as total solar irradiance, vertical ozone profiles and Earth's radiation budget in the JPSS-2 (2021) timeframe.

SUPPORTING RESEARCH PROGRAMS AND PROJECTS

Interagency Collaborative Research Programs and Projects

The National Earth System Prediction Capability

The National Earth System Prediction Capability (ESPC) is an ongoing collaborative project among the National Oceanic and Atmospheric Administration (NOAA), U.S. Navy, U.S. Air Force, Dept. of Energy (DOE), National Aeronautics and Space Administration (NASA), and the National Science Foundation (NSF).

The ESPC program will provide a more accurate global ocean and atmospheric forecast system with longer skillful forecast times at synoptic, sub-seasonal, seasonal, and inter-annual scales through integrating and coupling global atmosphere, ocean, ice, land and near-space forecast models into an operational suite of prediction systems that reduce errors relative to current modeling approaches. Additionally, it will continue to develop a national common modeling architecture to improve cross-agency collaboration, emphasizing more efficient computational architectures to allow real-time high-resolution operational prediction at these extended time scales. In FY 2015, predictability demonstration plans will be refined, and science workshops and early benchmark testing will be conducted. The long range program goal is to advance skillful forecasts (relative to averaged climatology) from the operational capability, currently 7-10 days, to 30 days and longer. Additional information can be found at: www.espc.oar.noaa.gov.

The United States Weather Research Program

The United States Weather Research Program (USWRP) is an interagency program for weather research and the transition of research to applications. The member agencies include NOAA (lead), NASA, the National Science Foundation (NSF), the Navy, and the Air Force. NOAA's Office of Oceanic and Atmospheric Research (OAR), through its Office of Weather and Air Quality, helps plan NOAA USWRP priorities, implements the program, and monitors progress. In FY 2015, the USWRP will include the following projects:

The Joint Hurricane Testbed (JHT). The NOAA USWRP provides support for the JHT, to evaluate and transfer into operations, promising hurricane observational methods, analysis and forecast applications developed in the research community. JHT also establishes and maintains an infrastructure to facilitate the modification and transfer of research applications into the operational computing, communication, and display environment. This testbed is located at NCEP's National Hurricane Center in Miami, Florida. USWRP will continue to support the JHT in FY 2015.

The Hydrometeorological Testbed (HMT). The NOAA USWRP has invested in research and transition of research to applications to improve quantitative precipitation forecasts through NOAA's HMT (http://hmt.noaa.gov/), led by the Physical Sciences Division (PSD) of NOAA/OAR/ESRL. The HMT conducts research and analyses to improve the understanding of extreme precipitation producing atmospheric phenomena and the development of prototype tools for flood and extreme precipitation forecasting. This support will continue as the HMT expands from the West Coast to a recently established field program in the southeast United States. This project is operated in collaboration with OAR's ESRL, National Severe Storms Laboratory (NSSL), and NWS's Weather Prediction Center. In 2014, the HMT at the Weather Prediction Center (HMT-WPC) conducted its fourth annual Winter Weather Experiment, focused on emerging short range microphysics-based snowfall forecasting techniques, as well as extending winter weather forecasts as much as a week in advance.

The Hazardous Weather Testbed (HWT). The NOAA USWRP provides support for the development, testing, and evaluation of new severe weather forecast and warning techniques for use by forecasters in the National Weather Service (NWS). The HWT also develops new decision support tools and allows for direct interaction between researchers and forecasters.

Understanding Air Chemistry. The NOAA USWRP supports projects on air quality modeling improvements that benefit operational air quality forecasting. New techniques for using air chemistry data in the models are developed and tested in collaboration with NOAA's Air Resources Laboratory (ARL) and the Earth Systems Research Laboratory's (ESRL) Physical Sciences Division and Global Sciences Divisions. This research contributes to improving National Weather Service smoke, dust, and volcanic ash predictions so people can act to limit the adverse effects on human, surface transportation, and aviation.

Social Sciences for a Weather Ready Nation. The NOAA USWRP funds social science research that enables NOAA to test research results that improve communication during dangerous weather situations. Topics addressed range from assessing NWS's flood forecasting products to understanding how the public understands tornado warning uncertainty to improve how the NWS communicates tornado warnings.

Joint Center for Satellite Data Assimilation

The Joint Center for Satellite Data Assimilation (JCSDA) is a USA partnership between NOAA, NASA, the U.S. Navy, and the U.S. Air Force dedicated to developing and improving our ability to exploit satellite data more effectively in environmental analyses and forecast models. The JCSDA is a collaborative effort that allows the development and testing of techniques required to assimilate the billions of satellite observations available daily to be shared by several agencies. This effort would otherwise be duplicated across the agencies.

The JCSDA has the following goals: (1) Reduce the average time for operational implementation of data from new satellite sensors to less than one year, (2) Increase the use and impact of current satellite data in numerical weather prediction models, (3) Advance common data assimilation infrastructure and methods, and (4) Assess the impacts of data from advanced satellite sensors on environmental predictions. By meeting these goals, the JCSDA achieves its objective of maximizing the nation's return on its investment in global observing systems and modeling, and providing improvements to the U.S. forecasting system (civilian and military), which in turn contributes to protecting citizens lives and property and the nation's global interests.

Wind Forecast Improvement Project

DOE and NOAA have recently completed positive efforts to improve wind forecasting for wind power and utility system operators through the Wind Forecasting Improvement Project (WFIP 1.0). The project targeted the Upper Midwest and West Texas, increasing the number of atmospheric observations in the region as input to the mesoscale models improved initial conditions and analyzing the resulting improvement in wind plant forecasts. The primary objective was to improve the shorter term forecast – thus lowering the risk to the wind industry on 0-6 hour forecasts (mainly ramps) and day-ahead forecasts. Wind plant operators utilize the foundation forecasts as input for individual wind plant operations. The benefits of these efforts not only include better

Wind turbines located in the North Site of the WFIP 1.0 Project Area. Photo Courtesy of WindLogics.

optimization of wind plant operations in the short term, but also reduction in spinning reserves in the day-ahead period, and valuable information to aid efficient wind plant siting and design for yearly and wind plant operations through the lifetime of the turbines. Thus improving the foundational model forecasts of wind plant inflow will lower cost to utilities of wind-derived electric power for all users, helping to enable wind as a cost-competitive energy alternative and leading to a larger penetration into the Nation's energy portfolio.

WFIP 1.0 demonstrated that, with added observational systems, capturing detailed data on wind speed and direction in the lower planetary boundary layer, initial conditions of the models could be improved leading to greater accuracy in predicting short-term ramp events with verified and consistent improvements within the 0-14 hour time frame. The observations included in the WFIP were wind profiling radars, sodars, lidars, and wind measurements from hub height anemometers and tall towers.

WFIP 1.0 was a success for developing public-private partnerships towards a common end goal – resolving a scientific challenge. For WFIP 1.0, wind energy companies shared meteorological observations collected on tall towers and wind turbines on wind farms with NOAA, who used the data for model verification and validation, as well as assimilation into its advanced, rapidly refreshed weather models. NOAA and DOE have for several years been encouraging renewable energy companies to sign data sharing agreements under which NOAA will maintain the proprietary nature of such data, while using them for research and operational weather forecasts to reduce risk for the wind industry.

DOE and NOAA will build on these efforts through a new effort, the Wind Forecasting Improvement Project in Complex Terrain (WFIP 2.0), which will develop further improvements to foundational atmospheric model physics in predicting wind ramp and day-ahead events for areas of complex terrain. WFIP 2.0 will focus on improving the physical understanding of atmospheric processes which directly impact the wind industry forecasts and incorporating the new understanding into the foundational weather forecasting models. The awardee, in partnership with DOE, NOAA, and a Balancing Authority (BA), will conduct a field campaign in an area of complex terrain to assess how physical processes alter wind speeds at hub heights.

From this research, the team will work to develop physical modeling schemes or atmospheric theories that can be incorporated in foundational weather models to improve wind forecasting. DOE headquarters will contribute funding for WFIP 2.0 and scientific expertise from four of national research laboratories, while NOAA will contribute scientific expertise in collecting atmospheric data and in executing weather models and forecasts. The FOA winner for WFIP 2 will be an integral part of the team and work in all these areas where possible. The WFIP projects are the first large-scale projects in a growing number of joint agency efforts under a Memorandum of Understanding on "Weather-dependent and Oceanic Renewable Energy Resources" signed by NOAA and DOE in January 2011. The agreement set up a framework for NOAA and DOE to work together on renewable energy modeling and forecasting.

NOAA/National Weather Service/Environmental Modeling Center

Continually improving the accuracy, timeliness, and accessibility to prediction services is a result of research and development (R&D) within the NWS, at other NOAA offices such as OAR, and externally from universities and private corporations. NCEP/EMC uses advanced modeling methods developed both internally and cooperatively with universities, the international scientific community, NESDIS, OAR, and other government agencies. EMC is a partner in the JCSDA, which is designed to accelerate the use of research and operational satellite data in NCEP operational models. The EMC conducts applied research and technology transfers and publishes results in various media for dissemination to the world meteorological, oceanographic, and climate communities. EMC also participates in ongoing interactive research programs such as NOAA's Hurricane Forecast Improvement Project (HFIP) and the WRF community model. Furthermore, EMC efforts in collaborative development resulted in improvements to mesoscale and global models, as well as advances in hurricane track forecasts, climate forecasts, and air quality forecasts.

NOAA/Hurricane Forecast Improvement Project (HFIP)

HFIP contributes significantly to NOAA's forecast services through improved hurricane forecast science and technology. HFIP provides the unifying organizational infrastructure and funding for NOAA and other agencies to coordinate the hurricane research needed to significantly improve guidance for hurricane track, intensity, and storm surge forecasts and accelerate the transition from research to operations. HFIP is in the fifth year of its mission to reduce track and intensity errors, increase the probability of detection and decrease the false alarm ratio for rapid intensification and extend the lead-time for hurricane forecasts out to seven days. HFIP approaches meeting those goals by improving existing numerical forecast systems through a regimen of revising, testing, refining and implementing promising technologies. HFIP developed a high-performance computing system housed at the NOAA Skaggs Building in Boulder, Colorado to support the significant computational demands of such an approach. With the computational infrastructure, HFIP focuses multi-organizational activities to research, develop, demonstrate and implement enhanced operational modeling capabilities, dramatically improving the numerical forecast guidance made available to National Hurricane Center (NHC). Participants include Federal laboratories and academic partners. Prior to each hurricane season, NHC reviews and selects a set of enhanced experimental guidance products they will evaluate operationally during the season. HFIP runs an experimental forecast exercise on the high-performance computing system in Boulder to provide the selected products to NHC forecasters.

HFIP's structured approach accelerates bringing promising technologies and techniques from the research community into operations. In particular, HFIP has successfully
* aligned focused research efforts within NOAA and with NOAA's interagency and academic partners;
* established a process to leverage outside research capabilities in support of project objectives (Federally funded grantees working within a community code repository);
* defined and implemented a solution (the seasonal, real-time experimental forecast system) to accelerate research into operational products; and
* established a high performance computing infrastructure and attendant protocols to support research-to-operations activities.

NOAA/National Environmental Satellite, Data, and Information Service

Center for Satellite Applications and Research (STAR)

STAR's mission is to create satellite data products using observations of the land, atmosphere, and ocean and transfer those products from research into routine operations. In addition, STAR supports the assimilation of the data from new satellite instruments into NOAA's numerical prediction models. STAR calibrates the Earth-observing instruments of all NOAA satellites to provide reliable measurements for assessing the current conditions on Earth in a timely manner, predicting changes in conditions, and studying long-term trends in the environment. STAR works to create products that monitor atmospheric, oceanic, and environmental hazards; enhance NOAA's infrastructure for remote sensing; reduce the risks associated with launching new, untested, and very expensive satellites and sensors; and expand its support to satellite data users.

Selected STAR Projects

Improving the Prediction of the Formation and Evolutioon of Derechos and Associated Micreburts. STAR is using a combination of several satellite products (i.e., vertical temperature and humidity lapse rates, convective potential, etc.) throughout the atmosphere to help predict the formation and evolution of strong convective storm winds that occur in derechos and their associated microbursts. A new algorithm, the Microburst Windspeed Potential Index (MWPI), is being tested using current Geostationary Operational Environmental Satellite (GOES) measurements. The GOES MWPI is being evaluated by NWS to assess its ability to improve forecasts of these events and is expected to be in operation by early 2016. Potential applications are assisting airports in planning for closures, rerouting of open ocean transit, and providing early warning to the general public.

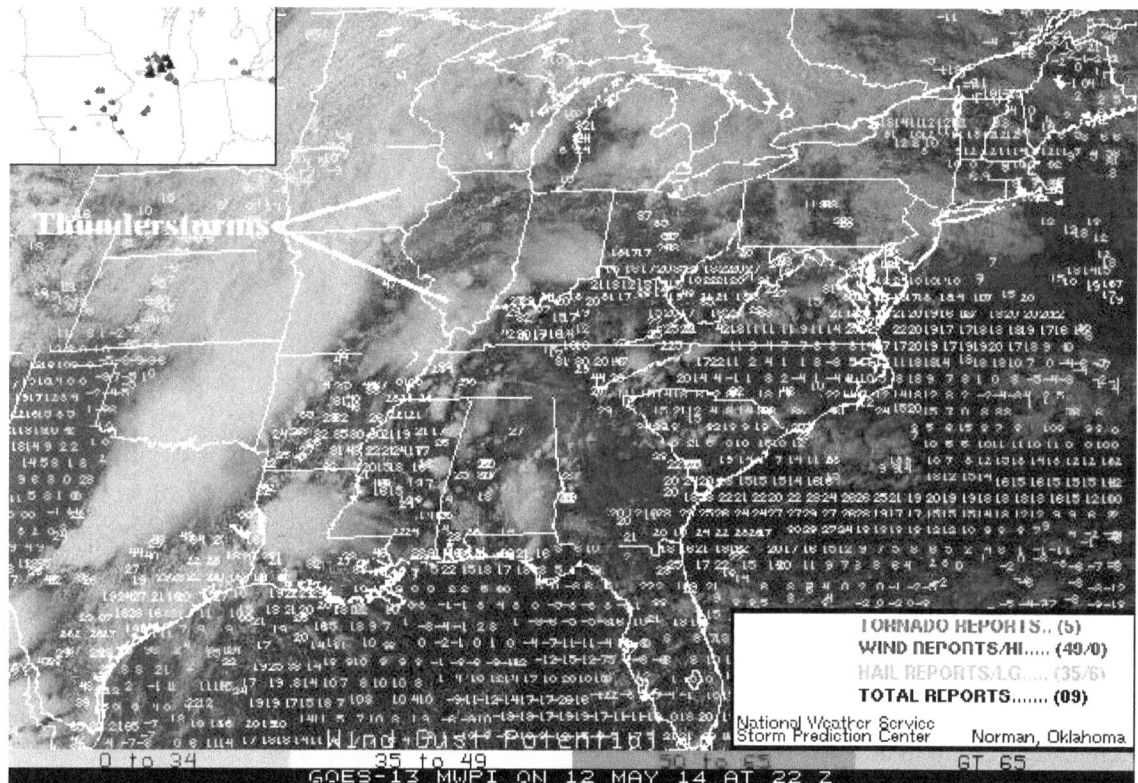

GOES Microburst Windspeed Potential Index (MWPI) product image at 2200 UTC 12 May 2014. The color-coded legend at the bottom of the image associates index values with microburst wind gust potential (in knots). Note the high concentration of severe wind reports (wind > 50 knots) over the upper Midwest in proximity to high index values (> 50, red color). Real-time experimental product images are available on STAR MWPI product page: http://www.star.nesdis.noaa.gov/smcd/opdb/kpryor/mburst/mwpi.html

Precipitation Estimation Using Satellite Observations. Precipitation estimation data from satellites provide a critical supplement to other sources of rainfall information for flood and flash flood forecasting, water resources applications, and a myriad of other uses. In many parts of the world, satellites represent the only reliable source of rainfall information. Data from GOES instruments sensing in the infrared and visible regions of the electromagnetic spectrum provide high-resolution, rapidly-updated rainfall information for hazardous-weather applications. More accurate estimates of rainfall can be derived from microwave-frequency instruments onboard POES and Suomi NPP, but their less frequent updating makes them more suitable for longer term water monitoring. In FY 2015, NOAA will continue to develop the algorithms for the next generation of NOAA's GOES. In addition, modifications to the current generation of algorithms will be explored in search of better ways to serve the users of these data.

Leading the Suomi NPP/JPSS Sensor Data Record (SDR) teams to ensure high levels of data maturity and quality. In FY 2012, STAR scientists successfully led the Suomi/NPP/JPSS Sensor Data Record (SDR) teams, ensuring achieving high levels of SDR data maturity and quality for global weather, water, climate and other environmental applications. The SDRs are the fundamental building blocks for all applications of its satellite data. The maturity and quality of the SDRs are critical for the success of the Suomi NPP/JPSS mission and its timely transition to operations. A high level of Suomi NPP/JPSS satellite data maturity allows weather forecasters to improve their forecast scores, modelers to make better predictions, and users to monitor the earth and disastrous events more accurately to benefit society. The Suomi NPP SDR data have already been tested globally with very positive feedback for numerous applications including weather forecasting, ocean monitoring, fire monitoring, power outage detection, and other applications to benefit the public. Future work in this area will continue to allow users and the general public to gain access to the Suomi NPP data for numerical weather prediction, ocean and environmental monitoring, and climate change detection.

NOAA/Office of Oceanic and Atmospheric Research (OAR)

The OAR Research Laboratories and Programs conduct an integrated program of research, technology development, and services to improve the understanding of the Earth's atmosphere and to describe and predict changes occurring to it. OAR programs, laboratories and their field stations are located across the country. Scientists study, monitor and model atmospheric and other processes that affect weather, air quality, water resources, and climate. By researching Earth's atmospheric system, OAR improves the understanding of weather and intense events such as hurricanes, tornadoes and severe thunderstorms in addition to seasonal droughts, and the next century's climate. OAR integrates these findings into environmental information products for weather support for aviation, hydrometeorology, emergency response, and climate. These products improve critical weather and climate tools used by the private sector, government agencies, decision-makers, and the public.

OAR Laboratories and Programs

Air Resources Laboratory (ARL)

ARL supports basic meteorological products and services through research and development in atmospheric dispersion, air quality, and boundary layer science. Primary research activities and programs are 1) using the ARL-developed Hybrid Single-Particle Lagrangian Integrated Trajectory (HYSPLIT) model to track the transport, dispersion and fate of harmful chemicals and materials; 2) providing the Real-time Environmental Applications and Display sYstem (READY) that allows users to access and display meteorological data and run trajectory and dispersion model products on ARL's web server; 3) improving air quality forecasting models to address health issues associated with ground-level ozone (O_3) and fine particulate matter ($PM_{2.5}$) for NOAA and other researchers; and 4) designing, developing, and deploying meteorological and turbulence instrumentation for use on land and in the air.

In FY15, ARL will:

- Continue to apply the HYSPLIT model to track and forecast, as needed, the release of radioactive material, volcanic ash, dust, and wildfire smoke, and other hazardous materials. ARL will continue to provide HYSPLIT-Mercury assessments for the Great Lakes Restoration Initiative.

- Continue collaborations with NOAA's Office of Response and Restoration which have resulted in the incorporation of the CAMEO/Aloha chemical source term model into the READY HYSPLIT interface. The end product is a one-stop tool for the emergency response community to use to model local to global scale events.

- Continue the development of a five- year re-analysis dataset, generated using the Community Multi-scale Air Quality model, to address O_3 and $PM_{2.5}$ for the benefit of the air quality and climate research communities.

Generate a new forecast product using data from the Joint Polar Satellite System to link ocean isoprene emissions to changes in ozone and aerosols that affect air quality and climate Prepare and deploy the Best Aircraft Turbulence probe as part of the Fluxes of Carbon from an Airborne Laboratory (FOCAL) system) for a study over the permafrost regions of Alaska.

Atlantic Oceanographic and Meteorological Laboratory (AOML)

AOML conducts mission oriented scientific research that seeks to understand the physical, chemical, and biological characteristics and processes of the ocean and atmosphere, both separately and as a coupled system. The Laboratory's research themes (oceans and climate, coastal ecosystems, and hurricanes and tropical meteorology) employ a cross-disciplinary approach, conducted through collaborative interactions with national and international research and environmental forecasting institutions. The work also provides reliable information based on oceanic and atmospheric measurements and analysis.

AOML scientists study atmospheric and oceanic processes that affect tropical cyclones, tornadoes, rainfall (drought), and climate impacts on weather. In addition to research, AOML serves as the operational data center for several programs that directly improve weather services through providing in situ observations from Argo profiling floats, surface drifters and the Pirata Northeast Extension. Research includes improved understanding and application development surrounding physical processes like the El Nino-Southern Oscillation (ENSO) phenomena, North Atlantic Oscillation (NAO), tropical Atlantic variability, the Meridional Overturning Cell, wind-driven gyres in the Atlantic, the global carbon cycle, and other climatically-relevant atmospheric compounds. Of particular note includes AOML research into tropical Cyclones and ocean observing technologies.

Ocean Observing Technologies. In addition to the many weather-related observing systems, AOML is dedicated to improving the development, deployment, and monitoring of oceanographic-related observing technologies and related data in support of meteorology forecasts. As part of this effort, AOML collaborates with the National Weather Service to gather and transmit in real-time marine weather observations from ship-based observations. The AMVER SEAS 2K, a real-time ship and environmental data acquisition and transmission system (developed and support by AOML), reports atmospheric and oceanographic data from Ship of Opportunity Program. AOML also manages the Global Drifter Program, the Argo profiling float data assembly center, and the XBT data assembly center that provide real-time oceanic observations to help improve weather forecast models.

Tropical Cyclone Research. To improve tropical cyclone track and intensity forecast guidance, AOML's Hurricane Research Division (HRD) uses in situ and remotely sensed data collected by aircraft, satellites, and buoys and computer model simulations of the inner core of tropical cyclones and their surrounding environment. An aircraft field program is used to gather datasets representing all stages of the storm's lifecycle; these datasets are used to both support operational needs and provide the cornerstone of HRD's research to advance the understanding and prediction of tropical cyclones. The observations are primarily collected during the hurricane season using two NOAA turboprop aircraft and a Gulfstream-IV jet operated by NOAA's AOC.

HRD's major objectives are the assimilation and evaluation of the new G-IV tail Doppler radar observations to improve forecast guidance, evaluation of the Global Hawk unmanned aircraft systems (UAS) observations to improve forecast guidance in collaboration with the NASA Hurricane Severe Storm Sentinel (HS3) program, and collecting observations specifically to evaluate the atmospheric and oceanic boundary layer physical parameterizations used in NOAA's operational Hurricane Weather Research and Forecasting (HWRF) model. As part of the OAR Sandy Supplemental supported research activities, HRD will evaluate and assess the benefits of using new and emerging observing technologies consisting of aircraft-deployed low altitude UAS and Doppler wind lidar (DWL) profiling systems to improve forecast guidance through evaluation of and improvements to the physical parameterization routines used in HWRF.

HRD is also the principal OAR component of NOAA's Hurricane Forecast Improvement Project (HFIP). Begun in 2009, the HFIP is a unified 10-year NOAA plan to improve the 1-day to 5-day forecasts for tropical cyclone activity, with a focus on rapid intensity change. Researchers at HRD, together with 2-3 researchers at OAR's Geophysical Fluid Dynamics Laboratory (GFDL), 5-6 researchers at ESRL, and 5-6 researchers at NCEP/EMC make up the NOAA core capability for hurricane R&D and play a major role in the HFIP. The *objectives* of the HFIP are to coordinate hurricane-related R&D within the NOAA entities mentioned above and to broaden their interactions with the outside research community in order to address NOAA's operational hurricane forecast needs.

In support of OAR's Sandy Supplemental supported High Impact Weather Prediction Project (HIWPP), HRD is collaborating with NCEP/EMC to accelerate the development of a fully two-way interactive moving nested, multi scale, non-hydrostatic modeling system based on EMC-developed Nonhydrostatic Multiscale Model on B-grid (NMM-B) in the NOAA Environmental Modeling System (NEMS) framework. Multiple HWRF-like nests will be designed to operate at about 3 km resolution within a basin-scale and global version of NMM-B to capture tropical cyclone inner core process and interactions with the large-scale environment, critical for improving track, intensity, rainfall and size predictions.

Earth System Research Laboratory (ESRL)

At NOAA's largest research facility, the Earth System Research Laboratory (ESRL), scientists (from four complementary divisions) study and monitor atmospheric and other processes that affect weather, air quality, water resources, and climate. By researching the dynamic Earth system, ESRL improves understanding of what causes today's weather, next month's hurricanes, next season's drought, and next century's climate. ESRL integrates these findings into environmental information products from weather support for aviation, fire weather, to methods that improve critical weather and climate tools used by the private sector, government agencies, decision-makers, and the public. ESRL research is so comprehensive that it is impossible to give each program or project equal mention. The following highlight a very small portion of products and services provided by ESRL.

ESRL provides long-term, high quality observations of atmospheric composition that are critical for understanding changes in the Earth system. These observations are used worldwide by scientists studying climate, air quality, and ozone-depletion. They form the core of international observing systems and are critical for delivering updates of key products such as CarbonTracker, the NOAA Annual Greenhouse Gas Index (a national climate indicator), methane climatologies, CO_2 trends (daily updates; a national climate indicator), the Ozone-Depleting Gas Index, GlobalView, and other long-term atmospheric composition data sets and products. They are used to set standards for GCOS, GEOSS, and WMO Global Atmosphere Watch. These observations stem from 6 global baseline observatories, ESRL-operated observing networks for stratospheric and tropospheric ozone, aerosols, surface radiation, ozone-depleting gases, and greenhouse gases, and a system of tall towers and aircraft observations across North America.

ESRL carries out process studies which improve understanding of the physics and dynamics associated with weather and other atmospheric phenomena. In FY15, ESRL will be the lead investigator on CalWater2, a broad interagency observing strategy with a major field campaign planned for early 2015 to study atmospheric rivers (ARs) and aerosol-cloud-precipitation interactions along the U.S. West Coast with the goal of improving the forecasts of extreme precipitation events. Facilities consist of the NOAA R/V Ron Brown and 3 research aircraft including the NOAA WP-3D and G-IV and the DOE G-1 in addition to the NOAA HydroMeteorological Network (HMT). An early start campaign was successfully completed in February 2014 with the NOAA G-IV consisting of 12 research flights in AR-related phenomena in the Eastern Pacific. ESRL is active in Arctic research. This year ESRL will submit a report on the NOAA Science Challenge Workshop on the Arctic (held in Boulder in May 2014) that includes a summary of the workshop findings with recommendations for NOAA actions for the next 5-10 years.

ESRL works with NOAA's National Weather Service, other federal agencies, and private companies to transition technologies into operations. In 2014 the number of NOAA patented Science of a Sphere installations surpassed 100 world-wide. One of ESRL's most notable areas of achievement is development and R2A of numerical weather prediction products. Over the past several years, ESRL has developed NWP models such as the Rapid Update Cycle (RUC) and Rapid Refresh (RAP) and transitioned them to NWS operations. The RUC was NOAA's operational North American forecast model for over a decade until it was replaced by the RAP. The RAP is NOAA's current hourly updating model that forecasts weather for North America on a 13km grid. The RAP was made for those who need frequent forecast updates of high-impact weather, including those in the aviation, energy, and severe weather forecasting communities. Version 2 of the RAP was transitioned to NWS operations in 2014 and continues to be updated. The RAP will be replaced by the High Resolution Rapid Refresh (HRRR) model which will represent a significant new level of capability in the numerical weather prediction arena. HRRR is a real-time 3-km resolution, hourly updated, cloud-resolving atmospheric model, initialized by 3-km grids with 3-km radar assimilation.

Over the next year, ESRL in the process of transitioning the following technologies to operations:

- The High Resolution Rapid Refresh numerical weather model is scheduled to be transitioned to NWS operations by September 2014.

- The Meteorological Assimilation Data Ingest System (MADIS) is being transitioned is scheduled for full operations at NWS in 2015. MADIS collects, integrates, quality controls, and distributes high-frequency weather data from both NOAA and non-NOAA sources.

- ESRL signed a Cooperative Research and Development Agreement with the private sector to conduct collaborative research and development pertaining to deriving and improving Global Positioning System/Global Navigation Satellite System neutral atmospheric measurements known as GPS_Met. ESRL intends to transfer the existing NOAA GPS_Met data acquisition and processing system technology by the end of the year.

ESRL develops, operates and manages high-performance computer systems that enable running many of NOAA's experimental weather models. These enhanced technologies speed up computing by 20-30 times and allow massive amounts of data to run at higher spatial resolutions providing increased weather forecast accuracy.

ESRL develops state-of-the-art forecast and decision-support systems to improve collaboration and decision-making between forecasters, emergency managers, and the public. These decision support systems provide decision-makers with additional tools to help them issue and communicate timely and accurate hazard information. Three systems currently in operational use include: Advanced Weather Interactive Processing System (AWIPS), The Meteorological Assimilation Data Ingest System (MADIS), Hazard Services, and Integrated Support for Impacted Air-Traffic Environments (INSITE).

ESRL research provides a scientific basis for efforts to improve air quality, and will continue to monitor long-lived gases in Earth's changing atmosphere to advance understanding of climate-related effects of short-lived pollutants, atmospheric particles, and their interactions with clouds. Additionally ESRL will continue collaboration with USDA/USFS to identify and quantify emissions of gases and aerosols from burning of different biomass materials for improvement of forecast models that provide air quality information to the public and to responders on-scene at wildland fires. They will continue participation in the Tropospheric Ozone Lidar Network (TOLNet) to further understanding of processes in the atmospheric boundary layer that affect ozone and aerosols.

In FY15, researchers will conduct intensive field studies in regions of oil and gas exploration and development to evaluate emissions of volatile organic compounds and greenhouse gases, their impact on downwind air quality, and their impact on climate. These studies include SONGNEX aboard the NOAA P3 aircraft and Fugitive Emissions aboard the NOAA DHC6 aircraft.

Geophysical Fluid Dynamics Laboratory (GFDL)

The Geophysical Fluid Dynamics Laboratory (GFDL) mission is to advance scientific understanding of climate and its natural and anthropogenic variations and impacts, and improve NOAA's predictive capabilities, through the development and use of world-leading computer models of the Earth System. GFDL scientists focus on model-building relevant for society, such as hurricane research, prediction, and seasonal forecasting, and understanding global and regional climate change. GFDL research encompasses the predictability and sensitivity of global and regional climate; the structure, variability, dynamics and interaction of the atmosphere and the ocean; and the ways that the atmosphere and oceans influence, and are influenced by various trace constituents. The scientific work of the Laboratory incorporates a variety of disciplines including meteorology, oceanography, hydrology, classical physics, fluid dynamics, chemistry, applied mathematics, and numerical analysis.

Research at GFDL improves our understanding of atmospheric circulations, ranging in scale from hurricanes to extratropical storms and the general circulation, with an emphasis on extreme weather events and the interplay between weather phenomena and climate variability and change at regional scales. High-resolution atmospheric models with comprehensive treatment of physical processes are central tools in this work.

Ongoing research will continue to explore a potential breakthrough in predicting seasonal hurricane activity: atmospheric models forced with observed SST can skillfully predict the interannual variability of the number of hurricanes in the Atlantic, showing that the random part of this annual Atlantic hurricane frequency (the part not predictable given the SSTs) is relatively small. Identifying the processes that are key to determining tropical cyclone frequency may allow us to extend dynamical forecast lead-time toward one year. GFDL will also continue to improve statistical methods for using coupled climate models to develop long-lead-time forecasts of seasonal hurricane activity, and experimental forecasts will continue to be updated on GFDL's website. Even longer lead-time predictions are targeted through a vigorous research program on decadal variability, predictability, and predictions. At the heart of that program are efforts to better understand physical processes that contribute to decadal variations and predictability in the climate system, particularly in terms of the role of the ocean as a driver of decadal-scale variations. Augmenting this research are decadal hindcast and prediction experiments yielding model-produced "predictions" for each year from 1961 to 2011. The ongoing analysis of these simulations seeks to evaluate whether predictive skill for near-term (decadal) climate forecasts is increased when starting from the observed state of the climate system, in addition to the predictive skill that arises from changing radiative forcing. Preliminary assessments of the predictive skill in these simulations reveal that they have skill that is at least comparable to other international efforts.

Complementing this effort is research to improve our understanding of the interactive three-dimensional radiative-dynamical-chemical-hydrological structure of the climate system, from the surface and troposphere to the upper stratosphere and mesosphere, on various time and space scales. Continued research in FY15 will focus on a better understanding of the response of global and regional climate to natural and anthropogenic aerosol emissions. GFDL's world-leading high-resolution models use the very latest numerical techniques to provide extremely energetic, realistic simulations of the climate system. Through their use in FY15, GFDL seeks to further scientific understanding of the role of the ocean and atmospheric constituents in climate variability and change, on regional to global scales. This research continues to exploit NOAA's high performance computing capability housed at DOE's Oak Ridge National Laboratory. Despite the relatively high computational cost, GFDL scientists believe it is critical to move to regional-scale high-resolution climate models to better understand the causes and predictability of decadal-scale climate fluctuations, as well as the role of the ocean in critical climate change issues, such as oceanic heat uptake.

Great Lakes Environmental Research Laboratory (GLERL)

The Great Lakes Environmental Research Laboratory (GLERL) develops and operates technology for scientific observations in the Great Lakes ecosystem. GLERL scientists acquire the data and develop information needed to improve our understanding of the Great Lakes ecosystem and support decision-making for improved resource management. Advancements in areas such as remote sensing, the miniaturization of sensors, and other data-gathering technologies have vastly increased the data gathering capacity of the scientific community. Additionally, GLERL tests and applies models for predicting the effects of natural and human-generated changes on the Great Lakes environment. Modeling of the atmosphere, lakes, seasonal changes in ice cover, and the ecosystem dynamics of the lakes adds to our understanding of how the Great Lakes basin changes over the course of months and years. GLERL research on regional climate projections is based on atmospheric and coupled hydrodynamics-ice-ecosystem models. Research tools are designed to examine the effects of climate on regional air temperature, precipitation, water levels, lake temperature and thermal structure, ice cover, and ecological changes and trends.

In FY 15, priorities for the Great Lakes Environmental Research Laboratory (GLERL) include continuing research, development, and transfer of remote sensing products from the Great Lakes CoastWatch Node. GLERL will focus on satellite synthetic aperture radar (SAR) and moderate resolution imaging spectroradiometer (MODIS) ice classification and mapping. GLERL will also focus on continued data collection and analysis for the advancement of ice-ocean ecosystem modeling in the Artic. GLERL's hydrologic modeling program will advance the GLERL hydrometeorological database, improve Great Lakes basin runoff and evaporation estimates, and continue to enhance the Great Lakes hydroclimate dashboard. GLERL will continue efforts on regional climate projection impact assessment and historic trends analysis. GLERL will assess the similarity and differences between model runs from the Coupled Hydrosphere-Atmosphere Research Model (CHARM) with model runs from other GCMs. GLERL will continue to develop and validate modeling systems that contribute directly to projection of the overall water budget and levels of the Great Lakes, as well as enhances understanding of relevant climate change processes on a regional scale.

National Severe Storms Laboratory (NSSL)

NSSL seeks to improve the accuracy and timeliness of forecasts and warnings of hazardous weather events such as thunderstorms, tornadoes, flash floods, lightning, winter storms, and their associated impacts. Moving research to operations is a core part of NSSL's mission. After more than a decade of development, testing and experimental evaluation, NSSL, in collaboration with the National Center for Environmental Prediction (NCEP), is bringing the Multi-Radar/Multi-Sensor (MRMS) system to NWS operations. MRMS consolidates and "cleans up" data from a wide variety of sources including radars, satellites, and surface observations to provide forecasters with the best-possible real-time analysis of the atmosphere at any location. This system has shown considerable improvement in radar-based rainfall estimates and provides forecasters with an ability to "mine" the data set to extract new types of aviation, severe and hydrologic weather information. By placing high-quality, remotely-sensed data in a three-dimensional framework accessible to NWS forecasters, MRMS will be opening the door to exciting new possibilities in improving severe weather warnings and forecasts.

Weather Radar Research. The NWS has recently completed the dual polarization hardware upgrade (developed by NSSL) to the WSR-88D to provide enhanced capability. NSSL is working to improve the WSR-88D software algorithms including the polarimetric Hydrometeor Classification Algorithm (HCA) by extending the types of precipitation automatically classified by the radar to winter weather precipitation types and hail size discrimination. NSSL is also actively engaged in a risk reduction activity for the Multifunction Phased Array Radar (MPAR) technology with the Federal Aviation Administration (FAA) and industry in designing and testing dual-polarized phased array radar panels required to determine the feasibility of using MPAR as the replacement technology for weather and aircraft surveillance radars (i.e. multi-function). Two important accomplishments occurred this past year: 1) Dual- polarization phased array radar panels were developed through engagement of industry partners (primary risk reduction element) and 2) "adaptive" scanning strategies were developed using the National Weather Radar Testbed (NWRT), allowing for more frequent updating from the phased array radar. In addition, the potential impact of the faster updates has been systematically assessed through experimental warning exercises with NWS forecasters in the Hazardous Weather Testbed. The major MPAR risk reduction activities for FY14 are to demonstrate multi-function capability on the NWRT and to continue evaluating the small dual-polarized phased array radar panel.

NSSL is also actively engaged in a risk reduction activity for the Multifunction Phased Array Radar (MPAR) technology. Since 2003, NSSL has improved the capabilities of the National Weather Radar Testbed (NWRT) to demonstrate the potential of PAR technology to simultaneously perform aircraft tracking and weather surveillance. Emphasis has been placed in developing proof-of-concept strategies for focused and tailored observations of weather; more recently, the NWRT software was upgraded to enable the demonstration of scheduling algorithms for multifunction operation. Case studies and Phased Array Radar Innovative Sensing Experiments held in the HWT are examining the importance of these data to understanding severe storm processes and improving the accuracy of severe thunderstorm and tornado warnings. With the Federal Aviation Administration (FAA) and industry, NSSL is involved in designing and testing dual-polarized PAR panels required to determine the feasibility of using MPAR as the replacement technology for weather and aircraft surveillance radars (see section 3, Aviation Services).

High Impact Hazardous Weather Research. NSSL is taking the lead on a groundbreaking project called Warn-on-Forecast (WoF). This research project aims to create computer-model projections that accurately predict storm-scale phenomena such as tornadoes, large hail, and extreme localized rainfall. The goal of WoF is to provide NWS forecasters with reliable guidance for the likelihood of each of these phenomena at any location up to an hour before they strike, enabling a new early-warning paradigm for extreme localized weather events.

NSSL is actively engaged in several lightning research projects, including three different approaches to the assimilation of lightning data into forecast models (e.g., nudging water vapor mixing ratio to force convection in the WRF-ARW, employing an operator to assimilate lightning data into an EnKF model, and investigating relationships between lightning bursts and changes in the intensity of inner-wall convection of hurricanes).

Revitalizig the Warning Decision Process. NSSL is working with the NWS to develop a new vision for the warning decision process, which continues to evolve as scientists and engineers work toward integrating the next-generation radar technology (e.g., rapid scanning phased array radar) and storm-scale numerical models to create a storm-scale ensemble prediction capability for the NWS. Within the next decade, NSSL envisions operational units using WoF methodology; for example, a forecaster will use thunderstorm-resolving computer models for severe weather warnings in the same way as he/she does today with the current Doppler radar systems. To make the best use of these technological advances, NSSL is collaborating with the NWS in the exploration and development of a new paradigm for hazardous weather forecasting known as Forecasting a Continuum of Environmental Threats (FACETs). FACETs is intended to move the NWS from a deterministic, binary, product-centric system to one in which a rapidly-updated, high-resolution stream of (probabilistic) hazardous weather information, fed by current and future scientific tools (e.g., WoF, PAR) can be optimized for user-specific decision making. As such, FACETs represents a fundamental change in the nation's hazardous weather forecasting system.

Office of Weather and Air Quality (OWAQ)

NOAA's Office of Weather and Air Quality (OWAQ) improves weather and air chemistry forecast information and products by funding, facilitating, and coordinating research. By working with scientists and academic partners, this research is transitioned into useful weather applications, watches and warnings to protect the lives and property of the American public and inform weather-sensitive U.S. industry. OWAQ provides outreach, linkages, and coordination between NOAA, other Government agencies, academia, the private sector, and non-profit research partnerships, particularly NOAA's Cooperative Institutes and Cooperative Science

Centers. OWAQ additionally supports social science studies to test research results that improve communication during dangerous weather situations.

In FY15, OWAQ will continue supporting projects through the U.S. Weather Research Program (USWRP) to improve high-impact weather forecasting and also improve NOAA's global modeling system from days to months through the Earth System Prediction Capability (ESPC) project. ESPC is a joint federal agency program that will provide a more accurate global ocean and atmospheric forecast system with longer skillful forecast times at synoptic, sub-seasonal, seasonal, and inter-annual scales through integrating and coupling global atmosphere, ocean, ice, land and near-space forecast models into an operational suite of prediction systems that reduce errors relative to current modeling approaches. OWAQ will also support testbeds and high-resolution numerical model improvements to enable better high-impact weather forecasts, including tropical cyclones and extreme precipitation. In addition, studies to simulate observations from proposed observing systems and use in weather forecasting models will help determine the impact on forecast skill, while projects on air quality modeling will contribute to improving National Weather Service smoke, dust, and volcanic ash predictions. OWAQ will also fund social science research that helps to understand how society uses and interprets weather information and how to improve the communication of weather information to the public.

Pacific Marine Environmental Laboratory (PMEL)

Meteorological research at PMEL focuses on air-sea interaction research in the Gulf of Alaska Bering Sea, and Chukchi Sea, as part of PMEL's Ecosystem-Fisheries Oceanography Coordinated Investigations (EcoFOCI) project, conducted jointly with NOAA's National Marine Fisheries Service/Alaska Fisheries Science Center. Financial support for the research is provided by NOAA, NSF, the North Pacific Research Board (NPRB), and the DOI/Bureau of Ocean Energy Management (BOEM).

PMEL's ocean climate research programs collect surface meteorological data from moored buoys and report in near-real time for ingest into global models. Data from PMEL's PIRATA and RAMA tropical observing systems in the Atlantic and Indian Oceans, and from PMEL's ocean climate stations at Ocean Weather Station Papa (Gulf of Alaska) and the Kuroshio Extension Observatory in the Northwest Pacific report surface meteorological data.

PMEL conducts research on atmospheric aerosols and their impact on air quality and climate. In 2015, PMEL is leading the Western Atlantic Climate Study - II (WACS-II). The purpose of the project is to characterize the chemical and meteorological processes that transform aerosols and their precursor gases and lead to perturbations to the region's radiation budget. Data will be used to improve simulations of radiative forcing of climate by ocean- and continentally-derived aerosols with a particular emphasis on African dust; characterize the emission, properties, and cloud nucleating ability of sea spray aerosol over a range of ocean conditions; and provide scientific information to support effective policy decisions.

NASA Supporting Research for Basic Meteorological Services

Research in Basic Meteorology and Atmospheric Science

The FY 2014 Budget will fund research competitively selected in FY 2013 through NASA's Research Opportunities in Space and Earth Sciences 2013 (ROSES-13) grant application solicitation. Many of the research activities carried out in FY 2014 will be tasks initiated in FY 2012 and FY 2013 based on the earlier ROSES-11 and ROSES-12 solicitations. Selections based on ROSES-11 and ROSES-12 are ongoing and are addressing diverse Earth Science research areas. NASA sponsored research continues to gain new insight into weather and extreme-weather events by the utilization of data obtained from a variety of NASA- and partner satellite platforms and hurricane field experiments. Major numerical weather prediction (NWP) centers both outside (European Centre for Medium Range Weather Forecasts (ECMWF)) and in the U.S. – NOAA/National Centers for Environmental Prediction (NCEP), NASA Global Modeling and Assimilation Office (GMAO), and the U.S. Navy – have shown notable improvements from the assimilation of Atmospheric Infrared Sounder (AIRS) data into their operational forecast systems.

NASA also has a long history of conducting airborne field campaigns in support of hurricane research (http://airbornescience.nsstc.nasa.gov/field/). Most recently, the Hurricane and Severe Storm Sentinel (HS3) Mission, a five-year Earth Venture Class Suborbital mission that was awarded in 2010, has been obtaining data from its base at the Wallops Flight Facility (WFF) on the coastline of Virginia during the hurricane seasons of 2012-2014 (https://espo.nasa.gov/missions/hs3/). This campaign uses two Global Hawk (GH) unmanned aircraft systems (UAS) with distinct payloads to address both over-storm and near-storm environmental issues. The HS3 Mission is designed to investigate some basic questions regarding changes in hurricane intensity:

1. What impact does the large-scale environment, particularly the Saharan Air Layer (SAL), have on intensity change?

2. What is the role of storm internal processes such as deep convective towers?

3. To what extent are these intensification processes predictable?

Short-term Prediction Research and Transition (SPoRT) Center

NASA encourages more rapid use of NASA's observations in operational weather predictions. The Short-term Prediction Research and Transition (SPoRT) center at NASA Marshall Space Flight Center is chartered to partially fulfill this research to operation transition activity. The SPoRT center has become NASA's primary research and operation transition interface with the National Weather Service for short-term (0 to 48 hours) weather predictions. Many NASA research data products are tested in near real time and disseminate to NOAA's weather forecast offices. NASA also funds external research proposals to collaborate with the SPoRT center to accelerate the transition of research data to at least one operational environment.

In FY2014, NASA continued the investment in SPoRT and the external research projects to accelerate the transition of research data to operational environments. Several proposals funded by ROSES-13 program element "NASA Data for Operation and Assessment" will continue the efforts in FY14. In FY2015, NASA plans to announce another run of NASA Data for Operation and Assessment ROSES element.

Technology Development

Technology investments are aligned with NASA's *Climate Centric Architecture* strategic document. A majority of the investments are centered on the National Research Council (NRC) Decadal Survey[1] activities, but many support NASA's foundational and climate continuity missions. Such investments focus on maturation of technologies to enable advanced space-based observations and modeling to improve understanding of the global integrated Earth system, including global and regional climate change. Earth Science Technology Program (ESTP) provides funding for instrument, component, and information technologies prior to mission formulation. Developing and validating technologies well in advance of a flight project help to improve acceptance and reduce costs. Projects are initiated each year through the ROSES solicitation, and the duration of each project is typically three years. For FY 2015, ongoing investigations will be managed in the Instrument Incubator, Advanced Information Systems Technology, Advanced Component Technology, and In-space Validation of Earth Science Technology areas.

NASA Applied Sciences Program

The Applied Sciences Program leverages NASA satellite measurements and new scientific knowledge to enable innovative and practical uses by public and private sector organizations. Projects, which are competitively selected through ROSES, are designed to discover and demonstrate new applications and facilitate adoption by non-NASA organizations. In FY 2015 the Applied Sciences Program will continue to initiate projects across a range of application areas, including health and air quality, water resources, disasters, and ecological forecasting. The Program's Capacity Building element sponsors programs and projects that improve the skills and capabilities of decision makers, community leaders, and resource managers in the United States and abroad related to accessing and applying Earth observations. NASA will continue to build such capacity through the SERVIR, DEVELOP, Gulf of Mexico Initiative (GOMI), and Applied Remote SEnsing Training (ARSET) program elements. The SERVIR program is a joint venture between NASA and the U.S. Agency for International Development (USAID), which provides satellite-based Earth observation data and science applications to help developing nations in Central America, East Africa and the Himalayas to improve their environmental decision-making. In FY2015, SERVIR plans to extend work into South East Asia. The NASA DEVELOP program fosters an interdisciplinary research environment for interns to work on applied science research projects with a local government partner under the guidance of NASA and partner science advisors. GOMI utilizes NASA Earth science assets to address regional priorities defined by the Gulf of Mexico Alliance (GOMA), a partnership of the states of Alabama, Florida, Louisiana, Mississippi, Texas and 13 federal agencies whose goal is to significantly increase regional collaboration to enhance the ecological and economic health of the Gulf of Mexico after

---

[1] NASA and its partners ask the NRC once each decade to look out ten or more years into the future and prioritize research areas, observations, and notional missions to make those observations. The last Decadal Survey was completed in 2007, entitled *Earth Science and Applications from Space: National Imperatives for the Next Decade and Beyond.*

hurricane Katrina hit the region. The ARSET project trains decision-makers and applied science professionals in the areas of Water Resources Management and Air Quality Applications to use NASA's Earth Science observations, tools, and models to support their decision-making.

National Science Foundation

To improve weather forecasts and public safety, the NSF supports basic research on observational systems, analysis techniques, and understanding of phenomena. NSF awards grants to single investigators or small collaborative groups working on specific topics, as well as larger interdisciplinary groups and the National Center for Atmospheric Research. Examples of recent award topics related to meteorological services include mesoscale predictability, boundary layer processes, and numerical modeling techniques. NSF also continues to support the analysis of data collected during recent major field campaigns dealing with tornadoes, hurricanes, winter storms, convection and chemistry, and the Madden Julian Oscillation. NSF has initiated funding on a large field campaign that will explore nocturnal organized convection in the US Plains states. NOAA will be an interagency partner in this effort. The Interdisciplinary Research in Hazards and Disasters program, which falls under the Science, Engineering and Education for Sustainability (SEES) portfolio in NSF, recently funded awards relevant to basic meteorological services, including the mitigation and response to hurricanes, tornadoes and flash flooding.

AGRICULTURAL AND LAND MANAGEMENT METEOROLOGICAL SERVICES

For purposes of this *Federal Plan*, Agricultural and Land Management Meteorological Services are those services and facilities established to meet the requirements of the agricultural industries and Federal, state, and local agencies charged with the protection and maintenance of the Nation's land areas. Meteorological services specifically tailored for wildland fire management are reported under the Wildland Fire Weather service category.

OPERATIONAL PROGRAMS INCLUDING PRODUCTS AND SERVICES

Interagency Programs

Joint Agricultural Weather Facility (JAWF)

In 1978 the World Board and the National Oceanic and Atmospheric Administration formed the Joint Agricultural Weather Facility (JAWF). Housed at USDA, JAWF monitors the weather and assesses its likely impact on crops around the world. Regular briefings by Board experts are an important information source for USDA commodity forecasters as well as for the Secretary of Agriculture and other top officials at the Department. Additional information on the JAWF is included in the U.S. Department of Agriculture narrative below and at http://www.usda.gov/oce/weather/.

U.S. Department of Agriculture

Office of the Chief Economist

The United States Department of Agriculture (USDA) Office of the Chief Economist (OCE) World Agricultural Outlook Board (WAOB) serves as the USDA focal point for economic intelligence and commodity outlook for U.S. and world agriculture. The WAOB coordinates, reviews, and approves the World Agricultural Supply and Demand Estimates (WASDE) report. The WASDE report provides USDA's forecasts of supply and demand for major U.S. and global crops as well as U.S. livestock. The WAOB maintains the integrity of this report by ensuring all information used to prepare the report is consistent, objective, and reliable. Because weather and climate have a significant impact on agricultural production, the WAOB employs meteorologists who specialize in preparing agricultural weather assessments, the group being collectively referred to as OCE/WAOB's Office of the Chief Meteorologist (OCM).

The primary mission of OCM staff is to routinely collect global weather data and agricultural information to assess the impact of growing season weather conditions on crops and livestock production prospects, keeping USDA commodity analysts, the OCE, and the Secretary of Agriculture and top staff well informed of weather impacts on crops and livestock worldwide. These assessments are integrated into USDA's monthly analytical process to develop monthly foreign crop estimates of area, yield, and production published in the monthly WASDE reports. In addition to providing routine agricultural weather assessments, OCM meteorologists prepare special assessments when extreme weather (e.g., droughts, heat waves, freezes, floods, and hurricanes) has been observed or is imminent. Alerts of anomalous weather conditions impacting agriculture around the globe are also routinely included in a daily report of agricultural developments that is sent to USDA policy makers each afternoon. When integrated with other data, these routine and special crop-weather assessments and analyses provide critical information to USDA decision-makers preparing crop production forecasts, formulating trade policy, and coordinating disaster relief.

While many of the group's core functions are internal, OCM staff routinely disseminates information to the public. Each morning, a written summary of current weather affecting agriculture in the United States is sent to the Secretary's office and posted on the OCE Web site: http://www.usda.gov/oce/weather/pubs/Daily/TODAYSWX.pdf. The meteorologists also routinely give public interviews, available not only through USDA's Information Office but also private sector media outlets. In 2012 and 2013, for example, OCM staff played a key role in providing assessments of the historic drought that impacted much of the Nation's agriculture by conducting interviews with national and international news outlets and disseminating unique products and assessments via the USDA homepage and blog.

In support of its mission, the OCM receives a full suite of meteorological data and products from the NWS for use in their assessments, including surface weather observations, radar data, satellite imagery, and model output. This activity is made possible through an agreement signed between the Departments of Agriculture and Commerce, outlining the creation of a Joint Agricultural Weather Facility (JAWF). JAWF was created in 1978 as an operational unit and is a cooperative effort between the WAOB and the U.S. Department of Commerce (DOC)/National Oceanic and Atmospheric Administration (NOAA)/National Weather Service (NWS)/National Centers for Environmental Prediction (NCEP)/Climate Prediction Center (CPC). As part of the agreement, the OCM serves as the focal point for dissemination of meteorological data and products to other USDA agencies, including weather data received from the Global Observing System, a worldwide network of nearly 8,000 meteorological reporting stations managed by the World Meteorological Organization (WMO). Additionally, the OCM obtains data from the NWS Cooperative Observer Program (COOP) to support domestic agricultural weather applications. In recent years, the NWS, National Hurricane Center (NHC), CPC, and the Hydrometeorological Prediction Center have supported USDA by providing an increasing number of their operational products in GIS-compatible formats. These NOAA GIS efforts have benefited the OCM significantly by increasing the speed and efficiency with which agricultural weather assessments can be prepared and enabling analysts to more accurately assess weather impacts on agriculture.

JAWF's flagship publication, the *Weekly Weather and Crop Bulletin (WWCB)*, is jointly produced by the OCM, the National Agricultural Statistics Service (NASS), and NOAA/CPC. First published in 1872 as the *Weekly Weather Chronicle,* the publication provides a vital source of information on weather, climate, and agricultural developments worldwide. The *WWCB* highlights weekly meteorological and agricultural developments on national and international scales, via numerous maps, charts, tables, and text products. In total, these products provide a comprehensive illustration of the weather and climate conditions affecting agriculture, benefiting

USDA decision makers and the agricultural community. Crop and weather information provided in the *WWCB* keeps crop and livestock producers, farm organizations, agribusinesses, state and national farm policy-makers, government agencies, and foreign buyers of agricultural products apprised of worldwide weather-related developments and their effects on crops and livestock. The extensive history provides a reference source that is rich in climate and agricultural information, which is essential for episodic-events monitoring and analog-year comparisons.

Although the main emphasis of the *WWCB* is on current growing-season weather conditions and agricultural developments in the United States, real-time agricultural weather assessments are also provided for foreign countries that are either major exporters or importers of agricultural commodities. While providing timely weather and crop information relevant to the monthly *Crop Production* and *WASDE* reports (issued by USDA/NASS and USDA/OCE/WAOB, respectively), the *WWCB* keeps the U.S. agricultural sector apprised of weather conditions impacting agriculture in foreign markets, which could possibly influence production decisions at the farm level.

In September 1994, the OCM published *Major World Crop Areas and Climatic Profiles* (Agricultural Handbook No. 664). Knowledge of historical weather and climate patterns and past agricultural production in major agricultural regions worldwide is critical to the success of JAWF's agrometeorological assessments, and this reference handbook provides the framework for assessing the weather's impact on world crop production by providing information on climate and crop data for key producing regions and countries. Coverage includes major agricultural regions and crops, including coarse grains, winter and spring wheat, rice, major oilseeds, sugar, and cotton. World maps show the normal developmental stage of regional crops by month. No longer issued in a paper format, an electronic version of the handbook was developed to provide periodic updates to the printed version as additional data become available. The *Major World Crop Areas and Climatic Profiles* book and other publications are available online at http://www.usda.gov/oce/weather/pubs/index.htm.

In the summer of 1999, the U.S. Drought Monitor (USDM) was developed to help improve drought assessments in the United States. The USDM is a collaborative effort between federal and academic partners, including the University of Nebraska-Lincoln National Drought Mitigation Center (NDMC), USDA, CPC, the NOAA/NESDIS/National Climatic Data Center, and the Desert Research Institute. Approximately 11 lead authors, two of whom work for the OCM, rotate the responsibility of preparing the USDM. Produced weekly, the USDM is a synthesis of multiple indices and impacts depicted on a map and in narrative form. The NDMC hosts the USDM on its Web site at http://droughtmonitor.unl.edu. The USDM, released each Thursday at 8:30 a.m. Eastern time, is a key source of information for briefing USDA top staff on U.S. drought developments. In recent years, the USDM has served as a trigger mechanism for several USDA programs and, because the USDM is prepared in a GIS, it is often overlaid on agricultural data to illustrate and quantify the spatial extent of drought affecting various agricultural commodities (**Figure 1**). In 2013, OCE entered into a cooperative agreement with the NDMC to improve U.S. agricultural drought monitoring capabilities and assessments by integrating USDA agricultural statistics and other relevant data sets into a GIS format to improve author's ability to incorporate various information types into their analytical process.

Similarly, the North American Drought Monitor (NADM) is a cooperative drought monitoring effort among drought experts in Canada, Mexico, and the United States. The NADM was initiated at a workshop in April 2002 and is part of a larger effort to improve the monitoring of North American climate extremes. Issued monthly since January 2003, the NADM is based on the end-of-month USDM analysis and input from scientists in Canada and Mexico. Major participants in the NADM program include the USDM collaborators – two of whom are OCM staff – as well as Agriculture and Agrifood Canada and the National Meteorological Service of Mexico. The NADM Web site is: http://www.ncdc.noaa.gov/temp-and-precip/drought/nadm/index.html.

A U.S. Drought Monitor Forum and a North American Drought Monitor Forum are held in alternating years. These meetings provide an opportunity for Drought Monitor authors, stakeholders, and members of the drought community to discuss the latest drought monitoring tools, drought analyses, and requirements. The most recent U.S. Drought Monitor Forum was held on April 15-19, 2013, in West Palm Beach, Florida, while a North American Drought Monitor Forum was recently held on June 17-19, 2014, in Toronto, Canada.

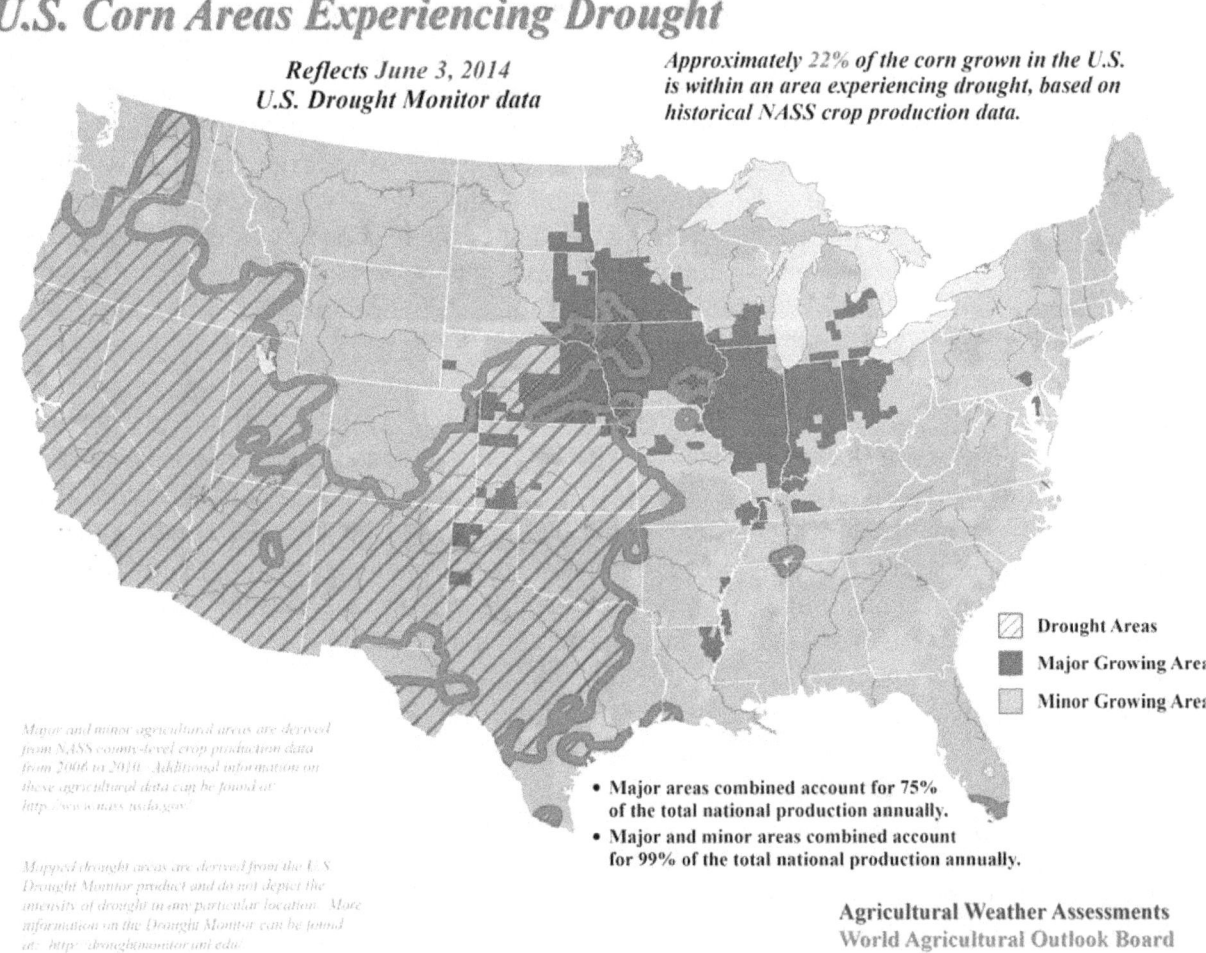

Figure 1. Percent of U.S. corn area in drought, according to the June 3, 2014, *U.S. Drought Monitor.*

OCE is one of several USDA agencies that have taken a lead role in the development of National Integrated Drought Information System (NIDIS), working closely with lead Federal agency NOAA and the Western Governors Association (WGA) over the years to address the specific needs of the agricultural community. The NIDIS builds upon existing drought monitoring tools and experiences, such as the USDM, to develop an early warning system that aids in drought preparation and mitigation. Recommendations for such an early warning system were outlined in a 2004 report from the WGA entitled *"Creating a Drought Early Warning System for the 21st Century: The National Integrated Drought Information System."* One of the early deliverables of NIDIS was the Drought Portal (http://www.drought.gov), which serves as the Government's multi-agency drought interface. The OCM is working with other USDA agencies to provide relevant drought information to the public via the Drought Portal.

As a result of the aforementioned activities with the NDMC, NOAA, and other agencies and organizations, OCE is a recognized leader within the USDA in the area of drought. In December, 2012, a Memorandum of Understanding was signed between USDA and NOAA to collaborate on projects of mutual interest, with particular emphasis on establishing new partnerships in the area of drought monitoring and mitigation. OCE enjoyed a lead role in developing the agreement and is currently working with NOAA to develop a work plan to identify near- and long-term deliverables and opportunities for collaboration. In 2014, USDA announced creation seven Climate Hub and three Sub-Hub locations. Under the direction of OCE's Climate Change Program Office, the Climate Hubs will deliver science-based knowledge, practical information and program support to farmers, ranchers, forest landowners, and resource managers to support decision-making related to climate change. Key partners in the networks include the public and land grant universities, Cooperative Extension, USDA researchers, the private sector, state, local and regional governments, the National Oceanic and Atmospheric Administration (NOAA), Department of Interior (DOI) regional climate change experts, and non-profits engaged in providing assistance to landowners.

Throughout its history, WAOB has successfully worked with international groups, including representatives of foreign governments and of multinational organizations. WAOB remains active in the World Meteorological Organization (WMO) Commission for Agricultural Meteorology (CAgM), and a WAOB meteorologist is currently serving on the Management Group of the WMO CAgM, helping guide Commission efforts to improve support systems for agrometeorological services. This position helps coordinate Expert Teams responsible for reviewing the operational applications of current agrometeorological data, analytical tools, and information delivery systems and making recommendations on the procedures, methodologies, and resources necessary to improve the capability for operational applications. Additionally, WAOB continues to support the World AgroMeteorological Information Service (WAMIS), a dedicated web server that hosts agrometeorological bulletins and advisories issued by WMO Members for the global agricultural community (http://www.wamis.org/). WAOB is collaborating with other WMO Members to expand WAMIS capabilities, including the development on a web-based GIS capability that will dynamically integrate weather and climate information.

Risk Management Agency

The Risk Management Agency (RMA) provides administration and oversight of programs authorized under the Federal Crop Insurance Act. RMA's Strategic Data Acquisition and Analyses (SDAA) unit works with Oregon State University's Parameter-Elevation Regressions on Independent Slopes Model (PRISM) Climate Group to develop and utilize spatial climate data sets to detect potential waste, fraud and abuse in the Federal crop insurance program and to assist underwriting in developing crop suitability mapping.

Department of Interior

Bureau of Land Management, Land Management Services

The Department of Interior's (DOI) Bureau of Land Management (BLM) utilizes air-resource-related (air quality, weather, and climate) information in order to manage public lands in a manner consistent with Congressional direction as expressed in the Federal Land Policy Management Act (FLPMA). FLPMA directs the BLM to periodically and systematically inventory resources through a land-use planning process and to manage public lands in a manner that protects the quality of scientific, scenic, historical, ecological, environmental, air and atmospheric, as well as other natural resources. The BLM also requires air-

Meteorological Station at Pinedale, WY

resource-related information to conduct environmental analyses under the National Environmental Policy Act (NEPA) for agency-initiated activities and land-use authorizations and to ensure compliance with pollution laws such as the Clean Air Act.

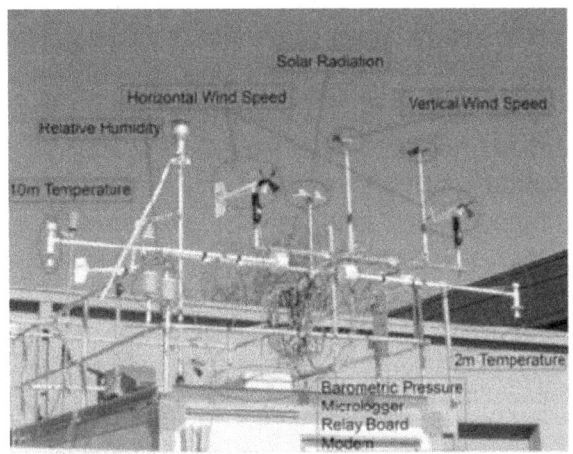

Meteorological System Components being tested before installation at a new monitoring site in Inigok Station, AK.

The BLM must therefore obtain, collect, and analyze air-resource-related information to (1) determine baseline conditions of air and atmospheric values on the public lands; (2) evaluate changes in baseline conditions (trends); (3) understand the extent to which other natural resources (vegetation, hydrology, wildlife, range, minerals, etc.) are influenced by atmospheric conditions so that informed management decisions can be made; and (4) to assist in developing models to predict future conditions; such as, atmospheric dispersion models to assess air quality impacts. The BLM obtains information of acceptable quality collected from existing monitoring networks operated by other agencies and programs whenever possible to promote efficiency and avoid duplication of efforts. Examples of these networks include the National Weather Service Cooperative Observer Network, Natural Resources Conservation Service (NRCS) SNOw pack TELemetry (SNOTEL) and Soil Climate Analysis Network (SCAN) networks, the National Interagency Fire Center Remote Automated Weather Station (RAWS) network, the Bureau of Reclamation Agricultural Weather (AgriMet) networks, the Community Collaborative Rain, Hail, and Snow (CoCoRaHS) network, the National Atmospheric Deposition Program (NADP), the U.S. Geological Survey National Streamflow Information Program (NSIP), and individual state climate offices.

SUPPORTING RESEARCH PROGRAMS AND PROJECTS

U.S. Department of Agriculture

Agricultural Research Service

The Agricultural Research Service (ARS) is the principal in-house research agency for USDA and conducts research on all aspects of agriculture including crop and livestock production, natural resources and sustainable agriculture systems, and human nutrition and food safety.

Weather-related research by ARS develops technologies to enable agriculture to cope with, and when possible benefit from, the effects of weather on crop and livestock production and natural resources and ecosystem services. Weather and climate data inputs and an understanding of weather-driven processes are critical to the development of crop growth and yield models, erosion models, nutrient cycling models, soil-plant-atmosphere process models, and many other simulation tools needed to provide continual advancements for agriculture. Such simulation tools are the foundations for the development of decision-support systems used for day-to-day decision making by producers and land managers, and policy-related decisions by strategic decision-makers. The Agricultural Model Intercomparison and Improvement Project (AgMIP) is a collaborative global effort to develop an ensemble crop yield estimation effort driven by the availability and use of weather data.

The interactions between genetics, environment (heavily focused on weather) and management practices (G x E x M) are under investigation as an approach to meeting the goals of sustainable agriculture as weather affects 1) the quantity and quality of desired yields, 2) the environment needed for agriculture and ecosystem services, 3) the economic viability of agriculture, and 4) the quality of life for rural populations and society as a whole. These goals are approached through a research strategy of: measure and understand, develop process-model prediction capacity, and then develop management technologies. Weather data and forecasting are integral to the research and applications of the research.

It is essential that crops and livestock be able to withstand weather stresses and losses from insect and disease exacerbated by the impacts of weather variations. Thus, the development of new crop and livestock varieties is a foundation of ARS weather and climate research. The interaction of water and high temperatures are of special interest.

The interactions of management systems and genetics are tempered weather. ARS weather research is pursuing the hypothesis that the development of basic understandings and process models of the interacting effects of crop varieties with specific management practices adapted to weather means and extremes offers promise towards meeting the challenges of expanded food production for a growing world population.

The ARS Long Term Agroecosystem Research (LTAR) network is distributed geographically to provide a coordinated outdoor laboratory environment encompassing regional and local weather differences. The geographic diversity and the ability to obtain weather data and data on weather-driven processes that can be compared over time and space are critically important strengths of the LTAR. Weather data is a foundation of LTAR measurements. ARS collaborates with domestic and international colleagues, including universities, NASA, NOAA, USGS, state and tribal organizations, and industry.

AVIATION SERVICES

For purposes of this *Federal Plan*, Aviation Services are those specialized meteorological services and facilities established to meet the requirements of general, commercial, and military aviation. Civil programs that are directly related to services solely for aviation and military programs in support of land-based aviation and medium- or long-range missile operations are included. Detailed aviation services/products for specific areas include, but are not limited to, ceiling and visibility, convective hazards, en route winds and temperatures, ground de-icing, in-flight icing, terminal winds and temperatures, turbulence, volcanic ash, and other airborne hazardous materials.

OPERATIONAL PROGRAMS INCLUDING PRODUCTS AND SERVICES

U.S. Code Title 49 Section 44720 (49 U.S.C. 44720) provides the basis for the Department of Commerce, through NOAA/National Weather Service (NWS), to provide "meteorological services necessary for the safe and efficient movement of aircraft in air commerce." Services are developed and provided at the request of the Federal Aviation Administration (FAA), which is designated as the Meteorological Authority for aviation weather services for the United States by the International Civil Aviation Organization (ICAO). In this capacity, FAA provides requirements for the provision of aviation weather services to NWS, which is designated as the National Meteorological and Hydrological Service Provider. The FAA is responsible for ensuring compliance with the services as defined and with maintaining ICAO Standards and Recommended Practices as specified in Annex 3, Meteorological Service for International Air Navigation.

The Department of Defense (DOD) service branches (U.S. Army, Navy, Air Force, and Marine Corps), and the Department of Homeland Security service branch (Coast Guard) provide their own aviation weather support. Each military service has its own meteorological support personnel except the Army, which is supported by the Air Force. Please refer to the Military Services section for details of military-unique aviation weather services.

Interagency Collaborative Operational Products and Services

National Volcanic Ash Operations Plan for Aviation

Under the auspices of the Office of the Federal Coordinator for Meteorological Services and Supporting Research (OFCM), the following agencies participate in the interagency Working Group for Volcanic Ash (WG/VA) and Committee for Aviation Services and Research (CASR): FAA, National Aeronautics and Space Administration (NASA), NOAA, U.S. Geological Survey (USGS), the U.S. Air Force, and the Smithsonian Institution. The WG/VA has prepared a National Volcanic Ash Operations Plan for Aviation. The purpose of the plan is to provide operational guidance by documenting the required procedures and information products of the government agencies responsible for ensuring safety of flight operations when volcanic ash has erupted into the atmosphere. This document also provides information on how the FAA, as the U.S. meteorological authority with regard to the ICAO, meets its obligations to the International Airways Volcano Watch, which is sponsored by the ICAO. There are several regional plans in addition to the national plan. Regional plans are currently in place for Alaska, the Pacific Northwest (Washington, Oregon), and the Northern Marianas Islands (draft framework). Future plans are being developed for Hawaii, California, and Puerto Rico/Eastern Caribbean. These plans are available on the OFCM web site at www.ofcm.gov. Regional plans typically also involve State and local agencies.

USGS. Through its five Volcano Observatories, the USGS is responsible for monitoring volcanoes in the United States and issuing notifications about volcanic activity as it waxes and wanes at individual volcanoes. USGS Volcano Observatories use a combination of ground-based, airborne, and space-based techniques to interpret

Volcanic ash hazards can be catastrophic to aviation operations.

precursory unrest and forecast expected volcanic activity (including when eruptions are not expected). Data and notifications of eruptive activity from USGS monitoring activities are supplied to FAA and DOD to provide warnings for pilots and aircraft operators and to NOAA/NWS to aid in its forecasting and tracking of ash clouds. Because of the proximity of Aleutian volcanoes to busy North Pacific air routes, the USGS's Alaska Volcano Observatory (AVO) has been and continues to be a world leader in the integration of volcano observatory operations with efforts to mitigate the risk from airborne volcanic ash to en route. USGS notifications and warnings about current volcanic activity throughout the United States are available to the public at http://volcanoes.usgs.gov/.

The eruption of Eyjafjallajökull in Iceland in the spring of 2010 and ensuing shutdown of European airspace focused attention on the global economic disruption that a volcanic ash cloud can have on the transportation of people and goods. USGS experts on the issue of airborne volcanic ash have been working with FAA, NOAA, and DOD colleagues, as well as with ICAO, to improve capabilities in mitigating the impact of the presence of volcanic ash in busy flight routes, both domestic and international. USGS also has established a new project that focuses exclusively on volcanic ash and brings together USGS efforts in research and development of new operational tools. One element of the new project is to collaborate with the NWS in Alaska on improving ash fall warnings for the public. The USGS has posted pages on its website devoted to practical guidance for dealing with ash hazards to transportation, communications, agriculture, water supplies, etc.; see http://volcanoes.usgs.gov/ash.

Recognizing that many potentially dangerous volcanoes have inadequate or no ground-based monitoring, the USGS recently evaluated volcano-monitoring capabilities and published "An Assessment of Volcanic Threat and Monitoring Capabilities in the United States: Framework for a National Volcano Early Warning System (NVEWS)" (available online at http://pubs.usgs.gov/of/2005/1164/). Results of the NVEWS volcanic threat and monitoring assessment are being used to guide long-term improvements to the national volcano-monitoring infrastructure operated by the USGS and affiliated groups. The most threatening volcanoes— those near communities and transportation infrastructure (ground and air) and with a history of frequent and violent eruptions—need to be well monitored in real time with an extensive suite of instrument types to detect the earliest symptoms of unrest and to reliably forecast behavior of the volcano. Waiting until unrest escalates to augment monitoring capabilities at these high-threat volcanoes puts people (including scientists in the field) and property at undue risk. Remote, isolated, or less frequently erupting volcanoes that nevertheless can pose hazards to air-traffic corridors require sufficient monitoring capability with ground-based instruments to detect and track unrest in real time so that other agencies responsible for en route flight safety can be kept apprised of the potential for explosive, ash-cloud-forming eruptions.

NASA. Through its fleet of satellite assets, NASA is able to rapidly generate and broadly disseminate imagery and data products on the location, heights, and characteristics of ash plumes and related hazards. These data products fuel a range of research and applications investigations that enhance our knowledge of solid Earth processes, atmospheric transport and composition, and the impacts that volcanic eruptions have on the Earth system. Although NASA does not have

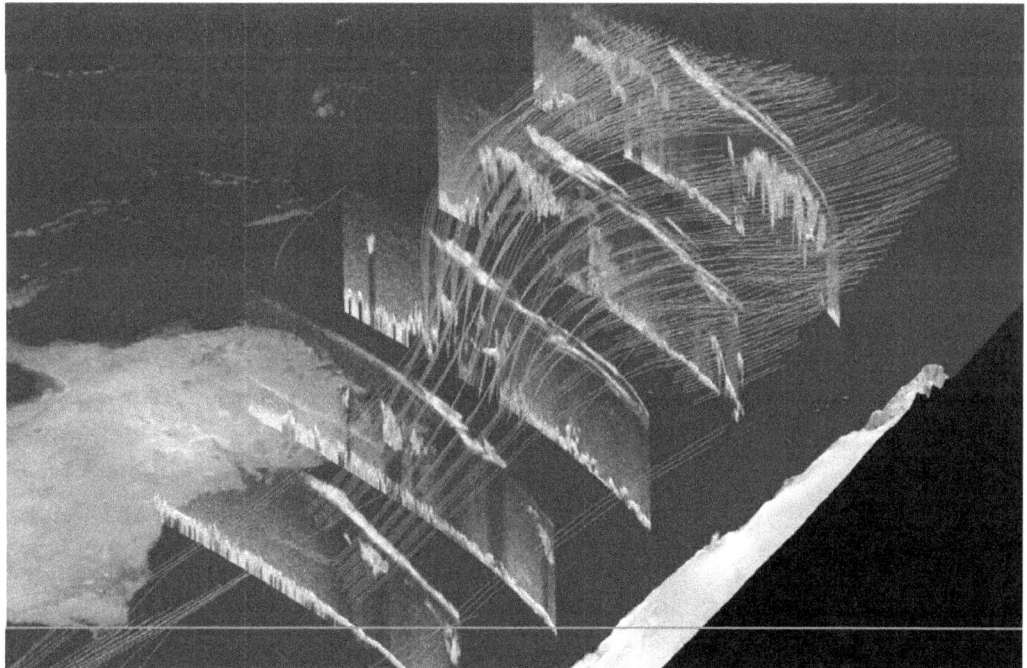

Sequence of CALIPSO 532 nm total attenuated backscatter curtains for 20 - 24 June 2011, depicting the passage of the Puyehue-Cordón Caulle volcanic ash cloud (yellow) near 8-12 km (20,000-40,000ft) across a large area from north Antarctica to Southeast Australia, Tasmania, and New-Zealand. The blue lines between the curtains illustrate forward model trajectories produced from the CALIPSO observations of the volcanic ash cloud. (Kurt Severance, NASA Langley Research Center)

operational responsibility for observation and analysis of volcanic gas and aerosol emissions, its fleet of research spacecraft provides data that are directly applicable to the societal hazards presented by these phenomena.

NASA's Earth Science Division primarily operates five on-orbit sensors that monitor volcanic ash, gases, aerosols, and eruptions. The five sensors currently on orbit are the Cloud-Aerosol Lidar and Infrared Pathfinder Satellite Observation satellite (CALIPSO), a joint mission between NASA and the French space agency CNES; the Ozone Monitoring Instrument (OMI)—a contribution of the Netherlands's Agency for Aerospace Programs (NIVR) in collaboration with the Finnish Meteorological Institute (FMI)—onboard the Aura satellite; the Moderate Resolution Imaging Spectroradiometer (MODIS) onboard the Terra and Aqua satellites; and the Suomi National Polar-orbiting Partnership (NPP) satellite with the Visible Infrared Imaging Radiometer Suite (VIIRS), Cross-track Infrared Sounder (CrIS) and the Ozone Mapping and Profiler Suite (OMPS) onboard. Suomi NPP is in partnership with NOAA. OMI and OMPS data, in which sulfate aerosol is used as a proxy for detecting volcanic ash, provide very accurate maps of the horizontal distribution of volcanic ash clouds, especially during the first several days following an eruption. The next generation of NOAA Polar Operational Environmental Satellites (POES), the Joint Polar-orbiting Satellite System (JPSS), will also deploy an OMPS. In FY 2013, NASA began making CALIPSO Cloud-Aerosol Lidar with Orthogonal Polarization (CALIOP) data and products available in near-real time for operational use. These data have been demonstrated to be of particular utility for improving trajectory model forecasts of volcanic ash, validating the location of volcanic ash determined by forecast models, and resolving the vertical structure and layering of ash clouds. The 30-meter High-resolution imagers on the Advanced Spaceborne Thermal Emission and Reflection (ASTER) instrument from the NASA Terra spacecraft can also be scheduled on a limited basis, and NASA-built sensors on NOAA Geostationary Operational Environmental Satellites (GOES) and POES also support volcanic ash monitoring. The NASA Science Mission Directorate, Earth Science Division, Applied Sciences Program, Disaster Area is currently responsible for developing satellite data applications for volcanic ash and is focused on extending research in which the Agency has employed these satellite measurements to improve the accuracy of trajectory model forecasts to improve aerosol dispersion (ash concentration) model forecasts for the Volcanic Ash Advisory Centers (VAACS). In addition, the NASA Jet Propulsion Lab (JPL) developed and tested small, dedicated Dragon Eye Unmanned Aerial Systems (UAS) in 2013 which penetrates volcanic plumes and coordinated these measurements with ASTER observations allowing scientists to compare sulfur dioxide concentration derived from the satellite with UAS measurements taken from within the plume. The NASA Aeronautics Research Mission Directorate (ARMD) also conducts volcanic ash research. Its efforts primarily concern the NASA ARMD Fundamental Aeronautics Program's Vehicle-Integrated Propulsion Research (VIPR) Project, which is a joint research project between NASA, USAF, USGS, FAA and the three largest engine manufacturers to determine propulsion system tolerances and thresholds for ingesting various concentrations of diffuse, persistent volcanic ash for increasing exposure periods. This purpose of this research is to improve the understanding of volcanic ash behavior and impacts in modern aircraft propulsion systems, to inform the development of operating procedures for operating in or near volcanic ash environments, and to develop improved aircraft engines and systems.

NOAA. NOAA/NWS is responsible for volcanic ash services in the United States. The program is currently managed under the Aviation Services Branch at NWS Headquarters in Silver Spring,

Maryland. Although the main focus has been and continues to be on the airborne ash hazards (mainly impacting aviation), there has been a move in the past several years to expand into an "all hazards" approach that incorporates both the NWS Public and Marine Services programs. The NWS is a co-lead in the development of regional volcanic ash response plans in the United States. Plans are currently in place for Alaska, the Pacific Northwest (Washington, Oregon), and the Northern Marianas Islands. Future plans are being developed for Hawaii, California, and Puerto Rico/Eastern Caribbean. These plans are available on the OFCM web site at www.ofcm.gov. As noted earlier, NOAA/NWS also operates the Anchorage VAAC and partners with NESDIS to operate the Washington VAAC—two of the nine such international centers.

NOAA/NESDIS is responsible for providing satellite data used in detecting and tracking volcanic ash in the atmosphere and is dedicated to providing timely access to global environmental data from satellites and other sources to promote, protect, and enhance the Nation's economy, security, environment, and quality of life. Many of the operational environmental satellites operated by NESDIS (see NOAA/NESDIS in Basic Services section) have channels available to help forecasters detect and track volcanic ash. The GOES-R and JPSS programs are joint NOAA-NASA programs that will provide more frequent, higher resolution imagery for the detection and tracking of volcanic ash beginning in 2015.

NOAA/OAR (Office of Oceanic and Atmospheric Research) Air Resources Laboratory (ARL) also participates to support NOAA's volcanic ash services. ARL will be conducting research and development to improve the volcanic ash application of the ARL-developed HYSPLIT transport and dispersion model that is used by the NWS/NCEP to support the U.S. VAACs. In particular, ARL will work to (1) improve the quantification of the initial ash mass and its vertical distribution, (2) develop a model evaluation database, and (3) work with other VAACs to promote standardization of volcanic ash dispersion products. ARL will collaborate with scientists at the NESDIS Center for Satellite Applications and Research (STAR) and the U.S. Geological Survey (USGS) Volcano Hazards Program. This work should lead to more globally harmonized dispersion guidance products and a better understanding of the dispersion model capabilities.

U.S. Air Force. Through its 2nd Weather Group at Offutt AFB, NE, the AF provides volcanic ash surveillance and analysis for DoD aviation operations worldwide. Analysts continuously monitor all active volcanoes, generating more than 3,500 bulletins per year. In addition to alert text bulletins, AF products include tailored satellite imagery, and graphical ash plume forecasts detailing coverage and concentration. The AF products ensure availability of information critical to DoD aviation, particularly outside the area covered by the Washington and Anchorage VAACs and when products from other international centers are not available.

Smithsonian Institution, Natural History Museum, Global Volcanism Program (GVP). GVP collects, catalogs, and disseminates information on ~1,600 volcanoes active in the last 10,000 years with a small staff working museum hours Monday to Friday in the Natural History Museum in Washington, D.C. The aviation community most depends on GVP for volcano names, locations, summit elevations, and a set of unique volcano numbers (VNUM's) for precise international communication about the source of eruptive plumes and related hazards. For example, Pacaya volcano in Guatemala (along the main approach to the Capital's airport) is designated on VAAC reports as "Volcano: Pacaya 1402-11." The numbers, or unique portions thereof, also provide a short-hand for labeling closely spaced volcanoes on maps. Each month GVP compiles the *Bulletin of the Global Volcanism Network* synthesizing information about volcano behavior, monitoring, and eruptions at variable time scales (months to years). The USGS and GVP jointly release the *Weekly Volcano Activity Report* summarizing eruptions and hazards during the passing one-week interval. Both reports appear on the GVP website (http://www.volcano.si.edu), which also features photos, eruptive histories, and other information, including the current list of data needed by the aviation community.

NOAA/National Weather Service

NOAA/NWS aviation weather projects support increasing and improving observation capabilities, forecast products and techniques, outreach and training, operational adaptation of applied research, and verification of forecast products. These projects have the goal of supporting the FAA in the safe and efficient flow of air traffic in the National Airspace System (NAS). In response to requirements from the FAA and the international community, aviation weather products issued by NWS span the globe.

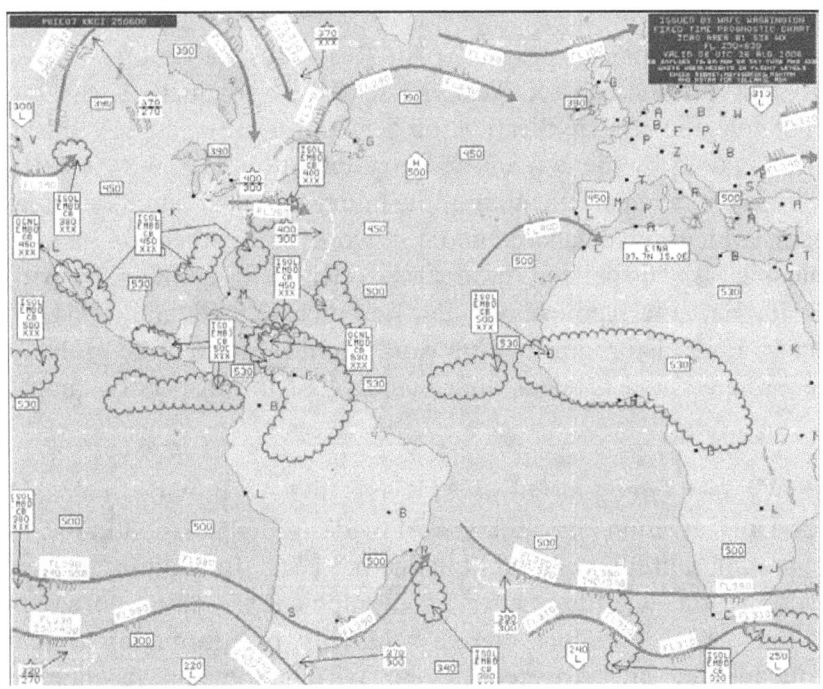

The Aviation Weather Center (AWC) has the responsibility, as part of the World Area Forecast Center, Washington, to provide global weather forecasts of significant weather phenomena. Presently, the AWC produces these High Level Significant Weather charts covering two thirds of the globe, both northern and southern hemispheres that are issued four times per day.

Under an international agreement through the ICAO, the United States meets its weather forecasting obligations to the aviation community through products and services of the World Area Forecast Center (WAFC) Washington, one of two global services, composed of three components of the National Centers for Environmental Prediction (NCEP): Aviation Weather Center, located in Kansas City, Missouri; NCEP Central Operations in College Park, Maryland, and the Telecommunications Operations Center (TOC) at NWS Headquarters in Silver Spring, Maryland. The Aviation Weather Center (AWC) prepares forecasts four times a day of globally significant thunderstorms, tropical cyclones, severe squall lines, moderate or severe turbulence and icing, and cumulonimbus clouds associated with these conditions. The forecast charts also include information on volcanic activity, radiological releases, jet streams, and tropopause heights. This information is made available by the WAFC Internet File Service (WIFS) to provide World Area Forecast System (WAFS) products and services, as defined by ICAO Annex 3, in an Internet web-based medium.

The AWC, along with the NWS Alaska Region's Alaska Aviation Weather Unit (AAWU) and the NWS Weather Forecast Office (WFO) in Honolulu, Hawaii, provides wind, temperature, and flight hazard (e.g., icing and turbulence) forecasts for flight planning and en route aircraft operations for the United States, the north Atlantic and north Pacific routes, and some routes in the southern hemisphere. This information is disseminated by AWC via WIFS in support of the World Area Forecast System (WAFS) for ICAO aviation data needs and in support of World Meteorological Organization (WMO) Region IV (North America) data exchange requirements.

Under an agreement with the FAA, NWS meteorologists are assigned to Center Weather Service Units (CWSUs) located in each of the 21 FAA Air Route Traffic Control Centers (ARTCCs). The CWSUs are currently staffed by 84 NWS meteorologists (4 at each of the 21 ARTCCs) to provide real-time support and decision assistance concerning weather impacts on air traffic. In addition to supporting the ARTCCs, the CWSUs provide meteorological support to en route centers, Terminal Radar Approach Control facilities, and airport towers. Because CWSU forecast support is embedded within the aviation mission, forecasters can focus on specific customer needs. In one example, a specialized marine stratus display system was developed to address the difficult issue of fog formation and dissipation in the San Francisco Bay Area. The San Francisco stratus system is used operationally by the CWSU forecaster, WFO aviation forecaster, FAA traffic managers, and airline meteorologists.

To operationally support the needs of aviation users today, the NWS WFOs prepare Terminal Aerodrome Forecasts (TAFs) up to eight times daily, with amendments as needed, for more than 660 public-use airports in the United States and its territories in the Caribbean and Pacific.

Thus, the NWS, through three Meteorological Watch Offices (MWO) and the two VAACs, provides large-scale, global aviation functions that can be sensibly centralized, while the NWS WFOs and CWSUs discharge local aviation functions. Additionally, NCEP's Environmental Modeling Center (EMC) supplies global gridded model data of temperature, winds, and humidity twice daily for flight levels from 5,000 to 45,000 feet.

NWS's Aviation Weather Services Program funds a broad range of initiatives designed to improve the delivery of aviation weather information to NAS users. These initiatives include the acquisition of aircraft-mounted water vapor sensors; development of software, tools, and training programs to enhance forecaster effectiveness; and development of products to improve weather information availability to the aviation community. NWS's NextGen Weather Program provides funds for the development and implementation of improvements to accuracy and accessibility of aviation weather information, aimed at meeting the goals of the Next-Generation Air Transportation System (NextGen).

Aviation Digital Data Service (ADDS)

In addition to satisfying global requirements for aviation weather, AWC hosts the Aviation Digital Data Service (ADDS), which makes available to the aviation community text, digital, and graphical forecasts, analyses, and observations of aviation-related weather variables. The ADDS serves as a platform for aviation weather products emerging from the FAA Aviation Weather Research Program and transitioning to operations. It has become an invaluable resource to users, especially in the U.S. general aviation community. The AWC's upgraded website infrastructure capitalizes on its FAA certification as a Qualified Internet Communications Provider, and the enhanced web presence provides increased levels of data reliability for users.

ADDS is a joint effort of NCAR Research Applications Program (RAP), Global Systems Division (GSD) of NOAA's Earth System Research Laboratory (ESRL), the NCEP AWC and the FAA. It is run operationally by NWS at the AWC and can be viewed on the internet athttp://aviationweather.gov/adds.

Volcanic Ash Advisory Centers (VAAC)

The United States, in agreement with ICAO, operates two VAACs as entities within NOAA. One of these, the Anchorage VAAC, is part of the AAWU located in Anchorage, Alaska, and works closely with the Alaska Volcano Observatory. The second VAAC, which is part of NOAA's National Environmental Satellite, Data, and Information Service (NESDIS) and NCEP, is located in College Park, Maryland. The VAACs monitor volcanic activity through satellite remote sensing, provide initial notification of a volcanic eruption upon detection, and forecast volcanic ash plume movement and evolution.

A recent change is the assignment of a small cadre of meteorologists to the FAA Air Traffic Control System Command Center (ATCSCC) to provide decision support to strategic planning and flow control for the NAS. These meteorologists work with a NWS Headquarters liaison to the ATCSCC to ensure continuity of support to the planners and to expedite analysis should questions or problems arise.

NWS International Obligations

National Weather Service meets international commitments through participation as consultants to FAA in ICAO groups, ad hoc task forces, and project teams. In addition, NWS leads or participates in a number of World Meteorological Organization (WMO) teams chartered by the WMO Commission on Aeronautical Meteorology. These teams include the Expert Team (ET) on Governance and Partnership (lead); the ET on Meteorological Services for Air Traffic Management (ATM) and Meteorological Information Exchange, which partners with ICAO's MARIE-PT; and the ET for Education, Training, and Competencies.

Federal Aviation Administration (FAA)

Timely and accurate weather observations and forecasts are essential to aviation safety and making the best use of aviation capacity. Pilots need to know the direction and speed of winds aloft in order to take advantage of tailwinds and minimize the effect of headwinds. They also need to know if there will be obstructions to visibility that restrict landings at their destination airport, and whether the runway is wet or dry and how that will affect braking action. Traffic flow managers and pilots use weather observations and forecasts to determine when they need to plan alternative routes to avoid severe weather. The FAA has the responsibility to collect and distribute aviation weather data – particularly hazardous weather to operators, pilots, and air traffic control and management.

Weather Systems

The FAA employs two categories of weather systems: weather sensors and weather processing/dissemination systems. FAA weather sensors include weather radars and automated surface observation systems that measure atmospheric parameters, such as surface temperature, prevailing wind speed and direction, relative humidity, cloud bases and tops, as well as wind shear and microbursts. These weather sensors provide real-time information to FAA weather processing/dissemination systems and to NWS centralized weather forecasting models. Weather processing systems organize, process, and distribute the sensor's observed data, matched with forecasts, and blends National Airspace System (NAS) operations with weather information. Data from multiple sensors feed numerical forecast model ensembles whose output can be disseminated and integrated in national and local processing systems to interpret broad weather trends affecting aviation operations. This information is then sent to air traffic controllers, traffic flow managers, dispatchers, and pilots in a seamless suite of weather tools that look backwards and forwards in time and along terminals, intended flight paths, flow corridors, sectors, and regions.

Weather Sensors

Weather sensors are divided in two portfolios: Wind-Shear Detection Services (WSDS), and the Automated Surface Weather Observation Network (ASWON).

Wind Shear Detection Services (WSDS)

The WSDS portfolio includes the Terminal Doppler Weather Radar (TDWR), the Weather System Processor (WSP), the Next Generation Weather Radar (NEXRAD), and Low-level Windshear Alert System (LLWAS). These systems automatically detect wind shear conditions near runways and approach/departure corridors and provide data to FAA weather processing systems to alert controllers, who can then warn pilots of gust fronts and wind shear in the vicinity of the airport. WSDS Work Package 1 (WP1) will address obsolescence of the legacy WSP) and LLWAS to ensure that Air Traffic Controllers will continue to receive the wind shear alerts necessary to maintain the safety of the NAS. Since these systems have been deployed, no major windshear-related incidents have occurred in the NAS.

Terminal Doppler Weather Radar

The most sophisticated windshear detection system is the TDWR. The primary mission of the TDWR is to enhance the safety of air travel through timely detection and reporting of hazardous weather conditions including windshear events, microburst, gust fronts, and thunderstorms in and near an airport's terminal approach and departure zones. There are 45 operational TDWRs serving 46 high-density airports with the most risk of windshear exposure. Using a three-dimensional pencil beam and fast update rate, TDWR produce detailed weather information across the entire terminal area at the surface and aloft throughout terminal airspace to controllers so they can issue warnings to pilots. TDWR weather data is transmitted to FAA automation systems. TDWR's unique products include high-resolution, rapid update of storm cells with height, precipitation, surface winds, windshear, etc. Activities are underway to address sustainment of the TDWR and the FAA is investigating whether to incorporate the TDWR functions into the NextGen Surveillance and Weather Radar Capability (NSWRC).

A Terminal Doppler Weather Radar

Weather System Processor (WSP)

WSP is an add-on weather processor enhancement to the Airport Surveillance Radar (ASR) Model 9 (ASR-9). The WSP was implemented as a low-cost, high-quality, wind shear detection system at medium and higher air traffic density airports not equipped with the TDWR. The WSP provides advanced Doppler weather radar performance with all the functions of the TDWR. The WSP is partitioned to the front-end radar (ASR) modification, and the back-end (WSP) development. The WSP coverage volume is considerably less that the TDWR but the ASR is better situated to detect the low-level shear events close to the airport. WSP detects and alerts on microbursts and wind shear, predicts arrival of gust fronts, and shows precipitation intensity on a graphical situation display for use by ATC without further meteorological interpretation. Activities are underway to address sustainment of these radars and the FAA is investigating whether to incorporate the WSP into the NSWRC.

Next Generation Weather Radar (NEXRAD)

NEXRAD is long-range weather radar that detects, analyzes, and transmits weather information for use by the ATC System Command Center, en route, terminal and flight service facilities. NEXRAD was developed under a joint program of the Department of Commerce (DOC)/NWS, Department of Defense (DoD), and the Department of Transportation (DOT)/FAA. NEXRAD radars are national critical assets that indirectly support most weather information users in the United States. These 160 systems are long-range, Doppler weather radars that detect and produce over 120 different long-range and high-altitude weather observation products and special products, including three-dimensional areas of precipitation by type, cloud cover by height, storm cells, winds aloft, turbulence, and icing. NEXRAD products and services are processed by FAA weather processing systems (discussed in next section). NEXRAD mosaics are used by en route and oceanic controllers to aid pilots in avoiding hazardous weather. The NEXRAD Dual Polarization (Pol) modification project has been deployed at all sites, and software improvements to further exploit this extensive weather capability are under development.

Dual Pol is an important upgrade to NEXRAD that improves detection of rain/snow mix, in-flight icing, hail, non-weather biological targets (bird strike concerns), and special DoD uses. It is expected to improve the forecasting of areas where in-flight icing will occur. Working with partner agencies, the NWS is investigating whether to incorporate planned long-range NEXRAD capabilities into the NSWRC.

Low Level Wind Shear Alert System (LLWAS)

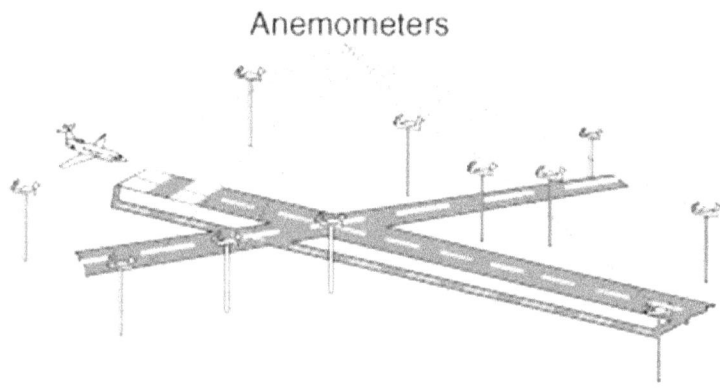

Anemometers

LLWAS uses an array of wind sensors located around the runway thresholds and along approach/departure corridors to measure surface wind direction and velocity. LLWAS compares wind velocity and direction, detected across the airport operations area, to determine whether hazardous wind shear events are occurring at, or near, the runways. LLWAS interfaces with TDWR and WSP at higher density airports where the systems are co-located, or operates as a stand- alone system at other airports that do not have a TDWR or WSP. LLWAS is undergoing service life extension and the FAA is investigating whether to incorporate the functionality of these sensors into the NSWRC.

Automated Surface Weather Observation Network (ASWON)

The ASWON portfolio includes a numbeer of surface sensors: Automated Weather Observing System (AWOS), Automated Surface Observing Systems (ASOS), Automated Weather Sensor Systems (AWSS), Stand Alone Weather Sensing (SAWS), Digital Altimeter Setting Indicator (DASI), Wind Equipment F-Series (WEF) Wind System, and AWOS Data Acquisition System (ADAS). These sensors measure weather parameters on the surface and report conditions to air traffic facilities and pilots. The terminal data they collect is important to pilots and dispatchers as they prepare and file flight plans. The overall surface data they collectively provide is vital for weather forecasting. Some elements of the ASWON portfolio are undergoing a technical refresh to keep these systems operating reliably.

Automated Surface Observing System (ASOS)

The ASOS, and other variants, such as the AWOS, AWSS and the SAWS system, have up to 14 individual sensors that measure surface weather data, including temperature, barometric pressure, humidity, type and amount of precipitation, cloud bases and amount of sky cover. These systems feed data directly to local air traffic control facilities and support automated broadcast of weather information to pilots and flight planning. They also provide regular, rapid updates for the NWS forecast models that predict future weather conditions including adverse weather. A technical refresh is underway to keep these systems operating reliably.

Digital Altimeter Setting Indicator (DASI)

The Digital Altimeter Setting Indicator (DASI) is a system that measures the atmospheric pressure and converts the measured pressure value into the actual sea level pressure based on the United States (U.S.) Standard Atmosphere. The value computed is known as the Altimeter Setting Indicator (ASI) value and is presented to the Air Traffic personnel in a digital format, e.g. 29.98, so they can inform pilots of the proper setting so the aircraft's altimeter will read the correct runway elevation at touchdown. The local DASI display panels provide a five-digit display of altimeter setting that is viewed on the Light Emitting Diode (LED) readout on the front. A keypad (or other means) on the unit enables the station altitude and a correction factor (if needed) to be entered into the system manually. Up to 10 remote displays may be accommodated.

AWOS Data Acquisition System (ADAS)

A software routine in ADAS receives lightning stroke information (latitude and longitude) system provides thunderstorm information (cloud-to-ground lightning reports) from the Automated Lightning Detection and Reporting System (ALDARS) out to 10 miles from the airport, and lightning information from 10 to 30 miles from the airport. It relays the stroke position to the ASOS and AWSS, which append the surface observation with the appropriate thunderstorm proximity indication. ALDARS appends the lightning information to the AWOS observation and sends the appended observation back to the AWOS for broadcast. Once the lightning data had been added to the observations, ADAS forwards them to NWS and WARP via WMSCR. The ADAS also disseminates weather it to area control facilities and FAA systems.

A Regional ADAS Service Processor (RASP), at each Continental United States (CONUS) Air Route Traffic Control Center (ARTCC), collects and bundles the automated surface observations from the Federal AWOS, ASOS, and AWSS within the ARTCC boundaries and distributes them to the local Weather Message Switching Center Replacement (WMSCR).

Some non-Federal AWOS are connected to WMSCR through FAA approved service providers. Only AWOS-III and IV are eligible for connection. They do not use ADAS.

Weather Processing Systems

Integrated Terminal Weather System (ITWS)

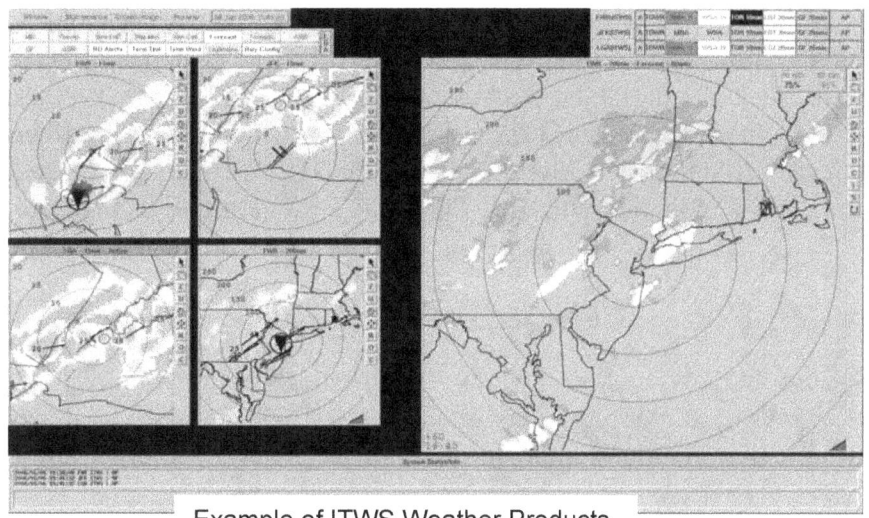
Example of ITWS Weather Products

ITWS generates automated weather products such as windshear alerts, gust front predictions (up to 20 minutes into future), storm cell intensity and direction of motion, terminal lightning information, detailed winds in the terminal area, and a one-hour storm forecast. ITWS receives weather data from automated surface observing sensors (ASOS and ADAS) and surrounding radars, (TDWR, ASR-9, ASR-11, and NEXRAD) and bundles it with NWS forecasts and value-added algorithms such as windshear prediction to provide real-time aviation weather information for terminal control facilities and pilots. Tower and Terminal Radar Approach Control (TRACON) controllers use the information to make more precise estimates of when runways should be closed and subsequently reopened. They also use the information to plan for efficient switches in terminal arrival patterns, which avoids inefficient maneuvering on taxi ways and on approach and departure, to accommodate a runway direction change as aircraft approach an airport. There are 34 ITWS Product Generators in-service at over one hundred sites (Air Traffic Control (ATC) Towers, TRACONs, Air Route Traffic Control Centers (ARTCCs) and the Air Traffic Control System Command Center (ATCSCC) providing common aviation weather information to 75 major airports. The FAA is targeting to implement the ITWS functionality in a NextGen Weather Processor (NWP) work package.

Corridor Integrated Weather System (CIWS)

CIWS gathers weather information along the busiest air traffic corridors to help air traffic specialists select the most efficient routes when they must divert traffic to avoid severe weather conditions. CIWS gathers convective weather information from radar and satellite images, blending it with NWS model data to provide an extrapolation of storm position and intensity 2 hours into the future. This process widens the common weather picture in between major metropolitan areas and expands en route weather situational awareness along busy jet-ways and air traffic corridors to help air traffic specialists select the most efficient routes when they must divert traffic flows to avoid severe weather conditions. CIWS has added a winter weather mode, convective initiation, and two-hour look-ahead aviation product to help pilots and flow managers better decide whether to fly over or around storms. The FAA is targeting to implement the CIWS functionality into the NWP.

Weather and Radar Processor (WARP)

WARP provides NEXRAD precipitation intensity data to controllers' displays. WARP compiles information from many FAA, NWS, and commercial weather sources for integrated display to ARTCC Traffic Management Unit (TMU) supervisors, sector controller's briefing terminals,

NAS operators of En Route Automation Modernization (ERAM), User Request and Evaluation Tool (URET), Advanced Technologies & Oceanic Procedures (ATOP), Dynamic Ocean Tracking System (DOTS), Alaska Flight Data Processor 2000 (FDP2K), and other automation/display systems and web users. WARP serves as a one-stop source for weather interpretation by the CWSU forecasting stations. WARP sustainment activities are underway and the WARP mosaic change is targeted to complete in 2015. The FAA is targeting to incorporate WARP functionality into the Common Support Services – Weather (CSS-Wx) and the NWP.

Aviation Digital Data Service (ADDS)

The Aviation Digital Data Service (ADDS) makes available to the aviation community text, digital and graphical forecasts, analyses, and observations of aviation-related weather variables. ADDS is a joint effort of the Federal Aviation Administration and the NOAA NCEPAviation Weather Center (AWC).

The success of ADDS (averages over 15 million hits a day in 2013) has allowed the FAA to showcase several important research initiatives, including the Current and Forecast Icing Products, the Graphical Turbulence Guidance, and the National Ceiling and Visibility Analysis. It is important to note that these products have been through thorough technical reviews and safety risk assessments before release on ADDS as operational products. Operational ADDS runs at the AWC in Kansas City, MO.

The FAAs Aviation Weather Research Program (AWRP) funds and directs the continuing development of ADDS as well as other experimental products being developed by the AWRP, while NOAA funds ADDS maintenance and operations. The results of the latest ADDS development efforts along with new experimental AWRP algorithm results can be viewed on http://weather.aero/, the Experimental ADDS website.

The Operational ADDS Web Site, currently running at AWC will be rehosted on the Weather and Climate Operational Supercomputer System (WCOSS) at NCEP Central Operations. This change will enable open source compatibility and easier access to data sources. The latest enhancements to AWRP capabilities in in-flight icing analyses and forecasts, turbulence nowcasts and forecasts, and national ceiling and visibility analyses, etc. will continue to be implemented onto ADDS. Additionally, mobile device compatibility, user portals for GA/135, 121 pilots, dispatch, Alaska, and passengers, and customized weather briefing capability are some of the main upcoming and long-term activities.http://aviationweather.gov/adds.

Weather Message Switching Center Replacement (WMSCR)

This system, with nodes in Salt Lake City and Atlanta, collects, validates, packages, and distributes alpha-numeric weather and Notices to Airmen (NOTAM) information nationwide. It is also a Service-Oriented Architecture (SOA) publisher of Pilot Report (PIREP) and altimeter data. In addition, WMSCR receives OMO data from ADAS for distribution to WARP and ITWS, as well as distributing Lightning Detection Data from ADAS to ITWS in the FAA Common Laser Dispatch Data file format. Initiatives in 2014/2015 include migration to a common platform at the Network Enterprise Management Center (NEMC) and continuing support of National Airspace Data Interchange Network (NADIN) II decommissioning. This functionality is targeted to be subsumed under a future CSS-Wx Work Package 2.

Weather Camera Program

The primary goal of the FAA Aviation Weather Camera (AWC) Program is to improve aviation safety and efficiencies by providing current visual weather information in the form of near real-time video camera images to aviation users in Alaska. The camera images are designated as an FAA supplementary weather product used for enhanced situational awareness and the images are made available free on the public website at http://avcams.faa.gov. The camera images provide pilots, dispatchers and Flight Service Station Specialists with up-to-date weather conditions at airports, mountain passes, and other strategic Visual Flight Rules (VFR) locations and enables them to make better informed decisions about whether or not it is safe to fly before becoming airborne. It provides pilots with the ability to "look before you fly" as opposed to the age-old procedure "fly out to take a look". New capability increases pilot situational awareness, enhances pilot flight decisions and when combined with other available weather information products, such as METARs, weather camera images become a powerful "go or no-go" aeronautical flight decision tool.

With the known benefits of the camera systems and the low costs of its operation there is a high level of interest to expand the service to the remainder of the NAS where shortfalls in weather related aviation safety and efficiencies are known or suspected. Past and ongoing aviation studies have identified additional safety and efficiency shortfalls in areas of the contiguous United States and Hawaii where safety and efficiency shortfalls are identified. Studies conducted by the FAA office of Aviation Safety Information Analysis and Sharing (ASIAS), and other FAA offices support the fact that weather cameras may be beneficial to aviation safety. In August 2013, the NTSB publically recognized the safety benefits of the weather camera system in Alaska and has forwarded a request to the FAA to install and maintain camera systems in Hawaii and the contiguous United States where cameras can address identified weather-related aviation accidents and safety shortfalls.

Flight Services

Flight Services collects and disseminates aeronautical and meteorological information, providing customized pre-flight and in-flight briefings to domestic and international general aviation communities, as well as to the military and commercial air carriers throughout the contiguous United States, Hawaii, Puerto Rico, and Alaska. These services are provided by flight service specialists to pilots via phone or radio, or pilots can access the information directly through online (internet-based) web portals. Flight service specialists interpret weather and aeronautical data to provide pilot weather briefings and flight planning services tailored for a particular flight. Alternatively, pilots accessing preflight information directly through a web portal are responsible for interpreting the weather and aeronautical information for their flight.

The FAA delivers these services through the following combination of contract vehicles and automation systems: Operational and Supportability Implementation System (OASIS), Flight Services for the 21st Century (FS21), and Direct User Access Terminal Service (DUATS).

Operational and Supportability Implementation System (OASIS)

OASIS is a flight service automation system operational at the 17 Flight Service Stations in Alaska since 2007 and is owned and supported by Harris Corporation. FAA Air Traffic Control Specialists operate OASIS and site level maintenance is performed by FAA Technical Operations personnel. OASIS provides integrated textual and weather graphics products, flight plan processing, emergency services, law enforcement, flight planning and regulatory information and other services as defined in FAA Joint Order (JO) 7110.10.

Flight Services for the 21st Century (FS21)

FS21 is the automation system operational at five Flight Service Stations located in CONUS operated by Lockheed Martin Corporation under the FAA's Automated Flight Service Station (AFSS) contract. The contract provides flight services to pilots throughout CONUS, Hawaii, and Puerto Rico. Lockheed Martin flight service specialists operate FS21 and are FAA-certified pilot weather briefers. FS21 provides integrated text and graphical weather products, flight planning and flight plan processing, emergency services, law enforcement, flight planning and regulatory information, and other services as defined in FAA JO 7110.10. The vendor also provides FAA-certified pilot weather briefers to operate the system.

Direct User Access Terminal Service (DUATS)

DUATS is an internet-based weather/aeronautical information and flight plan filing service that allows pilots to access information on-line to plan a flight, file a flight plan, and perform a self-brief without the aid of a Flight Service Specialist. The FAA contracts for this service with two separate vendors: Data Transformation Corporation (DTC) and Computer Sciences Corporation (CSC). DUATS provides aviation weather in both text (alphanumeric) and graphical formats according to standards defined in FAA JO 7110.10.

Contract Weather Observer (CWO) Program

The Office of Management and Budget (OMB) created the Contract Weather Observer program in 1995, for the FAA and NWS to provide oversight and program management activities. As of October 1, 2013, NWS transferred responsibility for training oversight, certification, and facility inspection of Limited Aviation Weather Reporting Stations (LAWRS), CWOs, and non-Federal Weather Observers (NF-OBS), to the FAA. The FAA continues to provide backup and augmentation to the ASOS and the NWS provide maintenance of the ASOS program.

In February 1996, a Service Standards Policy was developed and implemented by the FAA, NWS, and industry representatives for airports having an operational ASOS. The FAA then assumed sole responsibility for providing program management and supplying operational certified CWOs to provide backup and augmentation of the ASOS at Service Level A and B airports. At Service Level C airports, the FAA uses Air Traffic Control personnel to perform LAWRS duties, which provide backup for the ASOS.

Next Generation Air Transportation System (NextGen)

NextGen is an umbrella term for the ongoing, wide-ranging transformation of the United States' NAS to ensure that future safety, capacity and environmental needs are met. NextGen will fundamentally change the way air traffic is managed by combining new technologies for surveillance, navigation, and communications with workforce training, procedural changes, and airfield development.

NextGen requires efficient consolidation of large volumes of weather observations and forecast information for processing, dissemination, and integration into decision support system algorithms to produce the more sophisticated aviation weather products of how weather will impact NAS operations. NextGen core technologies will allow introduction of new NextGen operational improvements. Efforts under NextGen include: CSS-Wx, NextGen Weather Processor (NWP), ATM-Weather Integration Concepts, and NextGen Surface Observing Capability.

Common Support Services – Weather (CSS-Wx)

In conjunction with the deployment of the System Wide Information Management (SWIM) Enterprise Service, a common information publishing capability will be deployed that will include a first offering of aviation weather information. This subsumes the major functions ascribed to NextGen Networked Enabled Weather (NNEW) and will, over time, include additional information types – aeronautical information, flight information – as these move to new information protocols and formats. CSS-Wx extends the SWIM core services, and is being developed to enhance the collection and dissemination of aviation weather information and provide access to all users throughout the NAS. The FAA is currently performing investment analysis activities and in 2014 will make a decision on whether to implement CSS-Wx.

NextGen Weather Processor (NWP)

The NWP establishes a common weather processing platform that will enable the consolidation of the legacy FAA aviation weather processor systems and host new capabilities. As input, NWP uses information such as FAA and NOAA radar and sensors and NOAA forecast models. NWP will use sophisticated algorithms to create aviation-specific current and predicted weather, requiring no meteorological interpretation, for publishing via CSS-Wx. It will perform Weather Translation, which will enable the use of automated objective weather constraint information by decision-makers and Decision-Support Tools (DST). The FAA is currently performing investment analysis activities and in 2014 will make a decision on whether to implement the NWP.

ATM-Weather Integration Concepts

Currently NAS weather data is not well integrated into either manual procedures or automated decision-support systems. To support the predicted volume of future air traffic operations, improvements are needed. Unpredicted changes in weather are prime concerns because of the significant impact and disruption they create throughout the entire NAS. The current system does not respond well to unpredicted weather situations or to weather conditions that evolve differently than expected. This effort will address required improvements to support proactive planning operations rather than adjusting for impacts after the weather has changed. Activities in 2015 include technical studies for improvements to aviation weather information integration, transitioning an International Civil Aviation Organization (ICAO)-compliant Quality Management System (QMS) to operations, and harmonization of aviation weather requirements with the international community.

NextGen Surface Observing Capability (NSOC)

The existing surface weather observation sensor network is comprised of aging, stand-alone capabilities. Ongoing technical refreshes and Service Life Extension Programs (SLEP) can keep these sensors operating in the near to mid-term. While the current sensor network performs adequately, it is becoming increasingly costly to maintain. The NSOC will develop and evaluate methods to consolidate existing surface weather sensor networks, provide improved capability, and allow sensor outputs to be more universally available. In the near term, NSOC will explore existing automated winter weather sensing shortfalls in the ground-based weather sensor network. These shortfalls will be documented, socialized, and prioritized with key stakeholders including operators and project sponsors before being packaged into a weather work package. The work package will then be engineered and demonstrated, with an emphasis on the identification of potential operational and technical risks to successful integration into the NAS. System engineering activities will then commence to mitigate the risks, with the progress of the concept maturity documented via increasingly detailed design documents, evaluation matrices and reports and demonstrated through a series of increasingly challenging concept maturity technology demonstrations (CMTD).

FAA International Obligations

The FAA is the Meteorological Authority for the U.S. and establishes the requirements for meteorological services provided by the NWS to support international air navigation, as required by ICAO's Annex 3, Meteorological Services for International Air Navigation. As part of the requirements process the FAA, as a member of several ICAO meteorological groups, represents U.S. interests and strives for global harmonization of meteorological products and services. The FAA represents the Unites States on ICAO's World Area Forecast System Operations Group, ICAO's International Airways Volcano Watch Operations Group, ICAO's Aerodrome Meteorological Observation and Forecast Study Group, ICAO's Meteorological Warnings Study Group, and ICAO's Meteorological Aeronautical Requirements and Information Exchange Project Team.

SUPPORTING RESEARCH PROGRAMS AND PROJECTS

The National Airspace System of the Future

To address the growing demands on the NAS for the future, the 108th Congress and the George W. Bush Administration promulgated and signed into law VISION 100 Century of Aviation Reauthorization Act (P.L. 108-176). The Vision 100 Act calls for an integrated, multi-agency plan to transform the nation's air transportation system to meet the needs of the year 2025 and beyond, while providing substantial near-term benefits. The resulting Next Generation Air Transportation System (NextGen) initiative addresses critical safety and economic needs in civil aviation, while fully integrating national defense and homeland security improvements into the future NAS.

NextGen weather development activities will contribute to: (1) *Expanded Capacity* by providing air traffic managers the ability to better plan for predicted weather impacts on air travel, thus maximizing air space usage and optimizing flight routes; (2) *Improved Safety* by providing pilots and air traffic managers the ability to better assess and avoid hazards to air travel, such as severe turbulence, and (3) *Protection of the Environment* by enabling flight route optimization on the ground and in the air, thereby avoiding ground delays or holding patterns that require unnecessary jet fuel expenditure.

FAA Interagency Planning Office (IPO)

The FAA, NASA, and the Departments of Commerce, Defense, Homeland Security, and Transportation, along with the private sector and academic community, are working together with the Office of Science and Technology Policy to design and build the Next Generation Air Transportation System (NextGen). Under the direction from H.R. 3547 Consolidated Appropriations Act, 2014 to transition Joint Planning and Development Office activities to the FAA NextGen office, the FAA established the IPO in May 2014 to facilitate interagency collaboration, harmonize aviation-related research activities, and help effect technology transfer between the NextGen partners. The IPO worked with the agencies during FY14 to identify several potential technology development efforts that could enhance weather integration into national airspace operations and support emerging weather and air surveillance requirements.

FAA NextGen Office

Two principal FAA entities that report to the Assistant Administrator for NextGen are focused upon implementation of NextGen: the Office of NAS Lifecycle Integration and the Aviation Weather Division (AWD) under the Office of Advanced Concepts and Technology Development. The role of the Lifecycle Integration Office is to ensure that the plans for the several NextGen strategic thrusts, called NextGen Portfolios, are coordinated and integrated for efficient near- and medium-term implementation across the FAA. This includes the NextGen Weather research and development as pre-implementation activities in the NAS Infrastructure Portfolio. Common Support Services-Weather (CSS-Wx), discussed earlier, is a transformational program that will address weather dissemination within the Air Navigation Service Provider infrastructure. NextGen Weather pre-implementation is focused on improving weather observations, weather forecasts, and operational decisions based upon that improved weather information by integrating it into manual and automated decision support tools in the NAS. In addition to working with the AWD, this portfolio coordinates the investment analysis and acquisition of new weather systems and services within the FAA's Aviation Weather Services Directorate (AWSD), NOAA/NWS, and other agencies.

FAA AWG and AWSD Roles in NextGen Transition

The AWD and AWSD have important roles in the transition from today's aviation weather services to future CSS-Wx and NextGen weather forecast improvements, and other NextGen Weather Processing capabilities, as the FAA moves from air traffic *control* to air traffic *management* (ATM). In the NextGen system, most communications will occur as digital data, much of it transferred directly from computer to computer. Relevant information will be shared easily among system users through network-enabled information access.

The AWSD develops mission need and investment analysis for initial investment decisions for FAA aviation weather sensors, forecasting capability, dissemination systems, and integration of improved weather capability into the NAS. The focus is on NextGen, including collaboration with Single European Sky ATM Research (SESAR), with ICAO for advanced aviation weather standards, and with all the U.S. agencies involved in NextGen. This work addresses the high cost of weather to today's NAS, where weather is responsible for 70 percent of delays over 15 minutes and contributes to 24 percent of accidents and 34 percent of fatalities. Up to two-thirds of weather delays are avoidable, but despite a continuous flow of improvements available through aviation weather science and implementation solutions aimed at providing better weather information, weather continues to have significant impacts on aviation costs and safety. The purposes of this program are to reduce the number of weather-related aviation accidents; reduce aviation flight delays, diversions, and cancellations; improve the operational efficiency of the NAS; and harmonize ICAO standard with U.S. practices in aviation weather.

The NextGen program in AWSD is composed of three elements:

1. The Concept Identification and Development component generates, analyzes, manages, allocates, and validates requirements in the NextGen aviation weather portfolio. It focuses on the early stages of requirements from their inception/generation to the investment analysis and subsequent requirements decision. It develops transformational (NextGen mid-term and far-term capabilities), as well as evolutionary requirements (NextGen near-term capabilities). It formulates agreements between government and industry stakeholders on policies needed to meet requirements for airborne weather observations, including cost sharing, data access and distribution, data reporting frequency, aircraft equipage, and other technical issues. Finally, this program develops policies necessary for the allocation of roles and responsibilities in the provision of weather state information to meet requirements and U.S. commitments to ICAO.

2. The Global Harmonization component carries out FAA's role as the U.S. Meteorological Authority to ICAO. It promotes global harmonization through the development of ICAO Standards and Recommended Practices and manuals/guideline documents for surface and airborne observations/forecasts and global dissemination of aviation weather information that are supportive of NextGen. This work is accomplished through developing and presenting, to 12 ICAO planning, study, and operations groups, U.S. positions on issues arising from the ICAO Volcanic Ash program, the World Area Forecast System, the international Space Weather program, and amendments to ICAO Annex 3 and other guidance material to incorporate the NextGen concept of the 4-D Weather Data Cube.

3. The <u>System Performance</u> component develops metrics that provide a framework for enabling the FAA to measure the benefits of weather information for air traffic operations. It maintains standards for surface observations for the backup and augmentation of ASOS.

The development of NAS weather requirements under this program is an essential artifact of the following NextGen documents: the Weather Concept of Operations, the Mission Needs Statement for Weather (MNS-339), the Preliminary Portfolio Requirements, and the Final Portfolio Requirements. The requirements work in this program feeds later-stage activities, as defined by the FAA Acquisition Management System (AMS) lifecycle, of the RWI solution set.

The NAS Infrastructure Portfolio

The NAS Infrastructure Portfolio is a planning and development portfolio that includes elements to ensure NextGen operational weather capabilities utilize a broad range of weather improvements and technologies to mitigate the effects of weather in future NAS operations. This portfolio has two major NextGen weather elements: NextGen weather observation improvements and NextGen weather forecast improvements. The NextGen weather elements will address many weather problems including, but not limited to, rightsizing the observations network, transition of weather research to operations, development of weather impact metrics, development of weather decision support tools, integration of weather information into operations, weather processor architecture redesign and restructuring, and transition planning for legacy systems. The NextGen weather elements will conduct planning, prototyping, demonstrations, engineering evaluation, and investment readiness activities leading to an implementation of operational capabilities throughout the NextGen near, mid, and far terms.

A consistent and effective weather observation sensor network will be a cornerstone to improved NextGen weather capabilities. Currently the United States has fielded multiple weather surface sensor networks that vary in age up to 30 years. Ongoing technical refreshes and SLEPs can keep these sensors operating in the near to mid-term. However, as the demands of the NAS increase in the future, the present array of surface sensor systems will not be capable of delivering the required functionality. In addition, potential NextGen weather observation requirements might exceed current surface sensing capabilities (e.g., improved weather model initialization for increased weather forecast accuracy). Current surface observation systems also contain considerable overlap and waste that should be engineered out of the NAS. NextGen weather observation improvements will explore concepts for a next-generation surface sensing capability that can satisfy all current surface sensing requirements in a single system and be easily expandable to meet any future NextGen requirements.

The second NextGen weather element, NextGen weather forecast improvements, addresses the need to enable better weather decision-making and use of weather information in the transformed NAS. This need includes: (1) integrating weather information tailored for decision support tools and systems into NextGen operations, (2) implementing improved forecasts by transitioning advanced forecast capabilities from aviation weather research, (3) developing and using metrics to evaluate the effectiveness of weather improvements in the NAS, (4) developing probabilistic forecasts that can be effectively used in air traffic and traffic flow management, and (5) determining the most effective solution for a processor architecture to support these capabilities.

RWI will propose recommendations for near, mid, and far time frames, including a recommendation for transition of legacy systems.

Collectively the effect of the NextGen NAS Infrastructure Portfolio weather elements will result in aviation weather information no longer being just a stand-alone display, requiring cognitive interpretation and impact assessment, with limited ability to significantly mitigate delays. Instead, weather information is being designed to integrate with and support NextGen decision-oriented automation capabilities and human decision-making processes.

NOAA/National Weather Service

NOAA's commitment to the NextGen initiative has two core components: NextGen IT Services which will provide improved access to NOAA weather information in the formats required by NextGen users and NextGen Weather Science and Applications which will improve the accuracy, resolution and timeliness of weather information relevant to aviation users.

Specifically, NextGen IT Services will provide an information technology infrastructure comparable to those already employed by other government agencies and by industry which is flexible and extensible to keep pace with user's weather information needs in a rapidly changing technological environment. The greater and easier access to NOAA weather information for aviation decision-makers will facilitate better integration of this information into aviation users' decision-making processes and systems.

NextGen Weather Science and Applications will develop aviation weather information data sets, consistent in time, space and among weather elements, providing a complete picture of how weather will impact aviation across the NAS. This more accurate aviation weather information, achieved through higher resolution weather models, will improve air traffic managers' ability to fine-tune their assessment of the impact of the weather on airports and air routes to safely maximize available air space. The Program will also develop advanced aviation forecast generation techniques and tools to allow NWS meteorologists to generate aviation weather information faster and more accurately.

NOAA aviation weather information will be available to users of the NAS and to the FAA Air Traffic Management community through the FAA's Common Support Services – Weather Program (CSS-Wx). While NextGen IT Services implements common weather data services, CSS-Wx will also implement these common data services within the FAA enterprise, and will utilize the weather data and information from NOAA for use in weather impact decisions for aviation.

NOAA/Office of Oceanic and Atmospheric Research

Earth System Research Laboratory (ESRL)

Within NOAA's Office of Oceanic and Atmospheric Research (OAR), the Global Systems Division of the Earth System Research Laboratory (ESRL/GSD) evaluates aviation weather impact variables such as icing, turbulence, ceiling and visibility, and convective weather and develops decision tools for NWS forecast offices, FAA traffic managers, and commercial and civil aviation. Specifically, GSD has and will continue to develop capabilities to allow the forecaster to integrate, view, and manipulate observations from current and planned meteorological sensing systems using computer-assisted data display and synthesis techniques.

Improvements in weather and climate prediction are tied to developing and running finer scale weather forecast models and model ensembles on increasingly powerful High-Performance Computing Systems (HPCS). Central Processing Unit-based (CPU-based) HPCS computing is growing more expensive as hundreds of thousands of CPU cores are being combined to provide increased compute capacity. These large CPU-based systems require large and expensive facilities and infrastructure to provide sufficient power and cooling ($5M for construction and $1M annually to run). Given tight federal budgets, CPU-based systems are, therefore, increasingly unaffordable. GSD is exploring the use of Massively Parallel Fine-Grain computing (MPFG) systems, which can be 5-10 times faster than CPU-based systems, to run NWP models. By using MPFG systems, savings to NOAA for facilities space, infrastructure, and electrical costs can be substantial.

Plans for FY 15:

- Development of decision-support tools for use at FAA Center Weather Service Units
- Evaluate upgraded algorithms for detecting and forecasting aviation impact variables
- Run several Numerical Weather Prediction (NWP) models on (MPFG) systems to determine performance improvements over CPU-based systems
- Create product description documents and product schemas for the Request For Proposal, and subsequent documents, associated with the FAA's CSS-Wx implementation contract competition

National Severe Storms Laboratory (NSSL)

NSSL is participating in the effort to help NOAA's support for the NextGen initiative. Planning with the FAA and the NWS began in FY 2009 and is anticipated to continue for the next several years. NSSL is also working with the FAA's AWRP to develop weather radar applications that enhance the safety and efficiency of the aviation community and the NAS. Work is focused on both convective weather and winter weather, with special attention to treating all WSR-88D radars within the continental United States as a single network. As part of this effort, the Multi-Radar, Multi-Sensor (MRMS) system is being used to support intensive research directed toward polarimetric radar applications unique to aviation needs. Examples using WSR-88D dual-pol radar data include wintertime quantitative precipitation estimation, detection of icing (supercooled liquid water) conditions and winter precipitation type (snow, sleet, freezing rain, etc.) and data quality issues unique to FAA users. Work has also begun on a proposed prediction system will involve building an algorithm to predict the porosity of thunderstorms and thunderstorm complexes/systems by applying pattern recognition and geospatial data mining to forecast fields from convection-allowing numerical weather prediction models. The algorithm will first be applied to members of a convection allowing model ensemble to diagnose whether the modeling system has any skill at predicting porosity. Then, assuming the ensemble has sufficient skill for relevant forecast lengths, NSSL will explore methods for using the ensemble for predicting the likelihood that a region or specific flight corridor will be porous for airplane traffic.

NSSL is actively engaged in a risk reduction activity for the Multifunction Phased Array Radar (MPAR) technology with the FAA and industry in designing and testing dual-polarized phased array radar panels required to determine the feasibility of using MPAR as the replacement technology for weather and aircraft surveillance radars (i.e. multi-function). The NSSL continues to fund the development of 2 small-scale demonstrator phased array radar systems with research interests at the University of Oklahoma and MIT Lincoln Laboratory. These demonstrators will provide an evaluation of dual polarization phased array radar technologies on a planar and cylindrical platform. NSSL has also begun a multi-year collaboration with the FAA to upgrade the National Weather Radar Testbed (NWRT) to an active array dual polarization antenna and should be completed by FY17.

Federal Aviation Administration

Aviation Weather Research Program (AWRP)

The AWRP program will continue research into geophysical phenomena in the atmosphere and around airports that present hazardous conditions for aircraft operations. Among these hazards are in-flight icing, convective weather, turbulence, restricted ceilings and visibility, volcanic ash, liquid, freezing and frozen precipitation, etc. Additional work is being done to improve models and in developing advanced weather radar techniques.

Service Analysis

Weather Research for Service Analysis is a critical analysis of the various technologies for translating weather information into aviation constraints, and the application of the constraints to quantify operational impacts. This work is conducted as part of the New Service Needs as initial input into the Mission Analysis phase of the Acquisition Management System (AMS) lifecycle. Major areas of work include: Conducting research associated with the Air Traffic Management (ATM) Weather Integration Plan for the Next Generation Air Transportation System (NextGen); Providing critical analyses of the various technologies for translating weather information into aviation constraints, and the application of the constraints to quantify operational impacts; Determining which technologies address constraints and which address impacts; Proposing weather translation technologies adequate for use with such programs as Wake Turbulence Mitigation for Departure (WTMD), Tower Flight Data Manager (TFDM), En Route Automation Modernization (ERAM), Traffic Flow Management System (TFMS), Closely-Spaced Parallel Operations (CSPO), and Time Based Flow Management (TBFM).

Observations

Liquid Water Equivalence (LWE) Rate Measurement & Reporting-Surface Snow/ Ice

At airports in cold climates, a ground anti-icing/de-icing program is an essential pre-takeoff service. Decision making in a ground anti-icing/deicing program is the responsibility of the pilot. Pilots need to have the most accurate airport weather observation information available to support their ground anti-icing/deicing program decisions. Recent weather research has provided compelling evidence that snow precipitation intensity, which is currently determined by a visibility-based method, can be more accurately determined by measuring the LWE rate and providing this information as part of the airport weather observation. Pilots can use this enhanced information to more accurately determine their hold-over times. To capitalize on the federal government's investment in ASOS, a cost effective way to provide this LWE rate capability is to implement it in ASOS. The FAA is currently pursuing ways to provide this capability in ASOS. This improvement will enhance aircraft and passenger safety and support traffic flow management. The FAA is also pursuing development of a commercial LWE Rate System (see In-flight Icing for details).

Advanced Weather Radar Techniques (AWRT)

The AWRT research focuses on monitoring and improving the quality of information derived from the Weather Surveillance Radar 88D (WSR-88D), TDWR, and Canadian and Mexican weather radar networks. To properly initialize convective weather, turbulence, icing and numerical weather prediction products, it is critical that ingested weather radar data be of high quality. The AWRT research is focused on developing quality control techniques that ensure weather radar information is free from false weather returns caused by biological, clutter, or out of tolerance radars. Further research involves monitoring the benefits from the recently fielded WSR-88D dual polarization capability. Ongoing efforts towards improving the quality of Canadian weather radar data, which is being incorporated into various air traffic management tools, will continue.

The AWRT research also involves techniques for mitigating gaps in individual weather radar networks via seamlessly integrating all weather radar networks together to form high resolution, three dimensional mosaics. This work has produced the Multi Radar Multi Sensor (MRMS) capability. Operating at a thirty second update rate and 1 Kilometer (KM) resolution, MRMS provides high resolution mosaic input to several FAA AWRP efforts, as well as Automatic Dependent Surveillance-Broadcast (ADS-B), Unmanned Aircraft Systems (UAS), Traffic Flow Management System (TFMS), and air traffic modeling and simulation research efforts at the William J. Hughes Technical Center (WJHTC).

Forecasts and Modeling

In-Flight Icing

This research is aimed at developing improvements to in-flight icing diagnosis and forecasting with a goal of reducing the rate of aircraft icing related accidents and fatalities for aircraft operations in the NAS. Over the continental United States, the Current Icing Product (CIP) and Forecast Icing Product (FIP) have been developed to provide hourly updates of current and forecast in-flight icing conditions out to 12 hours. These products include probability and severity of icing conditions, as well as super-cooled large drop potential. CIP and FIP products are available to all NAS users of the ADDS website, which provides comprehensive user-friendly aviation weather graphics including icing, turbulence, and convection. Current research efforts include development of a current and forecast icing product for Alaska, and development of enhanced icing weather diagnosis and forecast algorithms using weather parameters including liquid water content, drop size, and temperature. The latter research effort called Model of Icing Conditions for Real-Time Operations (MICRO) will be used to produce future ice accretion products that will likely apply to a particular class of aircraft (prop, turboprop, jet, etc.) for use in the NAS by pilots and flight crews in avoiding hazardous icing areas. Planned future efforts also include development of a global (oceanic routes) forecast icing product. Research was completed in FY2011 to enhance efficiency and safety during winter terminal operations through the development of a prototype for Liquid Water Equivalent Rate System (LWES) specification for determining holdover times more accurate than those from currently used visibility tables. Results from this research are being incorporated into the Terminal Area Icing Weather Information NextGen (TAIWIN). The goal of the TAIWIN effort is to develop a comprehensive terminal area ground and in-flight icing weather product capability (for use by air traffic, flight crews, and ground de-icing operations) to provide terminal area icing weather information for operational decision-making.

Convective Weather

Convective weather is a critical area of research due to the considerable impact convective storms have on NAS operations. These storms contribute to delays, diversions and cancellations, along with well-documented effects on safety and capacity. The FAA's AWRP Convective Weather Research Program seeks to address this problem by: (1) exploring how thunderstorm information is used to make decisions and assessing the need for improved dissemination and communication of this information; (2) increasing the fundamental understanding of how thunderstorms form, behave, and impact the NAS; and (3) developing advanced thunderstorm analysis and predictive capabilities.

The following research projects are underway: (1) determine the accuracy of high-resolution models to forecast convective weather in the 2-8 hour time frame; (2) Improve the blending of different weather model types for prediction of convective storms in the 1-4 hour time frame; (3) develop an initial technology to estimate radar-like precipitation intensity and echo top height of storms beyond radar range by combining global lightning data, satellite data, and output from numerical weather prediction models with existing radar mosaics; (4) Develop and advance probabilistic convective weather hazard guidance for oceanic regions from 0 – 36 hours; and (5) identify and quantify the impacts of ramp closures due to lightning threats on NAS efficiency and safety of ground personnel, and determine if inefficiencies in lightning procedures (or the lack of procedures) can be mitigated with streamlined and consistent decision support guidance. In addition to the specific areas of research identified above, convective weather research will also support the initiatives of the Collaborative Decision Making (CDM) community and work to improve the prediction of other convective weather hazards such as hail, windshear, and turbulence.

Turbulence

This research has focused on producing real-time turbulence nowcasts and probabilistic forecasts of turbulence. The method utilized in meeting these objectives is a turbulence-forecasting capability in conjunction with two supporting sensor capabilities: one for in-situ detection of turbulence and the second for remote sensing of turbulence. The in-situ work has resulted in the deployment of an aircraft-based Eddy Dissipation Rate (EDR) turbulence detection algorithm on aircraft at United Airlines and Delta Air Lines. Current efforts include deployment at Southwest Airlines. The remote sensing work has targeted the use of data from the NEXRAD radar network. Data from the NEXRAD Turbulence Detection Algorithm, currently operational on NEXRAD, will be used as input for the production of a turbulence analysis (nowcast) product.

A separate effort, the Weather Technology in the Cockpit (WTIC) Turbulence/EDR Uplink Demonstration, is assessing the feasibility of using a low-cost device for the displaying of turbulence forecasts and EDR information in the cockpit for direct use by the flight crew. This work will provide a more complete Cost Benefits Analysis of the use of the turbulence data in the flight deck.

Ceiling and Visibility (C&V)

This research has focused on developing automated ceiling and visibility products to support current needs and future NextGen requirements for improvements in safety and terminal area traffic flow efficiency. C&V research will provide NAS users with improved detection and analysis of ceiling and visibility conditions in terms of the geographical extent, timing and duration of events and improved accuracy of forecasts. C&V can significantly disrupt terminal operations by degrading an airport's arrival rate and, in more severe cases, closing the airport entirely. The FAA is working with the NWS to more effectively and efficiently integrate numerical model output and C&V algorithms into the production of NWS terminal aerodrome forecasts (TAFs) and area forecasts (FAs). Similarly, some of these advanced, automated capabilities will eventually be provided directly to Air Navigation Service Providers (ANSPs) and NAS operational decision makers. Currently, a gridded national C&V analysis (CVA) uses frequently updated C&V data from airfield observations to display C&V conditions as an automated national gridded display available via ADDS. C&V forecast information will eventually be provided hourly in both probabilistic and deterministic form to support both human users and future automated decision support systems. Also developed under AWRP, is the Helicopter Emergency Medical Service (HEMS) display tool used by emergency helicopter crews to facilitate their No-Go decisions. New HEMS capabilities will include base map upgrades, C&V trending, user-supplied map datasets, storage, and retrieval of user preferences.

Volcanic Ash

The FAA continues to lead the U.S.'s efforts to ensure the nation's air transportation system maintains an excellent safety record for operations in air space contaminated with volcanic ash. Following the eruption of the Icelandic volcano Eyjafjallajökull in April 2010, the FAA participated in the ICAO Volcanic Ash Task Force. This task force primary mission was to develop a global safety risk management framework to determine safe levels of operation in airspace contaminated by volcanic ash and deal with the issues from large volcanic eruptions that disrupt global air transportation. That volcanic event, which centered in Europe, caused global disruption in the aviation community, resulting in cancelations and delays of flights.

The FAA is taking the lead, in cooperation with other U.S. agencies and international partners, in the following areas and initiatives:

- Development of a Concept of Operations for volcanic ash information in support of air traffic operations and management

- Development of a collaborative forecasting model to support the information exchange and collaboration between various Volcanic Ash Advisory Centers as well as

Meteorological Watch Office, Air Navigation Service Providers, and Airline Operations Centers in volcanic ash cloud analysis and forecasts

- Support an engine testing initiative for volcanic ash

- Other efforts supported by various U.S. agencies and international partners include:

- Continued research efforts to improve volcanic ash dispersion and transport models, which are key to providing improved forecasts of the location of ash clouds

- Improvements in volcano monitoring and the detection and reporting of volcano eruptions

- Initial documentation of best practices for the VAACs

The successful completion of these programs will result in improvements to the air transportation system, which will reduce the economic impacts from large volcanic eruptions and ensure no decline in safety.

Model Development and Enhancement (MD&E)

This AWRP research is targeted at developing and improving weather prediction models and data assimilation systems to better characterize the state of the atmosphere with the goal of providing superior aviation weather information to enhance NAS safety and capacity. These assimilation and prediction systems utilize all the latest weather observations and advanced supercomputers to create the most accurate and timely depiction of the future state of the weather. The future weather information output from these computer models serves as the underpinning for virtually all weather guidance beyond the first hour.

This research has been a collaborative partnership of the FAA, NOAA, NCAR, the Center for the Analysis and Prediction of Storms, the Air Force Weather Agency, and the Naval Research Laboratory to build the state-of-the-art Grid-point Statistical Interpolation (GSI) data assimilation system and Weather research and Forecast (WRF) modeling framework. A key result has been the 2012 implementation into NWS operations of the hourly-updated Rapid Refresh (RAP) weather prediction system. This system has yielded better wind forecasts and improved diagnoses and forecasts of weather hazardous to aviation, including en-route turbulence, convective weather, in-flight icing, and restricted visibility, over an expanded domain including Alaska.

Current efforts are focused on improvements to the RAP to enhance convective weather, turbulence, ceiling and visibility, and icing forecasts as well as continued development of the High Resolution Rapid Refresh (HRRR), which provides storm-scale resolution to capture convective activity at the cell level. The HRRR is expected to be operational at NOAA NCEP in FY 2016.

Weather Technology in the Cockpit (WTIC)

The purpose of this research program is to reduce the negative impacts of adverse weather conditions on the National Airspace System (NAS) and aircraft by developing, verifying, and validating recommendations to support airworthiness standards for enabling availability and improving the quality and quantity of meteorological (MET) information available to the aircraft. There are numerous research projects within the WTIC portfolio to achieve this overall project goal of recommending a [1]Minimum Weather Service that aid pilots in decision-making when

faced with adverse weather conditions. The portfolio of projects will perform the research necessary to address the overarching WTIC research questions (the specific FAR Part 91 and FAR Part 121/135 research questions will be delineated below). Instrumental in gaining answers to these research questions is research to resolve any shortfalls, gaps, and issues that were identified as gaps in MET information in the cockpit. A brief summary of several WTIC research projects are detailed in the following subsections.

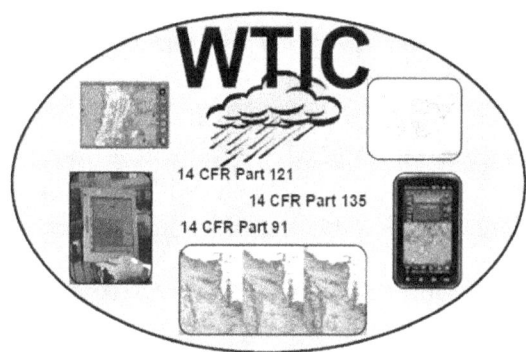

General Aviation(GA) Minimum Weather Service

The GA minimum weather service research effort is focused on the identification of Causal Factors for GA weather-related accidents and the gaps associated with weather information and presentation in the cockpit. The specific FAR Part 91 WTIC research will attempt to ascertain:

- Why has the weather-related accident rate for GA aircraft not decreased more with the commercial advances in MET cockpit technology and information? (i.e., obtain more insight into causes and locations of GA accidents and incidents)

- What is the minimum weather service for GA?

- What is the minimum MET information needed in the cockpit?

- What is the minimum quality of the MET information for the cockpit?

- What are the minimum rendering standards needed to enable correct interpretation of the minimum MET information in the cockpit and to enable consistent, safe, and effective decision-making?

- What are the shortfalls in pilot understanding and proper use of the information, and what are the pilot training needs to meet these shortfalls?

---

[1] The term "minimum weather service" as used in this document is defined as the minimum weather information content required in cockpits along with the associated parameters of that information, such as reliability, accuracy, update rates, and spatial resolution. The minimum weather service will also include rendering recommendations to reduce the likelihood of interpretation errors.

Research projects as part of the GA minimum weather service include defining information required to aid GA pilots in decision-making regarding adverse weather, reviewing the Aviation Safety Reporting System (ASRS) to identify and assess casual factors and trends of weather-related incidents, and conducting a causality analysis to identify, quantify, and assess gaps or issues with cockpit MET information. Additional research will include assessing the feasibility to develop agile, low latency, adverse weather cockpit alerts that require minimal cognitive processing by the pilot. The GA minimum weather service effort will also evaluate the utility of selected current MET products as decision support tools in high stress flight scenarios, and review issues associated with transition from Visual Flight Rules (VFR) into Instrument Meteorological Conditions (IMC) to determine decision making trends.

The Mobile Application project is researching MET information that provides benefits to the cockpit when presented on a mobile application. This project is also researching limitations of existing technology to support the concept of operation for this mobile application.

Results from Phase II of the WTIC Standardization Assessment project indicated statistically significant variations in pilot awareness of adverse weather events. Specifically, the recognition by pilots of METARs (surface observations) changing from indicating Visual Flight Rules (VFR) conditions to Instrument Flight Rules (IFR) conditions at their destination and potential alternate airports was far below expectations. Therefore, a third phase was deemed necessary. The third phase of the WTIC Standardization Assessment project will attempt to identify the elements in the presentations (such as colors, shapes, or a combination) that contribute to the performance variations and the design of those elements that provide most consistent safe performance and adverse weather event recognition.

The WTIC program is also performing research to develop and test scenario-based GA pilot web-based MET training courseware that are compatible with independent and instructor led training. This includes an associated set of comprehensive MET questions for the pilot written exam.

The WTIC program works closely with multiple RTCA Special Committees to collaboratively develop standards and guidance material associated with the delivery and presentation of weather information in the cockpit. The most extensive support is provided to RTCA SC-206, Aeronautical Information Services Data Link, and their efforts to develop recommendations for data link architecture and the dissemination of MET information between aircraft and ground users. The WTIC program also supports standards development by participating in the SAE G-10 effort to develop a standard symbology for the presentation of MET information in the cockpit.

Part 121 / 135 Minimum Weather Service

The Flight Deck Information research of the WTIC Minimum Weather Service for aircraft operating under FAR Part 121/135 research will attempt to ascertain:

- What are the current operational inefficiencies, by phase of flight, attributable to shortfalls of MET information in the cockpit?
- What are the potential NextGen operational inefficiencies attributable to shortfalls of MET information in the cockpit?
- What is the minimum weather service* for major aircraft categories in Part 121 and 135?

This WTIC Flight Deck Information research project includes research to provide accurate and timely wind information to flight management systems (FMSs), and quantify any predicted benefits of associated NextGen application programs (e.g., TBO, CLEEN, Interval Management). In addition, this minimum weather service effort will examine methods designed to improve characteristics of MET information such as format, latency, accuracy, etc., so that the information is ready for direct integration into select decision support tools (DSTs).

The WTIC program In Situ Turbulence project goal is the successful integration of objective turbulence information into NAS operations, both during the strategic flight planning process as well as the tactical process of re-routing aircraft during flight. The intent here is to provide the cockpit with current and forecast turbulence information in order to permit the pilot to be a more informed decision maker in terms of flight planning and cabin management. In order to achieve this goal the WTIC program will support the expanded deployment of the EDR algorithm throughout the airline industry (through government funded efforts and advocacy) in order to support meeting the required density of turbulence observations to support NextGen.

The WTIC program MET Information in Oceanic and Remote Regions Shortfall Analyses project will identify current operational inefficiencies, potential NextGen operational inefficiencies, and potential safety risks (current and NextGen) attributable to gaps of MET information in the cockpit in oceanic and other remote regions. The results of the research will partly address the WTIC research question of identifying current operational and NextGen inefficiencies attributable to gaps of MET information or technology in the cockpit. Previously accomplished WTIC research indicated benefits to providing cloud top information in oceanic and remote regions. This research is intended to identify gaps that are not fully resolved by providing cloud top information in the cockpit.

The WTIC program Adverse Weather Alerting Functions Use Case Analysis will identify Use Cases where a weather alerting function may address a weather information gap or assist in pilot decision-making relative to an adverse weather condition along their intended route of flight. The research will also include analyses to determine the concepts of an alerting function with performance characteristics that could provide benefit to pilots. The deliverables from this research will be used by the WTIC program to select the weather alerting functions to be prototyped and evaluated in a flight simulation demonstration.

Standards Support

The WTIC program works closely with multiple RTCA committees. The most extensive support is provided to RTCA SC206 and their efforts to develop recommendations for data link architecture and the dissemination of MET information between aircraft and ground users. WTIC program standards support also includes participating in the SAE G-10 effort to develop a standard symbology for the presentation of MET information in the cockpit.

Flight Deck Information

The WTIC Wind Diagnosis project is focused on defining elements of wind quality that impact NextGen application programs and performing simulations to assess those impacts based on varying wind quality.

The Cloud Top Heights - Human over the Loop Demonstration (HOTL) will assess the impacts to decision making in a collaborative environment of providing cloud top data in oceanic and data sparse areas.

The Mobile Application project is researching MET information that provides benefits to the cockpit when presented on a mobile application. This project is also researching limitations of existing technology to support the concept of operation for this mobile application.

Human Factors

The Standardization Assessment project is performing a quantitative assessment of the impacts of non-standardized MET presentations for General Aviation (GA) displays. This research consists of three groups of pilots flying a GA simulator with one of three common MET presentations. Researchers then measure variations in performance and multiple human factors parameters such as eye fixations.

A second human factors project is working with the National Aeronautics and Space Administration (NASA) to perform detailed callbacks on Aviation Safety Reporting System (ASRS) incident reports that identify weather or data linked MET information as a contributing factor.

The WTIC program is also performing research to develop inputs for improving pilot training relative to MET information and new MET technology.

Air/Ground Integration

Research in this structural element includes modeling generic versions of common data links (e.g. Immarsat) to assess their bandwidth, latency, and other quality factors related to their capability to disseminate MET information.

The Uplink of Advanced Weather Radar Mosaic project is assessing the potential benefits and the feasibility of up linking weather mosaics to the cockpit and to downlink sensor data from aircraft to enhance the quality of the mosaics.

CLIMATE SERVICES

For the purposes of this *Federal Plan*, Climate Services are specialized meteorological and hydrological services established to meet the requirements of Federal, state, and local agencies for information on the historical, current, and future state of the earth system. Climate services include observations, monitoring, assessments, predictions, and projections of the atmosphere, hydrosphere, and land surface systems.

OPERATIONAL PROGRAMS INCLUDING PRODUCTS AND SERVICES

NOAA/National Weather Service

National Oceanic and Atmospheric Administration's (NOAA) National Weather Service (NWS) provides climate services through the National Centers for Environmental Prediction's (NCEP) Climate Prediction Center (CPC), the NWS Climate Services Division (CSD), and more than 150 NWS regional and local offices nationwide. The NWS Climate Services Division (CSD) provides programmatic support to CPC and NWS local offices. In collaboration with other NOAA offices, CSD develops training, education, and outreach materials. CSD designs climate data mining and analysis tools, as well as local climate products for users of climate information. CSD fosters climate services ties with other NOAA offices, federal government agencies, academia, and external climate communities.

CPC delivers a broad range of products and services including climate monitoring and forecasts, as well as regional climate impacts information. The product suite spans time scales from a week to seasons, extending into the future as far as technically feasible, and covers the land, the ocean, and the atmosphere. CPC also identifies the important physical factors responsible for variability, anomalies, and trends in global and regional climate in order to improve these products. The NWS Weather Forecast Offices (WFOs) use these products to deliver climate services to local users., Together with NWS regional offices WFOs are responsible for collecting climate data, developing locally-relevant products, and conducting local climate studies in response to user needs. Applications include the mitigation of weather-related natural disasters and uses in agriculture, energy, transportation, water resources, and health. Additionally, WFOs issue daily and monthly climate reports for their areas of responsibility, providing localized information about temperature and precipitation records and extreme events such as droughts. WFOs serve as the local NOAA user interface for climate services, including outreach and education. They are also responsible for the integrity and continuity of the historical climate record for their area of responsibility.

NOAA/NESDIS/National Climatic Data Center

The National Climatic Data Center (NCDC) is the largest weather and climate data center in the world. NCDC receives, processes, archives, and disseminates surface, marine, upper-air, radar, satellite, and model output data. NCDC serves a large and diverse community, responding to more than one million information requests per year. Environmental data and information is made available through both the Internet and physical delivery of products and services. NCDC's climate data products support decision making in many sectors of the economy, including energy, transportation, agriculture, insurance, engineering, health care, and manufacturing. NCDC's data holdings and expertise also enable the private sector to make better business decisions, save money, improve their products, expand their businesses, and reduce their impact on the environment.

The State of the Climate in 2012 Report. The "State of the Climate" series has provided a detailed update on global climate indicators, notable weather events, and data collected by environmental monitoring stations and instruments since the report's inception in 1990. For the 13th consecutive year, NCDC scientists served as lead editors of the report, which was compiled by 384 scientists from 52 countries across the globe and published as a supplement to the *Bulletin of the American Meteorological Society.* For the second year, the report was accompanied by an online rollout at Climate.gov, providing easy access to key themes of the report for the public.

The report used dozens of internationally recognized climate indicators to track and identify changes and trends in the global climate system. Each indicator includes thousands of measurements from multiple independent datasets. The diverse array of international authors helped consolidate this wide range of data, using their expertise to more fully understand and communicate the state of the complex climate system. The report provides a valuable reference for the increasing number of professionals and consultants who use climate conditions and trends in their work.

NCDC Website and Home Page. NCDC enhanced its website and redesigned its home page by simplifying access to systems, data, and services. In addition, more efficient maintenance was implemented and Google Analytics was introduced to better analyze website traffic and anticipate customer needs. This enhanced redesign allows better navigation for major topical areas, including data access and climate information, as well as streamlines content and data access systems into logical groupings. Customers are now able to easily gather pertinent data and information for their needs and operations. Additional benefits from this redesign include simplified hardware and software support, enhanced security, and better use of personnel resources by streamlining maintenance.

Climate Data Records. In 2013, NCDC expanded its national inventory of operational Climate Data Records (CDRs) to 16 by transitioning five new CDRs from Research to Operations. The new CDRs include the atmosphere mean layer temperatures record, microwave imager and sounder records, and a global precipitation record. The CDR Program is responsible for generating and archiving operational climate data records for the atmosphere, oceans, and land. These CDRs provide authoritative, observation-based information on changes to the land, oceans, atmosphere, and ice sheets.

The 16 operational CDRs provide input to emerging climate prediction modeling as well as maintain the Nation's record of climate history. This includes the severity and frequency of drought, floods, and hurricanes. The CDRs are produced from decades of satellite data and used by industry, government, and research communities to detect, assess, model, and predict climate change. Decision makers value these long-term records for building effective strategies to respond to, adapt to, and mitigate the impacts of climate variability and change.

NCDC developed long-term, seamless, homogeneous records characterizing climate change and variation in order to produce CDRs. As new climate algorithms and sensors are developed, NCDC reprocesses the entire period of record to update the data. Further, NCDC is establishing practical approaches to information preservation, long-duration software, and algorithm maintenance. These approaches leverage the application of software engineering expertise and tailoring of existing standards and best practices. NCDC also devised new requirements to ensure that CDR software design and documentation can accommodate migration to future computing platforms and software languages.

Record-Setting Volumes of Data. NCDC archives data to preserve them for posterity from every corner of the globe, including land-based, marine, model, radar, weather balloon, satellite, and paleoclimate data. NCDC routinely incorporates, stewards, and provides access to these data to meet the needs of a wide variety of customers. In 2013, NCDC planned for vast increases in data requests and implemented information technology infrastructure to proficiently handle increases in data delivery. NCDC also upgraded its system, which retrieves data from the digital archives, to improve efficiency and reduce data delivery times for radar and satellite orders.

Throughout the 2013 fiscal year, users from all over the world downloaded NCDC's model, satellite, and radar data from both the Asheville, NC, and Boulder, CO, sites—totaling a record volume of 5.04 petabytes. For perspective, one petabyte of data is equivalent to the digital space needed for over 13 years of high-definition television. Satellite data access had the highest volume of data downloaded with 3,600 terabytes, followed by model (802 terabytes) and radar (317 terabytes) data.

NOAA/Office of Marine and Aviation Operations

NOAA Ship **Ronald H. Brown,** an oceanographic and atmospheric research platform, is the largest vessel in the NOAA fleet (274 feet). With its instruments and sensors, *Ronald H. Brown* sails worldwide, supporting scientific studies to increase our understanding of the world's oceans and climate.

The *Ronald H. Brown* typically supports research programs of NOAA's Tropical Atmosphere-Ocean (TAO) Project (real-time data from moored ocean buoys for improved detection, understanding and prediction of El Niño and La Niña). The primary objective for the ship is servicing some of the approximately 60 Autonomous Temperature Line Acquisition Systems (ATLAS) and current meter moorings in the central and eastern equatorial Pacific. The *Ronald H. Brown* works in cooperation with the Woods Hole Oceanographic Institute to conduct mooring recovery and deployment operations of the Stratus Ocean Reference Station, located under the stratocumulus clouds off Chile and Peru. The ship conducts meteorological and air-sea flux observations to document and establish the accuracy of the moored metrological observations and to observe oceanic and atmospheric variability. This region is of critical importance to climate predictability.

Ronald H. Brown conducts the PIRATA Northeast Extension (PNE) project, a joint project of Brazil, France, and the United States. The overarching goal is to improve knowledge and understanding of ocean-atmosphere variability in the tropical Atlantic. The ship will also be conducting the CalWater 2 project in the Pacific. The goal of this work is to develop and test methods to improve U. S. West Coast forecasts of damaging landfalling winter storms emerging from the data sparse Pacific Ocean. The project will involve multiple air and sea research platforms.

The WP-3D and the G-IV aircraft will take part in several projects for the Office of Oceanic and Atmospheric research. In October, the WP-3D will conduct the Arctic Flux project, which will look at the effects of newly ice-free regions of the Beaufort and Chukchi Seas on the Arctic weather. During the winter, the WP-3D and the G-IV will conduct the CalWater 2 project, along with the *Ron Brown*. The WP-3D will also conduct the Shale Oil and Natural Gas Nexus (SONG-NEX) project, a study of the emission of compounds that contribute both to climate change and air quality degradation. The aircraft will then conduct the Plains Elevated Convection At Night (PECAN) project, which involves a multiscale investigation of the factors that control the initiation and evolution of convective clusters and mesoscale convective systems over the U.S. Great Plains region.

United States Air Force

Air Force Weather Agency, 14th Weather Squadron

The AF, through the 14th Weather Squadron (14 WS), provides centralized climatology services and produces specialized weather-impact information for the DoD, Intelligence Community, and allied nations. The 14 WS provides decision-enabling products, allowing DoD command authorities to anticipate environmental impacts on all aspects of military operations worldwide. The 14 WS enhances the combat capability of the United States by delivering timely, accurate, and reliable environmental situational awareness to the AF, the Army, Unified Combatant Commands, the Intelligence Community, and the Department of Defense. The squadron produces a suite of standard and tailored products such as Operational Climatic Data Summaries, Engineering Weather Data, Wind Roses, and Wind Stratified Conditional Climatologies, among others, and provides frequency of occurrence and duration for mission limiting factors. In addition to climatology products, the 14 WS provides forensic weather research, climate analysis, and climate monitoring and prediction. The squadron continues to modernize the Warfighter climatology toolkit by integrating GIS visualizations on its Web site and developing GIS-based spatial and cloud climatologies. These technologies provide Warfighters the data needed to optimally plan and execute operations.

Long-range outlook products cover time periods from one to six months into the future, and include seasonal forecasts and hazard assessments. The AF produces 6-month graphical outlooks derived by using statistical methods, with an ensemble of Global Climate Model forecasts, considering the El Nino–Southern Oscillation (ENSO), North Atlantic Oscillation, Southern Annular Mode, etc. The AF is enhancing its 6-month outlook production capability by leveraging research conducted by the Air Force along with other national and international resources (including climatology work by NATO allies), as well as advances in understanding teleconnections (i.e., ENSO, North Atlantic Oscillation, Southern Annular Mode) in order to provide planners with a risk management tool to positively impact operations.

United States Navy

Rear Admiral Jonathan White is also the Director of Task Force Climate Change. In order to establish capabilities in accordance with the Arctic Roadmap, special equipment and possibly new Polar Icebreaker (PIB) ships will be needed. In 1965 the Navy turned over all ice breaking responsibilities to the Coast Guard. The Navy supports the Coast Guard in their requirements, and it makes good sense to have a modern, dedicated Arctic icebreaker to assist with disaster response and search and rescue. But at this point, we don't expect Navy ships to be operating in the ice zone. Our interest in the Arctic has been spurred by the fact that there is increasingly open water there, and that is where we will most likely be operating. The breaking up of the Arctic sea ice means that the ice is more subject to winds and currents, so it is certainly conceivable that a ship operating in open waters could quickly find itself surrounded by sea ice. Having a reliable sea ice forecast and a rescue icebreaker would make it much safer for surface ships to operate there.

The U.S. Navy is currently engaged in strategic planning to increase operational capabilities and infrastructure in the Arctic in future years. Within the next decade, it will be operating routinely in the Arctic with an appropriate presence that includes more than just submarines. Navy's global area of responsibility is growing as it adds new coastlines and a new ocean. With the fiscal challenges confronting the Navy, it must use cooperative multilateral partnerships to successfully build an Arctic-capable force.

Navy is investing in significant Arctic research. It is important that it improve its understanding of the environment for safety of operations, because the Arctic will remain a harsh and challenging environment even as it becomes more accessible and active. This line of effort includes an emphasis on research to better understand the changing climate and improve Arctic weather forecast capabilities, and on high resolution surveying of the seabed.

The U.S. federal fleet currently has two icebreakers, USCGC Polar Star, has been refitted, extending its operational life by perhaps another decade. The U.S. also has one medium icebreaker, the USCGC Healy, which is around 15 years old and is currently dedicated to scientific missions. Ice-hardening ships is expensive and impacts ship capabilities, like speed and turn-radius. We are still trying to determine the need for ice-hardened hulls.

The White House released a new National Strategy for the Arctic Region on 10 May 2013 reflecting a growing recognition that the Arctic is opening up for human enterprise and will increasingly become a strategic priority for the United States. The Office of Science and Technology Policy (OSTP) is leading the effort considering the larger policy issues associated with national requirements for icebreaking services in both poles. The Science and Technology Policy Institute is summarizing past reports and seeking updates from key Federal agencies including the USCG, NSF, NOAA, NASA, and DOT.

The Naval Meteorology and Oceanography Command has assets that can assist in this national effort. The command manages a fleet of six world-class ocean survey vessels and operates some of the world's finest oceanographic analysis and prediction computer models. Navy also partners with the National Oceanic and Atmospheric Administration and the U.S. Coast Guard to monitor polar ice movement at the National Naval Ice Center in Suitland, MD. This is in line with the National Arctic Strategy's third line of effort—building and maintaining cooperative relations with other Federal Agencies, as well as with Arctic nations and allies. The Navy will work closely with the U.S. Coast Guard and other Arctic nation sea services to ensure they can meet the same mission requirements in the Arctic as they do in, on, and above every other ocean in the world.

RADM White believes the presence of well-meaning naval forces acts as a stabilizing influence toward mutual prosperity and safe maritime activity. Climate change and energy are two key issues that will play a significant role in shaping the futures security environment. Working closely with U.S. departments and agencies, the Department of Defense has undertaken environmental security cooperative initiatives with foreign militaries that represent a way of building trust through sharing best practices on installations management and operations.

United States Geological Survey

The United States Geological Survey (USGS) is both a provider of and user of climate-related services, but the balance between these two activities is quite different from that of NOAA. The USGS provides climate data and models, like NOAA, that can be used by resource managers and policy makers to anticipate and adapt to climate change. The USGS climate and land use change science program, however, also has a strong emphasis on understanding relationships between climate change and hydrological, geologic, and biological processes.

USGS has expertise and numerous research projects and products that describe long-term changes in the Earth's climate. The climatic datasets developed by the USGS are based primarily on Earth surface and subsurface records, in contrast to the atmospheric records developed by NOAA. USGS climatic datasets are derived from ice sheets, glaciers, and permafrost; tree rings, landscape-scale phenology, and other biological data; and rock, sediments, and other paleodeposits, in addition to modern meteorologic, hydrologic, and remote sensing instrumentation. The USGS manages an international ice core facility in Denver, Colorado, and has numerous studies devoted to understanding the natural variability of climate and its attendant changes and to providing context for the development of mitigation and adaptation strategies. The Ice Core Laboratory is just one example of the services provided by the USGS to the scientific community engaged in understanding the Earth's climate history. Another good example of the climate services provided by the USGS is the Department of the Interior's network of regional Climate Science Centers. These centers are staffed by the USGS and have research and information delivery capabilities designed to respond to the needs of natural resource managers.

Department of State (DOS)

Stratospheric ozone depletion has been recognized as a critical health and environmental problem for almost three decades. Under Department of State (DOS) leadership, the United States worked to negotiate international agreements to phase out ozone-depleting substances, which are expected to lead to a recovery of the ozone layer by the middle of this century. In 2009 these treaties became the first to achieve universal ratification; 197 countries, including the United States, have ratified both the Vienna Convention and the Montreal Protocol. The State Department makes annual contributions to the Vienna Convention's efforts on scientific monitoring of the ozone layer.

The IPCC, which was established by the WMO and the United Nations Environment Program (UNEP), held its first session in 1988. This organization serves as an intergovernmental forum to assess scientific, technical, and socioeconomic information relevant for the understanding of

http://ipcc-wg2.gov/AR5/report/

http://mitigation2014.org

The IPCC's Fifth Assessment Report is comprised of the contributions from the three working groups: (1) The Physical Science Basis [Sept 2013]; (2) Impacts, Adaptation & Vulnerability [Mar 2014]; and (3) Mitigation [Apr 2014], as well as a Synthesis Report that will be released in Oct 2014.

climate change, its potential impacts, and options for adaptation and mitigation. In doing so, the Panel draws on the expertise of thousands of scientists and technical experts. The IPCC is currently organized into three working groups, which examine (1) the state of the science, (2) impacts, adaptation and vulnerability, and (3) mitigation. In 2013-14, the IPCC completed its Fifth Assessment Report, which is comprised of the three working group contributions listed above, as well as a Synthesis Report [see image]. In addition to preparing assessment reports, the IPCC also contributes to international negotiations through preparation and review of special reports and development of methodologies requested by the United Nations Framework Convention on Climate Change (UNFCCC).

DOS also plays a central role in the USG's work within the intergovernmental Group on Earth Observations (GEO), which is charged with the implementation of the Global Earth Observation System of Systems (GEOSS). Additionally, DOS also serves as the primary USG – and international – donor to the Global Climate Observing System (GCOS), which is intended to be a long-term, user-driven operational system capable of providing the comprehensive observations required for understanding Earth's climate system, how it is changing and what the impacts of that change will be.

In addition to its primary role in the organizations listed above, DOS is active in several relevant Federal interagency processes, including the Committee on Environment and Natural Resources and Sustainability (CENRS) of the National Science and Technology Council, the Subcommittee on Global Change Research (SGCR) and the broader U.S. Global Change Research Program (USGCRP), as well as US Group on Earth Observations (USGEO) subcommittee and its interagency International Working Group. . The CENRS was established in 1993 to coordinate domestic scientific programs. USGCRP was created in 2001 to "integrate Federal research on global change and climate change" across 13 Federal agencies (from 2001-2008, USGCRP was known as the Climate Change Science Program, or CCSP). In addition to the above, DOS responsibilities include, but are not limited to, international aspects of food security, disaster warnings and assistance, WMO (including the recently launched Global Framework for Climate Services) and UNEP activities (including the Climate and Clean Air Coalition), and international meteorological and Earth observing programs.

SUPPORTING RESEARCH PROGRAMS AND PROJECTS

Interagency Collaborative Research Programs and Projects

U.S. Global Change Research Program (USGCRP)

The U.S. Global Change Research Program (USGCRP)—known as the Climate Change Science Program (CCSP) from 2002–2009—is a Federal program mandated by Congress that coordinates and integrates global change research across <u>13 Federal agencies</u> to most effectively and efficiently serve the Nation and the world. Through interagency partnerships and collaborations with leading experts, USGCRP works to advance climate science and improve the understanding of how global change is impacting society, both today and into the future.

The vision laid out in USGCRP's *National Global Change Research Plan 2012–2021: A Strategic Plan for the U.S. Global Change Research Program* ("the 2012–2021 Strategic Plan") maintains a clear emphasis on advancing global change science, but it also calls for a strengthened focus on ensuring USGCRP science informs real-world decisions and actions. USGCRP's four strategic goals are to:

1. **Advance Science** - Advance scientific knowledge of the integrated natural and human components of the Earth system to understand climate and global change.
2. **Inform Decisions** - Provide the scientific basis to inform and enable timely decisions on adaptation and mitigation.
3. **Conduct Sustained Assessments** - Build sustained assessment capacity that improves the Nation's ability to understand, anticipate, and respond to global change impacts and vulnerabilities.

4. **Communicate and Educate** - Advance communication and education to broaden public understanding of global change and develop the scientific workforce of the future.

For more information, please refer to http://www.globalchange.gov/

NOAA/NESDIS/National Climatic Data Center

Explaining Extreme Events of 2012 from a Climate Perspective. For the second year, NCDC scientists collaborated with their colleagues across the globe to examine the causes of certain extreme weather and climate events. Scientists from NCDC served as part of the team of lead editors for the report, entitled "Explaining Extreme Events of 2012 from a Climate Perspective" and published as a supplement to the *Bulletin of the American Meteorological Society.* Overall, 18 different research teams from around the world contributed to the peer-reviewed report that examined the causes of 12 extreme events that occurred on five continents and in the Arctic during 2012. In addition to investigating the causes of these extreme events, the multiple analyses of four of the events allowed the scientists to compare and contrast the strengths and weaknesses of their various analytic methods. Despite the different strategies, there was considerable agreement between the assessments of the same events. By further developing the ability to put extreme weather and climate events into the longer-term context of climate change, NCDC is helping provide the public with the information needed to make decisions about effectively minimizing and preparing for the impacts of these events.

An Independent Record of Global Warming. In collaboration with researchers from the University of South Carolina, the University of Colorado, and the University of Bern in Switzerland, NCDC developed a compilation of temperature records based on data from ice cores, corals, and lake sediment layers that revealed a pattern of global warming from 1880 to 1995 comparable to the global warming trend recorded by thermometers. While the thermometer-based global surface temperature record provides meaningful evidence of global warming over the past century, it is critical to have independent analyses to verify that record, because it can be affected by such things as land-use changes, shifts in station locations, and variations in instrumentation. For this analysis, the team used environmentally sensitive proxies to compile a temperature record spanning 130 years that is independent from thermometer-based records. The research, which was detailed in "Global Warming in an Independent Record of the Past 130 Years," published in *Geophysical Research Letters*, resolved some of the uncertainty associated with thermometer records.

User Workshops and Forums. In 2013, NCDC collaborated with the Cooperative Institute for Climate and Satellites-North Carolina (CICS-NC) to host two workshops and two forums to engage climate data users and business leaders. One of these was the "Frost and Freeze Data and Impacts to the Agriculture, Construction, and Transportation Industry" workshop held in March. Interaction during the two-day workshop focused on informing users of NCDC data offerings for sector-specific climate information needs and providing relevant information to the industries that are the most vulnerable and susceptible to changes in freezing conditions. Another similar workshop entitled "Climate Data and Applications Workshop: A Focus on Precipitation" was held at NCDC in December 2013. This two-day workshop focused on the different types of precipitation data, from land-based to remotely-sensed observations, available from CICS-NC and NCDC, as well as examples of how the data are useful in various applications.

Additionally, NCDC and CICS-NC hosted the inaugural "Executive Forum on Business and Climate" in June in Asheville, NC. This forum brought together business and industry leaders, academic researchers, and climate science experts to collectively examine weather and climate science trends, observations and predictions, related business risks, impacts and opportunities, and current market trends. The forum engaged participants through a combination of interactive discussions, case studies, and scenario-planning activities. CICS-NC also partnered with the Center for Climate and Energy Solutions (C2ES) to host a follow-up forum, "Identifying Business / Industry Needs for Resilience Planning," in Washington, DC, in November. This second Executive Forum on Business and Climate served as a knowledge exchange seminar and networking activity built around discussions on climate-related risks and opportunities for private sector businesses.

Regional Surface Temperatures for Six Continents and the Arctic. NCDC scientists helped construct a one-of-a-kind synthesis and analysis of temperatures over the last 2,000 years for six different continents and the Arctic. This effort is part of a large international effort under the auspices of the Past Global Changes Programme of the International Geosphere-Biosphere Programme. These reconstructions cover the last millennium on annual-to-decadal timescales and extend into the first millennium of the Common Era, which began in 1 A.D., on annual-to-multidecadal timescales. To date, scientists have produced 2,000 year-long annual-to-decadal temperature time series for the hemispheres and the globe, but not at the continental scale.

For this project, NCDC scientists served as group leaders for the North American regional reconstruction and as part of the core writing team for the highly cited paper, "Continental-scale temperature variability during the last two millennia" in *Nature Geoscience*. The Center also served as the data hub for the entire effort, producing state-of-the-art proxy and reconstruction datasets for the Common Era. Overall, this effort greatly extended the record of past climate and explicitly characterized uncertainties in the reconstructions. With these data, scientists will be able to better understand extremes in annual to decadal temperature fluctuations over the past 2,000 years in a way that is not possible using the instrumental record alone.

Sea Surface Temperature Datasets. The Niño 3.4 index is an important measure of El Niño Southern Oscillation (ENSO) conditions, which NOAA's Climate Prediction Center uses to support ENSO monitoring and forecasting. The Niño 3.4 index is calculated based on two NCDC datasets: the Optimum Interpolation Sea Surface Temperature (OISST) dataset and the Extended Reconstructed Sea Surface Temperature (ERSST) dataset. In 2013, scientists found that monthly Niño 3.4 indices calculated from the two datasets differed by as much as 0.5°C in some ENSO events. Such large differences complicate the monitoring and prediction of current and future ENSO conditions, so NCDC scientists analyzed the datasets and computational methods of the index to identify the causes of the differences.

NCDC completed several experiments on the data involving the integration of satellite observations into ERSST and the analysis of bias adjustment methodologies in both OISST and ERSST. These analyses found that the major cause of the difference was the bias adjustment applied to the satellite sea surface temperature measurements used to produce the OISST dataset. By accounting for these differences, NCDC not only verified the quality of the Center's data analyses, but also provided an objective basis for users to select the appropriate sea surface temperature product.

NOAA/Office of Oceanic and Atmospheric Research (OAR)

The mission of climate research activities within NOAA/OAR is to monitor and understand Earth's climate system in order to predict the potential long-term changes in global climate as well as shorter-term climate variations that are of societal and economic importance. More information on OAR's climate research is available at: http://www.oar.noaa.gov/climate/.

To achieve this mission, climate research across OAR is structured to support the long-term goal of Climate Adaptation and Mitigation described in NOAA's Next-Generation Strategic Plan (NGSP). The NGSP identifies four Objectives under the Goal: (1) Improved scientific understanding of the changing climate system and its impacts; (2) Assessments of current and future states of the climate system that identify potential impacts and inform science, service, and stewardship decisions; (3) Mitigation and adaptation efforts supported by sustained, reliable, and timely climate services; and (4) A climate-literate public that understands its vulnerabilities to a changing climate and makes informed decisions.

To meet these Objectives, OAR's research is executed and delivered through a network of NOAA line offices, laboratories and cooperative institutes, programs, and university-based partnership activities. Broadly, OAR's laboratories and cooperative institutes contribute, both directly and indirectly, to all four Objectives. OAR's Climate Program Office (CPO) is the strategy lead for the NGSP Climate Goal, and it provides resources, programmatic oversight, and coordination to ensure NOAA's climate research activities proceed in an integrated and cost-effective manner. In these roles, CPO brings together and maintains relationships across NOAA laboratories, cooperative institutes, and university-based partners. OAR's laboratories and cooperative institutes work in tandem with CPO to ensure NOAA meets its aim of an informed society anticipating and responding to climate and its impacts.

Specific aims of research conducted under the Goal's Objective of improved scientific understanding include:

- Describe and understand the state of the climate through sustained atmospheric and oceanic observations and research related to global distributions, trends, sources and sinks of atmospheric constituents that are capable of forcing change in the climate of the Earth
- Understand and predict climate variability and change from weeks to decades to centennial timescales
- Conduct advanced mathematical modeling of the climate and Earth systems, including natural climate variability, anthropogenic climate change, weather and hurricane forecasts, El Niño prediction, and stratospheric ozone depletion to improve the prediction of climate phenomena
- Sustain the observing systems essential for climate, oceanographic monitoring, and data management
- Conduct physical process research to advance a seamless suite of information and forecast products
- Understand how decision makers use climate information to improve the ability of society to plan for and respond to climate variability and change.

NOAA/OAR Laboratories and Programs

Atlantic Oceanographic and Meteorological Laboratory (AOML)

AOML conducts mission oriented scientific research that seeks to understand the physical, chemical, and biological characteristics and processes of the ocean and atmosphere, both separately and as a coupled system. The Laboratory's research theme related to <u>oceans and climate</u> includes interdisciplinary scientific investigations of the physics of ocean currents and water properties, and on the role of the ocean in climate, extreme weather events, and ecosystems. The tools used to carry out these studies range from sensors on deep ocean moorings to satellite-based instruments to measurements made on research and commercial shipping vessels and autonomous vehicles, and include data analysis and numerical modeling as well as theoretical approaches. This includes stewardship of oceanic measurements relevant for climate, including custodianship of major data sets, development and deployment of new sampling methods, new analysis tools, and carrying out long-term consistent environmental measuring programs.

In an effort to better understand and forecast climate, the Atlantic Oceanographic and Meteorological Laboratory (AOML) contributes to the maintenance of the global Argo program (a global array of 3,000 vertically profiling floats), the global drifter program (which maintains a global array of more than 1500 surface drifters), the expendable bathythermograph program (the deploys more than 8000 XBTs each year and processes more than 14,000 XBTs each year). AOML collaborates with the Pacific Marine Environmental Laboratory (PMEL) to monitor decadal-scale changes in a suite of physical, chemical, and biological parameters over the full water column, such as carbon dioxide uptake, ocean temperature, and circulation (e.g. Global Repeat Hydrographic/CO2/Tracer (GO-SHIP) surveys in Support of CLIVAR and Global Carbon Cycle objectives: Carbon Inventories and Fluxes program). In addition AOML maintains time series observations of the Florida Current and Deep Western Boundary Current (Western Boundary Time Series program) and the Deep Western Boundary Current in the South Atlantic (the Southwest Atlantic Meridional Overturning Circulation Array); these programs contribute to the national and international programs designed to measure the meridional overturning circulation in the Atlantic Ocean which is the large scale ocean circulation that redistributes heat, fresh water and carbon north and south in the ocean. AOML manages these observational programs and conducts ocean circulation research on data collected through them. All of these programs are major contributors to the World Climate Research Programme's Climate Variability and Predictability (CLIVAR) Experiment and the Global Ocean Data Assimilation Experiment. Combined with satellites, these data provide a quantitative description of the changing state of the ocean and the patterns of ocean climate variability from months to decades, including heat, freshwater and carbon storage and transport.

In FY15, AOML plans to increase understanding of the ocean's role in climate through the stewardship of long term ocean observations, analysis of observations and models and numerical simulation of ocean circulation induced changes to climate. For example, the Physical Oceanography Division (PHOD) of AOML will continue to show that the variation of surface and subsurface ocean temperature and salinity is important to and linked with the Atlantic meridional overturning circulation (AMOC), hurricane intensification and fisheries productivity on climate time scales.

Air Resources Laboratory (ARL)

To provide a sound scientific basis for understanding climate variability and change, the Air Resources Laboratory (ARL) contributes to several land-based reference observing networks: the U.S. Climate Reference Network (measuring surface climate variables),the Surface Energy Budget Network, as well as the Global Climate Observing System Reference Upper-Air Network (GRUAN).

ARL also provides analyses of variability and change in the boundary layer, free troposphere and stratosphere, with an emphasis on atmospheric temperature, stability, and circulation and on clouds. ARL has long-standing and internationally-recognized expertise using radiosonde observations, often combined with satellite observations or climate model results, to understand global upper-air changes on diurnal to multi-decadal time scales. In addition, ARL has developed long-term datasets using model analysis and observed rainfall for air stagnation, and merged in-situ data and satellite observations for dust storm climatology.

In FY15, ARL will:

- Continue to provide maintenance and testing of existing technologies at the climate reference networks and evaluate new technologies;
- Continue to provide leadership for guiding the full implementation of the GRUAN;
- Continue using radiosonde observations to evaluate models of the global carbon cycle and to evaluate other observational approaches to identify boundary layer structures and variations;
- Continue analysis of the variability and trends in the climate system, interactions between climate and air chemistry, and the performance of operational and research models.
- Seek to remove or transfer stations associated with the US Regional Climate Reference Network, scheduled to be terminated in FY14.

Climate Program Office (CPO)

CPO was established to advance the agency's climate portfolio and to manage the competitive research program that funds high-priority science to advance understanding of atmospheric, oceanic, land-based, and snow and ice processes, and how they affect climate. CPO also supports multi-disciplinary research and assessments that help people better plan and respond to climate change and variability. CPO-supported research is conducted in regions across the United States, at national and international scales, and globally.

One of CPO's core missions is to develop and sustain a global in situ climate observing system. NOAA's instruments are a vital part of an international system to monitor Earth's climate and provide data for NOAA research, modeling, predictions, and forecasts. We leverage climate science capacity and capabilities to provide the nation with critically needed early warning systems that are timely and relevant for stakeholders at a range of scales spatially (local to global) and over time (weeks to decades).

CPO will continue to support modeling, research, analysis, and data collection aimed at gaining a better understanding of climate processes. These activities are essential steps toward delivering climate prediction and assessment products for risk management and decision-making. CPO will also sustain grant programs that advance the knowledge and capacity of decision makers to assess risk, prepare for, and respond to the impacts of climate variability and change.

Earth System Research Laboratory (ESRL)

The Earth System Research Laboratory (ESRL) continues to make unparalleled contributions to the scientific understanding of climate change through its global atmospheric monitoring program, field missions to study atmospheric processes, laboratory studies, modeling work.

ESRL's Chemical Sciences Division (CSD) climate research provides an improved predictive capability through a better understanding of the connections between emissions, atmospheric composition, and Earth's climate system. Research is focused on addressing two of the greatest uncertainties in current climate models: water vapor and aerosols (airborne fine particles), in addition to ongoing work on non-CO_2 greenhouse gases.

The ESRL Global Monitoring Division (GMD) conducts sustained observations and research related to global distributions, trends, sources and sinks of atmospheric constituents such as greenhouse and other trace gases, aerosols, ozone, a that are capable of forcing change in Earth's climate. This research and long-term observations are used worldwide to advance climate projections and provide scientific, policy-relevant, decision-support information to enhance society's ability to plan and respond to change.

Physical Sciences Division (PSD) strives to interpret the causes of observed climate variation, and advance the understanding, parameterization, and measurement of climate-relevant physical processes. This knowledge is used to improve climate models and forecasts and develop new climate products that better serve the needs of the public and decision-makers.

All ESRL Divisions place a high priority on communicating the results of their research in decision-support information products to underpin national and international decision-making. Their contributions include leadership and extensive participation in state-of-understanding national and international assessments on climate.

FY 15 Plans:

- Generate data archive of ship and aircraft observations from participation in the NOAA/DOE CALWATER2 field program in the Indian Ocean.
- Carry out the Shale Oil and Natural Gas Nexus (SONGNEX) investigation using NOAA's "flying chemical laboratory" aboard the NOAA P-3 aircraft
- Execute a complementary airborne study of Fugitive Emissions concentrating on methane and other GHGs aboard the NOAA DHC6 aircraft.
- Analyze data from the SENEX and SEAC4RS field missions to understand climate effects of emissions, atmospheric chemistry, and transport in the southeastern U.S.
- Develop new prototype systems for regional drought diagnosis and prediction.
- Develop large ensemble, multi-model, high resolution global modeling ability to resolve high impact extreme events: heat waves, droughts, heavy precipitation.

- Develop climate and weather guidance to reduce and manage drought and floods impacts on watershed restoration.
- Provide an integrated database for the International Arctic System for Observing the Atmosphere (IASOA). IASOA is a consortium of 10 sentinel Arctic Observatories that collect a wide range of detailed measurements relevant to weather and climate. This activity will support the Polar Prediction Project.
- Lead development of International Arctic science focus groups integrating information on the environmental state of clouds, radiation, surface-atmosphere exchanges, trace gases, black carbon and regional predictions in the Arctic region.
- Continue development of Arctic atmospheric observatories to obtain high-latitude climate data.
- Continue long-term observations of atmospheric composition, providing annual updates of CarbonTracker, the NOAA Annual Greenhouse Gas Index (a national climate indicator), methane climatologies, CO_2 trends (daily updates; a national climate indicator), the Ozone-Depleting Gas Index, GlobalView, and other long-term atmospheric composition data sets and products.

Geophysical Fluid Dynamics Laboratory (GFDL)

Over the last half century in general, and the last few years in particular, GFDL has demonstrated world leadership in pushing the boundaries of climate prediction and projections of climate change. Through direct participation in producing the Intergovernmental Panel on Climate Change (IPCC) 2007 Assessment and the Administration's Climate Change Science Program Synthesis and Assessment Reports, GFDL's premier climate science capacity and recent investment in computer model infrastructure allow NOAA to deliver essential climate prediction information at the regional and local level and provide an invaluable and unique opportunity for the Nation to make critical progress in global change science. GFDL has delivered model output to the IPCC Fifth Assessment Report (AR5). Four model streams have been specifically designed for the climate integrations needed to address the major climate science challenges. All four streams are based on CM2.1, a GFDL coupled climate model with an atmospheric model resolution of 200km and an ocean model resolution of 100km. CM2.1 was considered among the highest quality models used in the previous IPCC Assessment Report, AR4, and was used again in AR5. Nearly 200 terabytes of model output was submitted to the 5th Coupled Model Intercomparison Project (CMIP5) for use in AR5. The four modeling streams are:

1. Decadal prediction experiments to better understand physical processes.
2. Earth System Models (ESMs), emphasizing biogeochemical carbon-climate feedbacks through the use of a closed-carbon cycle.
3. Coupled climate models (CM3) with more realistic, self-consistent but complex atmospheric physics and chemistry than in CM2.1.
4. High-resolution atmospheric models at 50km and 25km resolution forced by future ocean states.

GFDL scientists continue to play an active role in AR5, and in some instances lead, in: model intercomparison and model simulation analysis activities; submission of model data, quality control and archiving; analyses of the data outputs and inferences about the climate system.

These activities will lead to a number of peer-reviewed publications based on the CMIP5 database of model output. GFDL has also made significant contributions, through high-resolution model experiments, to the North American Regional Climate Change Assessment Program (NARCCAP), and GFDL model experiments and science results were also used in the recently released USGCRP National Climate Assessment. Significant progress in model development is expected in FY15 as the four modeling streams outlined above are unified into a new coupled model called CM4. New very high-resolution climate and Earth System models are being developed and applied to address scientific questions regarding the impact of atmospheric composition and ocean eddies on the climate system as well as regional-to-local impacts of climate variability and change on marine ecosystems.

Great Lakes Environmental Research Laboratory (GLERL)

The Great Lakes Environmental Research Laboratory (GLERL) conducts climate research and modeling focused on the region of the Laurentian Great Lakes, which is shared by the USA and Canada. Investigation of impacts of climate change on hydrologic resources in the Great Lakes began during the late 1980s, and has been considerably overhauled in recent years. Hydrologic models with one-way driving of the surface by the atmosphere have been replaced by regional climate models with two-way coupling of the atmosphere to both land and water surfaces. The older generation of this type of model, the Coupled Hydrosphere-Atmosphere Research Model (CHARM) is being replaced by a newer version based on the Weather Research and Forecasting (WRF) model.

Of particular interest is the influence of climate and climate change on water quantity in the Great Lakes basin, and atmospheric and climatic research at GLERL are coordinated with hydrologic study, as well as research on the dynamics of motion and thermodynamics of the lakes and their ice. In addition, GLERL has expertise in the dynamics of nutrient loading in rivers, and on the interaction among these nutrients, water temperature and dynamics, and the biota of the lower and upper food web, leading to end-to-end capability in the physical, chemical, and biological aspects of this regional Earth system.

Pacific Marine Environmental Laboratory

The Pacific Marine Environmental Laboratory (PMEL) contributes to the advancement our understanding of the impacts of climate variability and change through a long-term observations program that supports robust climate research. Key observational components include:

- The Global Tropical Moored Buoy Array (GTMBA) was developed by PMEL in partnership with NOAA's Climate Program Office and multiple international partners. PMEL now maintains both the PIRATA and RAMA Arrays in the Tropical Atlantic and Indian Oceans, respectively. The TAO/TRITON Array, the original component of the GTMBA, has been transitioned from PMEL to NWS/National Data Buoy Center. Moored buoys in the GTMBA report daily values of surface meteorological parameters and subsurface temperatures and salinity values via satellite for use by the NWS Climate Prediction Center and climate researchers world-wide.

- PMEL contributes to the maintenance of the global Argo Float Array, through an international partnership under the direction of NOAA's Climate Program Office and in conjunction with U.S. Argo partners Scripps Institute of Oceanography, Woods Hole Oceanographic Institution, the University of Washington, and AOML. PMEL scientists deploy floats, support the addition of new measurement and increased depth capabilities of Argo floats, conduct research on improving quality control of the Argo data. PMEL and CPO are co-managing the development of Deep Argo floats, extending the depth range of Argo floats from 2,000 meters to 6,000 meters.

- PMEL scientists support the OceanSITES program of global moored ocean reference stations for climate, maintaining two moored platforms in the North Pacific, at the Kuroshio Extension Observatory (32.3°N 144.5°E) in the NW Pacific Ocean and at Ocean Station Papa (50°N 145°W) in the NE Pacific Ocean.

- In conjunction with the International CLIVAR program, PMEL conducts ocean research on a global scale to monitor decadal-scale changes in a suite of physical, chemical, and biological parameters over the full water column, such as carbon dioxide uptake, ocean temperature, and circulation.

- PMEL climate researchers are active in the Arctic region, using atmospheric models to predict long-term sea-ice extent and providing information on potential impacts to ecosystem managers and local communities on Arctic change. PMEL scientists, working with the National Marine Fisheries Service, also conduct research on the impacts of ocean climate change on marine resources, conducting research on ocean acidification and on the impact of ocean and climate change on North Pacific fishery and marine mammal populations.

U.S Department of Agriculture

National Institute of Food and Agriculture

National Institute of Food and Agriculture (NIFA) funding supports research projects that collect, analyze and utilize short and long-term weather and climate data as a base of information for the projection and prediction of climatic trends related to environmental impacts on agro-ecosystems, forests, and rangelands and the development of adaptation and mitigation strategies for natural resources and production management. Research, education and extension projects address the contribution of human activities, soil management, and crop and animal production to atmospheric greenhouse gases. Broader areas of study involve climate variability, carbon, nitrogen and water cycling, and their role in global change. The impact of changes in UV radiation and ozone level studies also fit into this broad global category.

Historical climate changes are derived from data gathering and modeling studies, enabling prediction of future crop production and irrigation needs. NIFA funding supports studies on the impact of climate and weather on food, feed, and fiber production and on natural resource protection and utilization. These studies relate to forest and agricultural plant growth, rangeland productivity, cropping system selection, livestock production practices and natural resource management.

Climate impacts on human, environmental and plant and animal health is also well represented in studies of both micro-and macro-climatic change. These involve studies dealing with the climatic impact on air quality, water quality and quantity and agriculture waste management related to agricultural practices and forest and urban development. Studies on climatic impact on nutrient cycling and carbon sequestration and emission are supported with NIFA funds. Research is also being supported to quantify the impact of climate change on the incidence and severity of drought, extreme heat and cold, emerging plant and animal diseases and pests, new invasive species, shifting biodiversity and ecosystem services, and adaptation to these events.

The Agriculture and Food Research Initiative (AFRI) has funded projects on a wide variety of weather and climate related research in collaboration with other U.S. federal agencies. AFRI's Climate Change Challenge Area currently focused research areas are, regional climate studies in agriculture and forestry, plant breeding, animal health impacts, and mitigation and adaptation in agriculture and forestry. Other AFRI areas of research related to weather and climate change include organic agriculture, carbon cycling, agro-ecosystem modeling, biofuel feedstock production and economic consequences of adaptation and mitigation strategies. Emerging areas of research include the impact of climate on environmental markets, food safety, and soil health.

United States Geological Survey

Water Resources Monitoring. The effects of warming temperatures on water resources are among the most certain and costly consequences of climate change in some regions of the United States, and water resources information should be an important component of (or closely linked with) a new national climate services program. The USGS is the nation's water resource monitoring agency, with one or more offices in every State and a National stream and groundwater monitoring network that supports water management efforts at State and local levels, as well as vast western water resources managed by two other Department of the Interior bureaus: the Bureau of Reclamation and the Bureau of Land Management (BLM). The USGS has a long-established hydroclimatology research program and is the primary Federal science agency for water resource information. The USGS monitors the quantity and quality of water in the Nation's rivers and aquifers and develops tools to improve the application of hydrologic information, including the effects of a changing climate. This broad, diverse mission cannot be accomplished effectively without the contributions of the Cooperative Water Program with the 50 states. For more than 100 years, the Cooperative Program has been a highly successful cost-sharing partnership between the USGS and water-resource agencies at the State, local, and Tribal Nation levels.

The USGS carries out research in climate change, regional hydrology, the carbon cycle, coastal erosion, and glaciology. The Water, Energy, and Biogeochemical Budgets (WEBB) program is studying processes controlling water, energy, and biogeochemical fluxes at five small research watersheds in the United States. This program includes research on the effects of atmospheric and climatic variables on watershed processes. There are also a number of ongoing studies to characterize trends in hydrologic data and to relate these trends to climatic variables. Researchers are also using global and regional climate models to enhance understanding of the potential effects of climate change and climate variability on U.S. land and water resources.

Glacier Monitoring. As part of its glaciology program, the USGS maintains an observation program on three benchmark glaciers representative of different climatic zones of the western United States, one in Washington, one on the south coast of Alaska, and one in the interior of Alaska. At each glacier, the program measures the winter snow accumulation, summer snow and ice ablation, air temperature, and runoff in the glacier basin. Beginning in 1959, this is the longest such record in North America. Analysis of this record is providing a greater understanding of climate variability and its effects on water resources of the western United States. The record clearly shows the effects of changing winter precipitation patterns associated with atmospheric conditions in the northeast Pacific Ocean, including El Niño-La Niña events and the Pacific Decadal Oscillation.

To augment its glacier monitoring efforts, the USGS is using National Systems data to measure fluctuations of glaciers in Alaska, Washington, and Montana. Mountain glaciers are ideal subjects for these systems because they are remote, have an appropriate space scale, and require infrequent but repetitive observations. The observations have established a baseline of regional glacial conditions. The resulting archive of ongoing observations is being used to determine recent trends in glacier size and terminus location. In addition, techniques have been developed to generate derived products that provide critical glacial parameters, including DEMs, equilibrium line altitudes, and ablation rates. These products are being incorporated into a glacial runoff model of the South Cascade Glacier, Washington, where they are proving to be a valuable source of otherwise unavailable data.

Snow and Ice Studies. The USGS, in cooperation with BLM, is using a variety of remote sensing data to monitor the rapid wastage of the piedmont lobe of Bering Glacier, Alaska. Landsat, Radarsat, ICESat, and Ikonos observations show that Bering Glacier is retreating rapidly and thinning in an accelerating retreat from an advanced position that resulted from a major glacial surge in 1993-95. The satellite data and ground-based observations have been combined to determine the surface flow velocities and calving rates of the glacier and to monitor the expansion of Vitus Lake and Berg Lake—two large lakes whose boundaries include the glacier terminus. The rapid change in glaciation is having a large impact on nearby terrestrial and aquatic ecosystems.

Geological Investigations. The USGS has traditionally led the USGCRP projects devoted to understanding cycles in the Earth's climate, abrupt climate change, ecological thresholds, and climate change in polar regions. These cycles are generally underrepresented in the available data from meteorological records. By combining paleoclimatic and instrumental data, USGS scientists have contributed substantially to understanding how past and current changes in the Earth's climate affect water, land, and biological resources.

Biological Indicators of Climatic Trends. The National Phenology Network, Breeding Bird Surveys, and Amphibian Research and Monitoring Initiative are examples of USGS biological science programs that provide national coverage and uniform protocols for reporting the occurrence of biological events that can augment analyses of changes in the physical climate system.

USGS Role in National Climate Services. The USGS has several additional programs and projects that would directly support and complement an interagency national climate services program. The USGS develops global and regional climate models and contributes to international programs of model development (e.g., the WCRP and phase 3 of the Coupled Model Intercomparison Project [CMIP3]). It targets model applications to investigate past, present, and future climate change (mean and variability) and how those changes influence and are influenced by surface systems. Applications include surface and subsurface hydrology, terrestrial and marine ecosystems, glaciology, and wildland fire research. USGS model simulations are often targeted to specific natural resource response and management questions rather than production runs for activities such as the IPCC assessments. Its modeling activities are flexible, with quickly implemented experimental designs that are often modified through an iterative process involving cross-disciplinary researchers.

The figure right depicts some of the potential areas of USGS collaboration and support toward a coordinated interagency national climate services program.

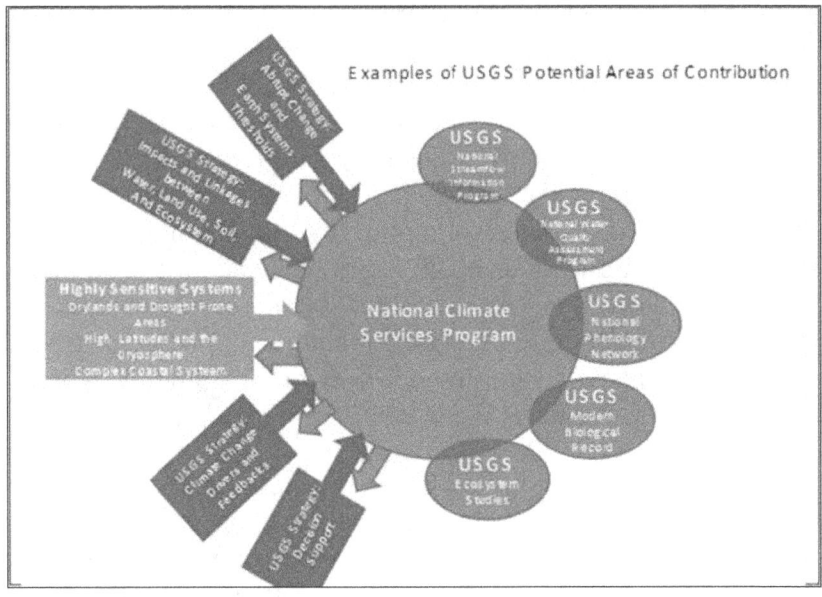

National Aeronautics and Space Administration

Science Mission Directorate, Earth Science Division

The 2010 National Space Policy states that the National Aeronautics and Space Administration (NASA) "shall conduct a program to enhance U.S. global climate change research and sustained monitoring capabilities, advance research into and scientific knowledge of the Earth by accelerating the development of new Earth observing satellites, and develop and test capabilities for use by other civil departments and agencies for operational purposes." The NASA program on global climate change research is comprehensive, encompassing continuous interactions between satellite mission development and formulation, satellite data analyses, Earth system modeling, new technology, and state-of-the-art scientific knowledge for applications. Scientific research and data analysis are conducted through competitive peer-reviewed opportunities offered through Research Opportunities in Space and Earth Science (ROSES) announcements.

NASA's climate research program is unique because it encompasses the development of observational techniques and the instrument technologies needed to implement them; laboratory testing and demonstrations from an appropriate set of surface-, balloon-, aircraft-, and space-based platforms; development and operation of satellite missions and production and dissemination of the data products resulting from these missions; research to increase basic process knowledge; incorporation of observations and research results into complex computational models that can be used to more fully characterize the present state of the environment and predict the future evolution of the Earth system; and development of partnerships with other national and international organizations that can use the generated information in environmental forecasting and in policy, business, and management decisions.

NASA is the largest funding contributor to the 13-agency United States Global Change Research Program (USGCRP) and provides the bulk of the global observations and research by the USGCRP. NASA delivers a substantial portion of the observations and research that forms the basis for international scientific assessments of climate change and other subjects such as ozone.

Satellite Missions. Satellites provide critical climate change measurements via global coverage, frequent sampling in both space and time, and near-uniform accuracy and stability. NASA initiated, and in selected cases has sustained for more than a decade, many global, high accuracy, well-calibrated data records, such as total solar irradiance at the top of the atmosphere; Arctic Ocean sea ice extent and thickness; Antarctic and Greenland mass change and other characteristics; global sea level and global ocean surface vector wind; global ocean near-surface chlorophyll-*a* concentration; global land use and land cover; ozone in the stratosphere; and, global precipitation, including water vapor, rainfall and snow. These critical climate data records are a foundation for national and international studies of global and regional climate change, sea level rise, and study of the earth as a system.

NASA has 10 on-orbit satellite missions related to climate services, as defined by this Federal Plan: Aqua, Aquarius/Satélite de Aplicaciones Científicas (SAC)-D, Aura, Cloud-Aerosol Lidar and Infrared Pathfinder Satellite Observation (CALIPSO), CloudSat, Global Precipitation Measurement (GPM), Ocean Surface Topography Mission (OSTM), Quick Scatterometer (QuikSCAT), Suomi-National Polar-orbiting Partnership (S-NPP), and *Tropical Rainfall Measuring Mission* (TRMM). Table 1 lists the climate services themes supported by these missions. On February 27, 2014, GPM was launched successfully from Tanegashima Space Center, Japan, and recently completed on-orbit checkout, transitioning to nominal mission operations on May 29, 2014. GPM is the last of the foundational missions that the Decadal Survey by the National Research Council[1] assumed would be a precursor to Decadal Survey missions.

Table 1. Correlation of NASA Earth Science Division operating satellite missions with OFCM themes of climate services.

Satellite	Launch Date	Climate Services Theme
TRMM	Nov 1997	Climate variability and change; water cycle
QuikSCAT	Jun 1999	Climate variability and change
Aqua	May 2002	Atmospheric composition; carbon cycle; climate variability and change; water cycle
Aura	Jul 2004	Atmospheric composition
CALIPSO	Apr 2006	Atmospheric composition; water cycle

[1] National Research Council (2007) *Earth Science and Applications From Space: National Imperatives for the Next Decade and Beyond*, 428 pp.

CloudSat	Apr 2006	Climate variability and change; water cycle
Aquarius	Jun 2011	Climate variability and change; water cycle
OSTM	Jun 2008	Climate variability and change; water cycle
Suomi NPP	Oct 2011	Atmospheric composition; carbon cycle; climate variability and change; water cycle
GPM	Feb 2014	Climate variability and change; water cycle

A daunting challenge in supporting the complexity of global and regional climate change science is the huge number of biological, chemical and physical variables that must be measured nearly simultaneously globally. To address this challenge, NASA has engineered constellations of satellites flying in close formation. For example, NASA's Aura, CALIPSO, CloudSat, and Aqua satellites, together with the Japanese Global Change Observation Mission 1–Water "Shizuku" (GCOM-W1), are called the "A-Train" constellation and produce an unprecedented quantity of data for atmospheric chemistry and composition. The time separation between the front and rear of the A-Train is 11 minutes, less than the lifetime of most clouds; this important feature allows researchers to utilize multi-satellite observations to examine processes related to cirrus cloud formation in large-scale models.

NASA has 4 satellite missions in development (Table 2) related to climate services, as defined by this Federal Plan, for launch from 2015 to 2020. Soil Moisture Active Passive (SMAP) is a Tier-1 Decadal Survey mission. Cyclone Global Navigation Satellite System (CYGNSS) is the first in the Earth Venture-class series of rapidly developed, cost-constrained small space missions, and is referred to as EVM-1 (Earth Venture Mission-1). Tropospheric Emissions: Monitoring of Pollution (TEMPO) is the first in the Earth Venture-class series of instrument missions of opportunity that will fly as hosted payloads on commercial or government satellites. Surface Water and Ocean Topography (SWOT) is a Tier 2 Decadal Survey mission that was accelerated by the Climate Initiative to provide insight into the movement and distribution of Earth surface waters including both freshwater and oceans.

Table 2. Correlation of NASA Earth Science Division missions in development with OFCM themes of climate services.

Satellite	Planned Launch	Theme
SMAP	2015	Climate variability and change; water cycle
CYGNSS	2016	Water Cycle
TEMPO	2019	Atmospheric composition; climate variability and change
SWOT	2020	Climate variability and change; water cycle

NASA continues with the pre-formulation studies of Decadal Survey and other mission concepts such as Pre-Aerosol, Cloud, and ocean Ecosystem Mission (PACE), Climate Absolute Radiance and Refractivity Observatory (CLARREO), Active Sensing of CO_2 Emissions over Nights, Days, and Seasons (ASCENDS), Aerosol-Cloud-Ecosystem (ACE), GEOstationary Coastal and Air Pollution Events (GEO-CAPE), Hyperspectral Infrared Imager (HyspIRI), *Lidar Surface Topography* (LIST), Precision and All-Weather Temperature and Humidity (PATH), Gravity Recovery and Climate Experiment II (GRACE-II), Snow and Cold Land Processes (SCLP),

Global Atmospheric Composition Mission (GACM) and Three-Dimensional Tropospheric Winds from Space-based Lidar (3-D Winds). Additionally, the Earth Science Technology Office (ESTO) makes significant investments in technology development activities directly addressing the priorities outlined in the Decadal Survey including the tiered missions as wells as Earth Venture through development of observation systems and information systems technologies.

The following activities will be undertaken or accomplished in FY 2015.

- SMAP will launch in early FY2015, complete its initial on-orbit checkout, and transition to operations.

- CYGNSS will complete its Critical Design Review (CDR) and enter system test phase (Phase D).

- TEMPO will complete its Key Decision Point – C (KDP-C) review and enter development.
SWOT will conduct its Phase B.

NASA Science Research

NASA's Earth Science Research Program addresses complex, interdisciplinary Earth science problems in pursuit of a comprehensive understanding of the Earth system. The Research Program pioneers the use of both space-borne and aircraft measurements through its implementation of several elements: multi-disciplinary research and analysis, interdisciplinary research, airborne science, modeling and data assimilation, and enabling capabilities. Research and analysis (R&A) emphasizes the development of new scientific knowledge, including the analysis of data from NASA satellite missions and the development and application of complex models that assimilate these science data products and/or use them for improving predictive capabilities.

The following research activities will be undertaken or accomplished in FY 2015:

- Continue expanding the supercomputing capacity to more than 2 quadrillion floating point operations per second (petaflops) for weather and climate modeling using the combination of CPUs and multi-core accelerator technologies such as Intel Phi and NVIDIA GPU processors.

- Increase the number of science data products delivered to Earth Observing System Data and Information System (EOSDIS).

- Continue operation of ground networks for measuring time evolution of atmospheric trace constituents, including gases and aerosols.

- The Earth Venture Suborbital-1 (EVS-1) missions are conducting the following research (observation investigations were mostly completed in FY14, some will continue into FY15, analysis of scientific data will continue in FY15); proposals for EVS-2 are being reviewed and selections will be made in late FY14/early FY15:

  o The Hurricane and Severe Storm Sentinel (HS3) EVS-1 investigation uses two Global Hawk unmanned aircraft systems (UAS) around and in hurricanes. One Global Hawk payload is optimized for direct observations of the hurricane itself.

The other payload includes sensors, which measure important parameters in the region surrounding the storm to help improve our understanding of tropical cyclone genesis and behavior. HS3 results will also improve our ability to predict storm path and intensity. GPM will be providing much higher quality data than previously available on rain structure in tropical cyclones in all ocean basins. The surface-wind monitoring ISS-RapidScat instrument to be launched to the International Space Station in 2014 will also provide valuable information on surface winds in storms.

o The Airborne Microwave Observatory of Subcanopy and Subsurface (AirMOSS) EVS-1 investigation uses an ultra-high frequency synthetic aperture radar to measure high spatial resolution soil moisture to provide insight into the vegetation activity and the uptake of carbon dioxide; the goal being to reduce uncertainty of global NEE (Net Ecosystem Exchange) estimates. The radar, deployed on a Gulfstream 3, operates over nine diverse biomes to provide information on a wide variety of climates and ecosystems. FY 15 will be the final year of a three year airborne campaign.

o The Deriving Information on Surface conditions from Column and Vertically Resolved Observations Relevant to Air Quality (DISCOVER-AQ) EVS-1 investigation uses airborne and ground-based measurements of trace gases and aerosols, correlated with satellite remote sensing measurements, to improve our ability to characterize and predict air quality. In order to continue efforts to characterize a variety of urban environments, a fourth measurement campaign is planned for Denver, Colorado during summer 2014.

- The second Earth Venture Instrument selection (EVI-2) will be made in late FY14 and the selected investigation will begin its formulation in FY15. The release of the third Earth Venture Instrument solicitation (EVI-3) will be made in FY15. The selection of the commercial geostationary satellite for the TEMPO (the selection for EVI-1) will also occur in FY15.

- The Earth Venture Mission-2 (EVM-2) CYGNSS project has made significant progress. The science product algorithms are proceeding well and testing has started using the selected nature model. The satellite observatories and constellation has completed the preliminary design and is beginning Engineering Model development. The PDR (Preliminary Design Review) and associated KDP-C were successfully completed resulting in agency confirmation and authorized entry of the project into its implementation phase. The launch service selection has completed with the selection of the Orbital Pegasus Vehicle, which will launch the entire constellation out of Florida.

- Through the Making Earth System Data Records for Use in Research Environments (MEaSUREs) Program, NASA is continuing its commitment to expand understanding the Earth system using consistent records. Emphasis is placed into linking together multiple satellites into a constellation, developing the means of utilizing a multitude of data sources to form coherent time series, and facilitating the use of extensive data in the development of comprehensive Earth system models. In FY 2014, the most recent selection of MEaSUREs projects made data products available and will continue to do so throughout the following five years. These data products include multi-sensor water vapor, temperature, and cloud climate data records, global precipitation climatology development, global water surface storage dynamics, and the cloud-boundary layer.

Department of Energy/Office of Science

DOE's Climate and Environmental Sciences Division (CESD) within the Office of Science focuses on a predictive, systems-level understanding of the fundamental science associated with climate change and DOE's environmental challenges; both key to support the DOE mission. CESD supports an integrated portfolio of research ranging from molecular to field scale studies with emphasis on the use of advanced computer models and multidisciplinary experimentation. As discussed next, CESD supports three research activities and two national scientific user facilities:

Atmospheric System Research (ASR)

The ASR activity seeks to resolve the two major areas of uncertainty in climate change projections: the role of clouds and the effects of aerosol emissions on the atmospheric radiation balance. Research from the ASR program results in improved physical formulations leading to state-of-the-art science related to clouds, aerosols, radiation, and precipitation. The program is geared to observe and advance understanding of the atmospheric system in a holistic, comprehensive fashion that addresses the full range of interrelated climatic processes. The anticipated end result is that climate models will have reduced uncertainty and improved climate simulation capability so that climate models can be used with increased confidence in decision- and policy making.

Climate and Earth System Modeling

Climate and Earth System Modeling in CESD focuses on development, evaluation, and use of Regional and Global Climate Modeling (RGCM), the development of Earth System Models (ESM), and Integrated Assessment Models to determine the impacts and possible mitigation, of climate change.

Achieving greater detail about uncertainty and future variability of the earth climate system is critical for decision-makers. There is a need to ascertain shifts in major modes of climate variability and climate extremes, to detect and attribute regional manifestations of climate change, and to conduct ever more thorough model validation. All these goals of the RGCM program remain significant challenges. This program also provides support for national and international climate modeling research and assessments. An understanding of the model biases seamlessly feeds back to the ESM program.

RGCM activities are organized into several distinct but coordinated components:

- The **Program for Climate Model Diagnosis and Intercomparison** (PCMDI) develops improved methods and tools for the diagnosis and intercomparison of climate and Earth system models. It provides major facilities for archiving climate model output, including frequently analyzed variables such as those used for the IPCC Assessment Reports. PCMDI makes such model output readily accessible to the climate modeling community.

- The **Climate, Ocean and Sea Ice Modeling Project** (COSIM) continues to develop the ocean model POP (Parallel Ocean Program), its hybrid-coordinate successor (HYPOP), and a sea ice model (CICE). COSIM is also developing a new Community Ice Sheet Model (CISM), designed for use at high spatial resolutions and at high latitudes. The scientific thrust of this work is to understand the role of oceans and ice in climate change, including (1) future sea-level rise caused by thermal expansion of the ocean and by melting of land ice; (2) stability of the high-latitude ocean thermohaline circulation; and (3) the unique high-latitude marine and ice ecosystems that reside along the ice edge and how they respond to changes in sea ice extent, including consequences for carbon and sulfur uptake and exchange.

- Multi-century simulations using the Climate Change Simulation Model (CCSM) are conducted by the DOE Climate Change Project at the National Center for Atmospheric Research (NCAR). Analysis of CCSM simulations provides insights into how natural and anthropogenic forcings affect the coupled climate system.

Research from the ESM program results in improved state-of-the-science dynamically coupled models for understanding future variability and predictability of the climate system. Significant scientific challenges need to be addressed, such as future changes in major modes of climate variability, climate extremes in a changing climate, detecting and attributing the regional manifestations of climate change, and carbon-cycle interactions with climate. Improved climate information at high spatial and temporal resolution is of immense significance to society and decision-makers.

Climate change is real, its effects are more immediate and profound than previously anticipated, and old questions (are humans the cause?) are yielding to new ones: What are the impacts? Who and what will be most vulnerable? What can we do about it, and how can we prepare? Against this backdrop, and with an eye toward: (1) regional and local scale insights; (2) quantitative predictions at the decadal, annual, and even shorter time scales; (3) policy-making, planning and decision-support tools; (4) impacts, adaptation, and vulnerability studies; and (5) highly integrated analyses spanning energy, environment, and economic security, new or vastly improved Integrated Assessment Models will inform some of the most significant U.S. energy and other infrastructure decisions and investments of this century. These models shape our fundamental understanding of climate change: the drivers of climate change; its pace and consequences; the implications and role for energy systems of the future; changes in availability of natural resources, food, and water; and shifts in global economies, vulnerabilities and overall national security.

Environmental System Science

The Environmental System Science activity in CESD seeks to advance a robust predictive understanding of energy-derived byproducts in terrestrial ecosystems extending from the bedrock to the top of the canopy and from molecular to global scales. This activity focuses on understanding the role of terrestrial ecosystems in a changing climate[2] and the role of subsurface biogeochemical processes in the fate and transport of heavy metal and radionuclide contaminants

---

[2] See http://science.energy.gov/ber/research/cesd/terrestrial-ecosystem-science/.

in subsurface systems.[3] DOE is responsible for what has been described as the largest, most complex, and diverse collection of environmental remediation challenges in the nation. While some of the problems are tractable and require only time and money to resolve, a large fraction of them cannot be resolved with existing knowledge and technology. The need for solutions to these challenging environmental remediation problems drives the Environmental System Science program.

Future climatic changes will almost certainly affect many important organisms and processes in terrestrial ecosystems, and these ecosystems provide society with a host of essential goods and services. The Environmental System Science program's research is directed at obtaining and then disseminating scientific knowledge of the most important effects of climatic change on ecosystems so that society can understand the ecological implications of climatic change and then plan for those changes. While the program focuses on U.S. terrestrial ecosystems, much of the knowledge gained has global applicability.

The terrestrial biosphere is a major factor influencing the transport and concentration of atmospheric greenhouse gases including carbon dioxide. Current limitations of our understanding of carbon cycling through terrestrial ecosystems account for significant uncertainties in projections of future climate scenarios. This program seeks to identify critical carbon cycle pathways, provide quantitative explanations for those pathways and integrate the resulting process understanding into coupled carbon-climate models.

CESD National Scientific User Facilities

Two scientific user facilities—the Atmospheric Radiation Measurement Climate Research Facility (ARM) and the Environmental Molecular Sciences Laboratory (EMSL)—provide the scientific community with technical capabilities, scientific expertise, and unique information to facilitate science in areas of importance to DOE's mission. ARM is a multiplatform facility that supports research for addressing the major uncertainties of climate models: clouds and aerosols. It provides the national and international research community unparalleled infrastructure for obtaining precise observations of key atmospheric phenomena needed to advance the understanding of understanding atmospheric

Radar Wind Profiler and radio acoustic sounding system (RASS), ARM site at Barrow, Alaska.

process and improve climate models. The facilities and capabilities of EMSL are available to the general scientific and engineering communities to conduct research in the environmental molecular sciences and related areas.

---

[3] See http://science.energy.gov/ber/research/cesd/subsurface-biogeochemical-research/.

Within DOE, ARM's major clients are the ASR, RGCM, and ESM programs of CESD (described above). The primary ARM objective is improved scientific understanding of the fundamental physics related to interactions between clouds, aerosols, and radiative feedback processes in the atmosphere. In addition, ARM has enormous potential to advance scientific knowledge in a wide range of interdisciplinary Earth sciences. ARM was the first climate change field research facility to operate cutting-edge instrumentation on a long-term continuous basis and at both fixed and varying locations around the globe. ARM field research sites are designed to study the effects of aerosols, precipitation, surface flux, and clouds on global climate change. The fixed sites are located in three diverse climate regimes representing mid-latitude, polar, and tropical environs (i.e., the Southern Great Plains and the North Slope of Alaska in the United States and the Tropical Western Pacific). With its aerial measurement capability and mobile ground facilities, ARM provides the world's most comprehensive continuous observational capabilities for obtaining atmospheric data specifically for addressing the major scientific uncertainties in climate change.

Each ARM site uses a leading-edge array of cloud-, aerosol-, and precipitation-observing instruments to record long-term continuous measurements of atmospheric and surface properties. ARM also provides shorter term (months rather than years) measurements with its two mobile ground facilities and aerial measurement capability. The combination of high temporal resolution at discrete locations makes ARM observations uniquely suited for studying local cloud processes, many aspects of which remain poorly represented in climate models. The resultant data are available through the ARM's data archive. These data are used as a resource for over 100 journal articles per year, which represent tangible evidence of the ARM's contribution to advances in most areas of atmospheric radiation and cloud research. Additional programmatic information is available via the ARM homepage on the Internet. In 2013, ARM initiated its strategy to improve climate predictability by combining High resolution earth system modeling; ARM facility observations, and Ecosystem research by initiating and enhancing its NSA infrastructure complex at Oliktok Point, Alaska. This will include the installation of a third ARM Mobile Facility (AMF3) that includes its Restricted Air Space (RA-2204). ARM plans to utilize tethered balloon and Unmanned Aerial Systems (UAS) in order to address the climate changing issues in the Arctic.

EMSL offers users access to more than 60 major systems, including many one-of-a-kind analytical instruments for studying atomic to molecular to larger-scale processes, a supercomputing platform and associated computational chemistry software, and the in-house scientific expertise to help obtain high quality results in a timely fashion. By co-locating multiple types of capabilities and scientific expertise, EMSL serves as an ideal place for research teams interested in integrating theory with experiment, as well as a place to conduct a wide range of single-investigator studies.

Detailed scientific knowledge of the physical, chemical, and biological processes occurring at the most fundamental levels is necessary to discover and fully utilize breakthroughs in areas such as hydrogen as a new energy source, improved catalysts and materials for industrial applications, insights into the factors influencing climate change and carbon sequestration processes, new approaches to managing legacy wastes such as radionuclide and heavy metal contamination, and making bioenergy sources a reality. The complex nature of DOE's energy, science, and environmental missions demands a wide range of leading-edge experimental and computational capabilities to enable scientists to conduct fundamental and multidisciplinary research using multiple experiment and computational approaches that will lead to scientific advances to help address the DOE missions. EMSL provides these leading-edge experimental and computational capabilities to the scientific community.

DOE's Next-Generation Ecosystem Experiments Project

The new scanning precipitation radar at the Barrow, Alaska, ARM site.

The Arctic is undergoing a system-wide reorganization in response to an altered climate. The mechanisms responsible for this change have been unpredictable and difficult to isolate due to a large number of interactions among individual components of the system. The Next-Generation Ecosystem Experiments (NGEE) project will quantify the complex physical, chemical, and biological behavior of terrestrial ecosystems in Alaska. The project will focus on interactions that drive ecosystem-climate feedbacks through greenhouse-gas fluxes and changes in surface energy balance associated with thawing permafrost and threshold-dominated permafrost degradation and thermokarst formation. Research sites will be located along a bioclimate gradient that spans tundra and shrub-tundra transition zones on the North Slope and Seward Peninsula.

The vision for the NGEE project is to deliver a high-resolution terrestrial system model for coupled thermal, hydrological, geomorphic, biogeochemical, and vegetation processes as needed to predict the evolution of a warming Arctic landscape and its feedback to the global climate system. This vision includes field observations, laboratory experiments, and modeling of critical and interrelated water, nitrogen, carbon, and energy dynamics and the important interactions, from the molecular scale to the landscape scale, that drive feedbacks to the climate system.

National Science Foundation (NSF)

The National Science Foundation funds basic climate research, modeling, and process studies. This research portfolio includes support to individual investigators and to groups such as the Center for Multiscale Modeling of Atmospheric Processes (CMMAP) and NCAR. With DOE as a partner, NSF funds NCAR to develop, maintain, and support the Community Earth System Model (CESM), a fully-coupled global climate model that provides state-of-the-art computer simulations of the Earth's past, present, and future climate states. As part of a cross-directorate and interagency effort, NSF and USDA are funding a third round of awards with a focus on decadal and regional prediction using Earth System Models. The NCAR/Wyoming Supercomputer is now fully operational and is available for a variety of research topics, including climate modeling studies, partially satisfying the ever growing need in the climate community for new computing resources..

The NCAR-Wyoming Supercomputing Center provides advanced computing services to scientists studying a broad range of disciplines, including weather, climate, oceanography, air pollution, space weather, computational science, energy production, and carbon sequestration. It also houses a landmark data storage and archival facility that will hold, among other scientific data, unique historical climate records. Courtesy: University Corporation for Atmospheric Research

EMERGENCY RESPONSE AND
HOMELAND SECURITY SERVICES

For purposes of this *Federal Plan*, Emergency Response and Homeland Security Services are those specialized meteorological services and facilities established to meet the requirements of Federal, state, and local agencies responding to natural disasters and security incidents. This category includes the use of atmospheric transport and diffusion (ATD) models for predicting the dispersion of airborne toxic substances; it also includes natural disaster monitoring and prediction services and the transport of water-borne toxic substances not included in basic services. For example, numerical weather prediction models used to forecast the path, intensity, and storm surge of landfalling tropical cyclones are part of basic services. Downstream models of the effects of a landfalling tropical cyclone on the infrastructure and population of a particular populated area could be included in this service category.

OPERATIONAL PROGRAMS INCLUDING PRODUCTS AND SERVICES

Federal Emergency Management Agency

The mission of the Federal Emergency Management Agency (FEMA) is to support U.S. citizens and first responders to ensure that as a Nation we work together to build, sustain, and improve our capability to prepare for, protect against, respond to, recover from, and mitigate all hazards. In carrying out its role, FEMA works with the Federal scientific community and meteorological agencies to ensure that appropriate risk information for hazards, vulnerabilities, and consequences is used to execute this mission.

FEMA's interagency collaboration and support is key to disaster impact assessments and plans.

FEMA's Planning Division administers the National Hurricane Program, which conducts hurricane evacuation studies, provides evacuation decision-making training, and provides a range of hurricane evacuation decision support tools to State and local government emergency management officials to develop their hurricane evacuation plans. Under an Interagency Agreement with the National Weather Service's National Hurricane Center (NWS/NHC), the NHC builds and utilizes its SLOSH (Sea, Lake and Overland Surges from Hurricanes) storm surge model as the hazard analysis basis for Hurricane Program studies, training, and decision support to State and local governments..

It is critical for FEMA to identify, develop, and/or utilize the most appropriate meteorological information to calibrate its preparedness, response, and recovery activities to build and deploy emergency management capability, and to design and implement mitigation measures which reduce the con–sequences from emergencies and disasters. These interests extend to national standards for geographic information systems (GIS) used for delivery of meteorological products and services by other agencies. As administrator of the National Flood Insurance Program (NFIP), FEMA publishes Flood Insurance Rate Maps for all flood-prone communities, which serve as the official demarcation for flood risk.

As part of its National Hurricane Program, FEMA also licenses, operates, maintains and distributes the HURREVAC software decision-support tool to approximately 20,000 local and state government coastal emergency managers annually. HURREVAC combines real-time NHC forecast data with evacuation clearance times to guide emergency managers as to when to execute hurricane evacuations to ensure evacuation is completed before the arrival of hurricane conditions. Additional information can be found at http://www.fema.gov/region-iii-mitigation-division/national-hurricane-program

FEMA actively supports the OFCM-sponsored Working Group for Disaster Impact Assessments and Plans: Weather and Water Data (WG/DIAP) and the WG/DIAP's efforts to develop and implement the National Plan for Disaster Impact Assessments which outlines the interagency procedures to coordinate and support the collection of perishable data after major storms. *(see related COASTAL Act narrative on page 2-6)* These data have applications in post-disaster mitigation activities, the NFIP flood hazard analysis, the FEMA National Hurricane Program hurricane, and other FEMA risk analysis activities, such as the Multi-Hazard Loss Estimation Methodology (HAZUS). The National Hurricane Program is the principal FEMA contact point for most meteorological matters, while the FEMA Risk Analysis Division is the primary contact for NFIP flood risk analysis.

U.S. Coast Guard

The U.S. Coast Guard (USCG) directly supports all Department of Homeland Security (DHS) missions. It provides DHS the broad authorities, capabilities, and partnerships necessary to accomplish its tasks in the maritime domain.

Coast Guard has three broad roles: Maritime Safety, Maritime Security, and Maritime Stewardship. Each USCG role is composed of several statutory missions. Each of the eleven statutory missions is listed below under the appropriate role.

- The Air Force Weather Agency (AFWA) provides backup to several NOAA operational centers, maintaining capabilities (for example) in severe weather, aviation, volcanic ash dispersion, and space weather, available if and when needed to support emergency response and homeland security.

- Maritime Safety Missions
  - Marine Safety
  - Search and Rescue

- Maritime Security Missions
  - Drug Interdiction
  - Migrant Interdiction
  - Defense Readiness
  - Ports, Waterways, and Coastal Security

- Maritime Stewardship Missions
  - Living Marine Resources
  - Marine Environmental Protection
  - Other Law Enforcement (Fisheries)
  - Aids to Navigation
  - Ice Operations

Most USCG missions support more than one role. For example, the Aids to Navigation mission primarily supports the Maritime Stewardship role by preventing pollution from vessel groundings and collisions, while facilitating the movement of people and goods. But this mission also supports our Maritime Safety Role by preventing accidents, injuries, and deaths. These interwoven roles and complementary missions call for Coast Guard personnel and resources that are similarly multi-mission capable. This characteristic of USCG people and platforms—their ability to perform multiple missions—brings greater effectiveness, insight, and agility to bear in any situation. It is a fundamental source of the Coast Guard's strength.

Support for meteorological operations and supporting research is detailed in other sections of this plan.

Interagency Modeling and Atmospheric Assessment Center

The IMAAC provides a single point for the coordination and dissemination of Federal dispersion modeling and hazard prediction products that represent the Federal position during actual or potential incidents involving hazardous atmospheric releases. Through plume modeling analysis, the IMAAC provides emergency responders with predictions of hazards associated with atmospheric releases to aid in the decision making process to protect the public and the environment.

The IMAAC is led by the Department of Homeland Security (DHS) and supported by seven other federal departments and agencies: Department of Defense (DoD), Department of Energy (DOE), Department of Health and Human Services (HHS), Environmental Protection Agency (EPA), National Aeronautics and Space Administration (NASA), National Oceanic and Atmospheric Administration (NOAA), Nuclear Regulatory Commission (NRC). The IMAAC Technical Operations Hub is operated by the Defense Threat Reduction Agency (DTRA).

Decision makers and first responders need timely and accurate plume predictions to help guide emergency response decisions. The IMAAC provides a suite of plume modeling tools that incorporate meteorological, geographic and demographic data, as well as hazardous material information, to predict the transport and potential downwind consequences of biological, chemical, radiological/nuclear, and natural releases. The IMAAC experts are available 24/7 to produce detailed quality-assured model predictions, utilize observations and field measurement data to refine analyses, and assist decision makers in product interpretation.

The IMAAC produces both technical analyses and briefing products tailored for communications to non-technical decision makers. The IMAAC plots show hazard areas, affected populations, potential casualties and/or fatalities, damage estimates, and health effect, public protective action and worker protection levels. The IMAAC utilizes NOAA National Weather Service's meteorological observations as well as variety of other available numerical weather prediction models (NWP) for the plume products. During IMAAC activations, the IMAAC Technical Operations Hub will consult with NOAA to determine the preferred higher-resolution forecast model for the event and will utilize that model if possible.

The IMAAC has responded to numerous real-world events, including chemical fires and train derailments, in-situ burns from the Deepwater Horizon oil spill, and sulfur dioxide volcanic emissions in Hawaii.

NOAA/National Weather Service

National Tsunami Hazard Mitigation Program (NTHMP).

The National Weather Service (NWS) has oversight responsibility for the NTHMP. The mission of the NTHMP is to work with communities in vulnerable U.S. coastal areas on preparedness activities to respond to tsunami events. In response to the destructive Indian Ocean Tsunami (December 2004), the U.S. Tsunami Warning Program, including the NTHMP, was upgraded and expanded to enhance the monitoring, detection, warning, and communications capabilities designed to protect lives and property for all U.S. communities at risk.

NOAA Weather Radio (NWR).

NWR is used as a reliable means of communicating weather-related warnings directly to the public. NWR is a nationwide network of radio stations broadcasting continuous weather information directly from the nearest National Weather Service office. NWR broadcasts official Weather Service warnings, watches, forecasts and other hazard information 24 hours a day, 7 days a week. Working with the Federal Communication Commission's (FCC) Emergency Alert System, NWR is an "All Hazards" radio network, making it a single source for comprehensive weather and emergency information. In conjunction with Federal, State, and Local Emergency Managers and other public officials, NWR also broadcasts warning and post-event information for all types of hazards – including natural (such as earthquakes or avalanches), environmental (such as chemical releases or oil spills), and public safety (such as AMBER alerts or 911 Telephone outages).

FEMA Integrated Public Alert and Warning System (IPAWS) and Wireless Emergency Alerts.

The NWS pushes watches, warnings, advisories, and special statements in Common Alerting Protocol (CAP) format to the FEMA IPAWS Alert Aggregator. Warnings which pose an "imminent threat" to the general public pass through IPAWS to commercial wireless carriers and are broadcast as a 90 character Wireless Emergency Alert (WEA) message on WEA-capable cell phones. The message notifies the recipient of the alert type, expiration time, a brief description of the action to be taken, and name of the alert originator (i.e., NWS). Numerous wireless carriers provide the CMAS/WEA service and more continue to come online. The number of WEA-capable devices on the market has dramatically increased since 2013. In fact, at least one major carrier reports their entire line-up of cell phones on the market as being WEA capable.

The NWS is actively collaborating with commercial wireless industry as well as Federal, emergency management, and academic partners in multiple Federal Communications Commission (FCC)-tasked working groups which have been tasked to identify possible future improvements to WEA.

The NWS and FEMA have made NWS CAP messages available over FEMA IPAWS and multiple NWS dissemination systems for repackaging and redistribution by alert distribution partners. The NWS also continues working with FEMA, the FCC, broadcast industry, and related partners on solutions which would provide compatibility between NWS-produced CAP and EAS signals broadcast by NOAA Weather Radio (NWR), so that NWS-CAP could be used to activate EAS.

Interagency Activities.

In partnership with the Department of Homeland Security (DHS), NWS forecasters provide meteorological support for response to terrorist acts and other homeland security concerns, as well as accidental releases/spills of hazardous chemical, biological, or radioactive materials or other environmental events. NOAA has a meteorologist permanently located at FEMA Headquarters as a liaison to provide real-time decision support to FEMA and other national government decision makers. In addition, the liaison (along with additional NWS meteorologists when necessary) will support activations at the National Response Coordination Center as a result of any major disaster or emergency. Similarly, NOAA has a liaison deployed to the DHS National Operations Center.

NWS meteorologists provide forecasts in response to Incidents of National Significance such as the space shuttle Columbia recovery effort, volcanic eruptions in Iceland, Sandy, and after major tornado events such as those in Tuscaloosa, Alabama and Joplin, Missouri during the spring of 2011. In addition, the NWS deploys a national cadre of specially trained Incident Meteorologists (IMETs) to provide onsite support for large wildfires and other homeland security concerns, as well as for accidental releases/spills of hazardous chemical, biological, or radioactive materials.

NOAA/National Ocean Service

Coastal Oceanographic Applications and Services for Tides and Lakes (COASTAL).

The COASTAL program focuses on non-navigational applications of CO-OPS observing systems, data, and products for ecosystem restoration and management. COASTAL also provides decision-support tools to aid managers and restoration practitioners to plan for both current and future coastal conditions, and to anticipate and mitigate natural hazards. Real-time water level and meteorological information is critical for emergency managers to make decisions related to evacuation and warnings for coastal communities as well as to produce storm surge predictions.

The Storm QuickLook product in particular incorporates water level and meteorological information measured at National Water Level Observation Network (NWLON) and Physical Oceanographic Real-Time System (PORTS®) stations. Storm QuickLook bulletins (figure to the right) are posted for tropical cyclones that affect the United States coastline, but have also been created for the Deepwater Horizon oil spill and, in FY 2011, for elevated water levels in the Mississippi River. These bulletins provide near real-time, continuously updating oceanographic and meteorological data measured at affected water level stations and are displayed on the CO-OPS website and on the NOAAWatch (the NOAA All-Hazard Monitor) page. FY 2014 goals for this product are to enhance displays and make the product more dynamic. Also, 6-minute interval GOES transmission capability supports the NWS storm surge warning program when expected water level elevations are predicted or observed during coastal storms and hurricanes. COASTAL also provides aid to Tsunami

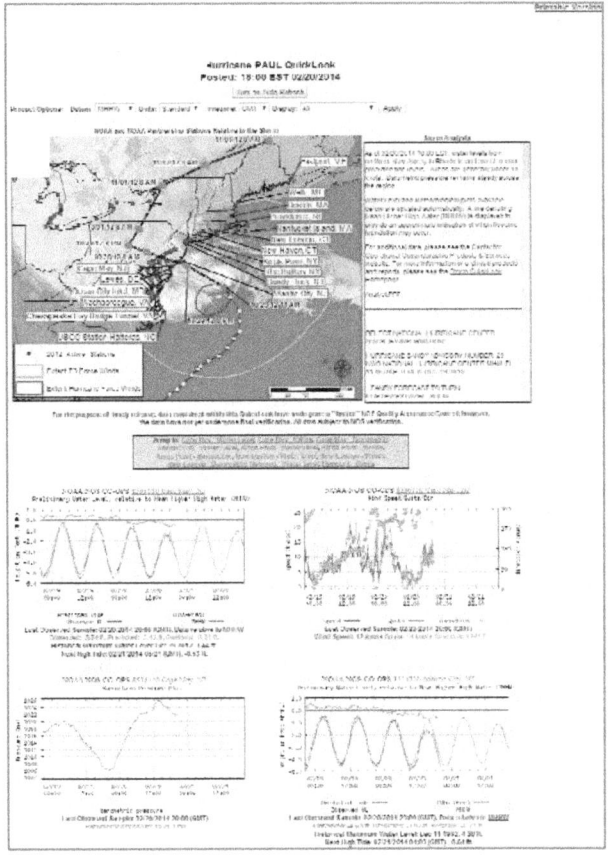

Warning Centers, by supplying one-minute water level observations at coastal NWLON stations, as well as 15-second data as requested after a tsunami event. These data are used to assess the impact and damage of a tsunami. For more information see http://tidesandcurrents.noaa.gov/coastal.html.

NOAA/Office of Marine and Aviation Operations

Among the NOAA/OMAO airborne observing systems fleet, the King Air (N68RF), AC-695A Commander 1000 (N45RF) and the Twin Otters (4 aircraft) can perform damage assessment flights following natural disasters. They provide high-resolution photographs to the public via the Internet. These photographs are extremely useful to local, county, and state government personnel, as well as to emergency managers and to the public at large, as they go about the business of assessing the damage and the nature and magnitude of the relief effort that will be required in the region.

United States Air Force

Air Force (AF) weather personnel enable military decision-makers to anticipate and exploit the weather for air, ground, space, cyberspace, and intelligence operations. As applies to this section of the federal Plan, the AF provides meteorological services and support in the form of data, information and products required by Air Force, Army, and other military operations in order to support the Nation's emergency response and homeland security efforts. Different AF organizations support various aspects of the Nation's homeland security efforts:

- The AF, through the 1st Weather Group, produces environmental products used to support both the daily and emergency response and homeland security operations of the U.S. Northern Command (USNORTHCOM) and the North American Aerospace Defense Command (NORAD). The AF's 2nd Weather Group produces fine-resolution model and forecast products for use by the Defense Threat Reduction Agency (DTRA) in its hazard dispersion modeling and related emergency planning and response efforts. The 2nd Weather Group also produces space weather products used to support both the daily and emergency response and homeland security operation of USNORTHCOM and NORAD.

- The AF produces both CONUS and worldwide geospatial representations of current and forecast weather used to support the National Geospatial-Intelligence Agency (NGA) and US Army Corps of Engineers in their homeland security missions.

- The AF provides backup to several NOAA operational centers, maintaining capabilities (for example) in severe weather, aviation, volcanic ash dispersion, and space weather, available if and when needed to support emergency response and homeland security.

- The AF's overseas Operational Weather Squadrons (the 17th OWS in Hawaii and the 21st OWS in Germany) may support emergency planning and response operations in their respective areas of responsibility. The Joint Typhoon Warning Center (JTWC), operated jointly with the Navy in Hawaii, routinely works with its sister U.S. hurricane forecast centers to issue forecasts and warnings to protect U.S. assets and interests across the Pacific and Indian Ocean basins.

United States Navy

Humanitarian Assistance and Disaster Response

Navy METOC personnel support Humanitarian Assistance and Disaster Response (HA/DR) operations aboard U.S. Navy and Coast Guard vessels. Military oceanographic survey vessels designated as T-AGS are continuously forward deployed and conduct hydrographic, oceanographic and acoustic surveys in all the oceans of the world. These ships have modern full ocean depth multi-beam and single-beam sonar systems for accurately measuring bottom depths and features, towed side-scan sonar systems for acoustic imaging of bottom features and navigation hazards, ocean current profilers, sub-bottom profilers for measuring stratification of seabed sediments, and over-the-side devices that collect physical ocean parameters such as temperature and salinity with depth. These ships are capable of hosting a number of roll-on/roll-off systems, including systems for collection of seismic data and unmanned underwater vehicles. The ships are also equipped with C-band communications to send data directly back to the Naval Oceanographic Office for immediate processing. Three of the ships are complemented with hydrographic survey vessels — smaller craft for shallow water hydrographic collection. We have been operating unmanned underwater vehicles (UUVs) for well over a decade. Their inventory consists of propelled vehicles, such as various models of the Remote Environmental Monitoring Units (REMUS) UUVs for collection of sonar, sub-bottom and optical data in addition to buoyancy controlled, high endurance UUVs, such as the Slocum glider and wave-powered Sensor Hosting Autonomous Remote Craft (SHARC) vehicles, which collect and transmit ocean and atmospheric data in real time for satellite transfer back to our operational production centers. Navy uses UUVs for direct support to operation missions. As part of our hydrographic capability, they also employ unique jet-ski variants called Expeditionary Survey Vehicles or

ESVs. Outfitted with single-beam and side-scan sonar, in addition to accurate Global Positioning System navigation, ESVs can be rapidly deployed and conduct hydrographic survey into the surf zone, to places where traditional systems can't go. This data is extremely useful for supporting expeditionary warfare and enabling missions like humanitarian assistance and disaster response. Weather forecasts and surveys of anchorages and piers provided commanders with the status of port accessibility, ensuring accessibility to the port critical to allowing food, water, medical supplies, materials, and other support to arrive by sea.

Partnerships and Stability Operations

Partnerships provide access to data sources, insight into emerging science and technology, and help all parties leverage limited budgets by reducing redundancy and pooling capabilities. Naval operational oceanography has a long history of executing cooperative military hydrographic and oceanographic surveys, as well as data, product and subject matter exchanges with international partners. Their ability to provide emerging partners with improved capabilities in the areas of meteorology, oceanography and hydrography is unique and is highly regarded by combatant commanders and naval component commanders as an outstanding tool for strategic engagement. In turn, their partners provide access to information vital to forward-deployed naval operations.

Navy METOC also maintains strong interagency partnerships. The integration of modeling efforts among the Navy, National Oceanic and Atmospheric Administration (NOAA), and U.S. Air Force, NUOPC provides an unparalleled global modeling capability that can be adapted by individual agencies for specific applications. Another strong example is their partnership with NOAA and the U.S. Coast Guard in operating the U.S. National/Naval Ice Center, providing ice analysis and forecasts to support safety of navigation for public, commercial and DoD use in the Arctic, Antarctic, Great Lakes, and all other large bodies of water affected by ice.

U.S. Marine Corps

Chemical Biological Incident Response Force

The CBIRF METOC Analyst serves to provide essential meteorological and hazardous prediction information in support of CBIRF`s real world and training operations coupled with providing daily and weekly weather forecasts and climatological briefs to the corresponding staff to support planning operations and typical battalion operations. Other duties include: research and implement new weather forecasting techniques or tools as well as new plume modeling software or tools; responsible for coordinating and maintaining liaison with local, state, and federal agencies such as the Defense Threat Reduction Agency (DTRA) or Joint Task Force Civil Support (JTF-CS) in order to enable appropriate sharing of information and a combined arms response to Chemical Biological Radiological Nuclear Explosive (CBRNE) incidents that may occur within CBIRF`s purview; and work with the Battalion Chemist to develop plume modeling products that are disseminated in response to training or real world CBRNE events.

U.S. Geological Survey

The USGS mission provides for "the classification of the public lands and the examination of the geological structure, mineral resources, and products of the National Domain." The USGS serves the Nation by providing reliable scientific information to describe and understand the Earth; minimize loss of life and property from natural disasters; manage water, biological, energy, and mineral resources; and enhance and protect our quality of life. Among its broad responsibilities and efforts are identification, assessment, and monitoring of potentially hazardous areas; development of capabilities to predict the time, place, and the severity of hazardous geologic, hydrometeorologic, biologic, and chemical conditions or events; and dissemination of the findings and their implications, including the provision of technical and scientific advice to public officials. The USGS also maintains Bureau-wide efforts intended to educate the public about natural hazards.

The USGS has been delegated the Federal responsibility to provide notification and warnings for earthquakes, volcanoes, and landslides. In addition, USGS data-collection networks provide real-time information needed by other agencies to issue forecasts and warnings related to a variety of hazards. For example, the USGS seismic network supports NOAA tsunami warnings; the USGS streamgage network supports NOAA flood forecasts (see Hydrometeorological and Water Resources Services); the USGS geomagnetic observations support solar storm forecasts (see Space Weather Services); USGS biologic monitoring of wildlife diseases enhances assessments of potential human pathogens such as the H1NI influenza virus; and USGS geospatial and remotely sensed information supports a broad spectrum of disaster-response activities and operations from an "all-hazards" perspective.

The USGS established a secondary reception station for NOAA GOES at the USGS Earth Resources Observation and Science (EROS) Center in Sioux Falls, SD. Three new satellite antennas and an existing antenna at EROS are used in support of this effort. Three of these antennas support communications with the GOES East and GOES West satellites, along with a hot spare. The remaining antenna is designated for a DOMSAT link, which is used for data dissemination. EROS also receives streamgage data in real time from the GOES satellites and is making these data available to USGS and other stakeholders. The receive station at EROS serves as a backup to the primary site [station] at Wallops Island, Virginia, which otherwise would represent a single point of failure in this vital data collection and dissemination system.

Beyond network operations, the USGS has the expertise and infrastructure to acquire, assess, disseminate, or preserve information that can be derived from the study of geological, hydrological, meteorological, chemical, or biological conditions before, during, or after an imminent or declared disaster or emergency. These capabilities can be tapped through mission assignments, interagency agreements, or third-party contracts, as provided by law and regulation.

Department of Energy

Field Site and National Laboratory Meteorological Services

In support of its national security, scientific research, and environmental stewardship missions, DOE has established and maintained operational meteorology programs at its field sites, national laboratories and offices. It has also managed various atmospheric research projects in support of emergency response and homeland security services among other activities. With respect to these programs, field sites and national laboratories collect quality-assured meteorological data, develop site-specific climatology from these measurements, use these data for various applications, and provide local site weather forecasting.

Emergency response and homeland security services require characterization of atmospheric processes that determine fate and transport of hazardous material releases. Meteorology programs directly contribute to the protection of public health and safety, and the environment, by accurately measuring and characterizing relevant local atmospheric processes to establish real-time and forecasted consequences.

DOE sites benefit from well-managed meteorological program support to 24/7 operations (e.g., work-force safety under severe weather and emergencies) and national defense programs.

Field Site and National Laboratory Operationao Meteorological Support

Operational meteorology programs provide customized meteorological monitoring services and local site weather forecasts for national defense projects, and for nuclear safety and emergency response programs. Each operational program is primarily focused on supporting emergency response, and protecting the safety of workers, public and the environment.

Operational meteorology programs are established at eight national laboratories, three reservations, Pantex, and Waste Isolation Pilot Plant. Some of these programs incorporate 24/7 severe weather watches to reduce risk of potential adverse impacts on facilities, construction projects, property, and workers:

- Two sites have received a National Weather Service (NWS) Storm Ready certification and other sites are pursuing this certification;

- Lightning protection and safety initiatives are integral elements of facility and worker safety systems supported by operational meteorology programs at eight locations;

- State-of-the-art instrumentation, including vertical profilers, sound detection and ranging instruments, and sonic anemometry, are employed in site monitoring networks; and,

- Data from some sites are part of the NWS database through the Meteorological Assimilation Data Ingest System.

The DOE Meteorological Coordinating Council, established in 1994, provides a forum for continuity of meteorological activities at DOE sites. Information on this council can be located at www.orise.orau.gov/emi/dmcc.

Many DOE reservations and national laboratories are located in areas with heterogeneous surface characteristics (e.g., land-water interface, mountain-valley morphology, forests) and experience complex localized weather conditions and resultant airflow trajectories. At these locations, diagnostic and prognostic characterization of local three-dimensional wind fields is required to determine trustworthy consequence assessments. Accordingly, multi-location comprehensive meteorological monitoring systems are essential at these facilities. Oak Ridge National Laboratory (ORNL) has produced a comprehensive characterization of the wind fields in the Oak Ridge, TN area that is being used to fine-tune its ATD modeling while other sites have extensively studied site-specific atmospheric flow phenomena to calibrate ATD capabilities.

Finally, several DOE sites use NOAA's Weather Research Forecasting mesoscale model as a component of its emergency response prognosticative capability.

Nuclear Regulatory Commission

The Nuclear Regulatory Commission (NRC) Office of Nuclear Security and Incident Response coordinates NRC responses to nuclear facility emergencies through the activation of the NRC Operations Center. The Protective Measures Team dose analysts in the NRC Operations Center rely on the Radiological Assessment System for Consequence Analysis (RASCAL) model to assess offsite consequences in the event of a radiological accident at a NRC-licensed facility.

The NRC also maintains an interest in the effects of transport and dispersion of airborne hazardous and nonradioactive materials on the safe operation of nuclear facilities and uses SAFER Real-Time to assess protective action levels for making recommendations in the event of a non-radiological accident at an NRC licensee.

SUPPORTING RESEARCH PROGRAMS AND PROJECTS

NOAA/Office of Oceanic and Atmospheric Research (OAR)

Atlantic Oceanographic and Meteorological Laboratory (AOML)

AOML conducts mission oriented scientific research that seeks to understand the physical, chemical, and biological characteristics and processes of the ocean and atmosphere, both separately and as a coupled system. While the focus of the laboratory is on improving understanding through research, AOML houses a large component of the global ocean observing system, including Argo profiling floats and surface drifters. AOML also has maintained the long-standing expendable bathythermograph (XBT) program, which has a long-standing collaboration with the National Weather Service and the United States Coast Guard to develop and support the AMVER program. The AMVER program (Automated Mutual-Assistance Vessel Rescue AMVER reports allow the U. S. Coast Guard to track a vessel's position. The AMVER program relies on ships to submit four types of reports: (1) Sail Plans; (2) Position Reports; (3) Arrival Reports and (4) Deviation Reports, when necessary. The U. S. Coast Guard updates their database with the position information from these reports, which allows them to identify vessels in the vicinity of a ship in distress. AOML mains the software and processes all reports.

Air Resources Laboratory (ARL)

NOAA's Air Resources Laboratory (ARL) develops and improves dispersion and air quality models and collects research-grade atmospheric measurements. Our programs support NOAA, broader US, and some international emergency response programs with an emphasis on chemical, nuclear, smoke, dust, and volcanic ash events. Our key NOAA customer is the National Weather Service, and we interact extensively with the National Ocean Service Office of Response and Restoration. ARL also provides a broad selection of web based assessment tools. ARL's primary emergency response model platform is the Hybrid Single Particle Lagrangian Integrated Trajectory (HYSPLIT) model and has been made operational at NOAA's National Centers for Environmental Prediction (NCEP). HYSPLIT also serves as the national dispersion forecasting capability in several other countries. HYSPLIT is the major product employed in the operations of the Regional Specialized Meteorology Center (RSMC) set up as a joint undertaking of ARL and NCEP under the auspices of the World Meteorological Organization (WMO). The

WMO/RSMC is the source of dispersion products in the event that a radioactive plume crosses international boundaries.

ARL provides meteorological support for the Department of Energy at the Idaho National Laboratory (INL) in Idaho Falls and at the Nevada National Security Site (NNSS) north of Las Vegas. ARL meteorologists staff the DOE Emergency Operations Center during drills and emergencies, such as accidental toxic chemical releases and wildfires. ARL operates an additional mesonet in East Tennessee that collects ozone and basic meteorological information. ARL operates an urban research meteorological network within the National Capital Region called DCNet that collects standard meteorological data and also measures characteristics of atmospheric turbulence. DCNet provides critical data and insights that improves dispersion predictions of airborne hazardous materials. While DCNet is a research network, its observations are used by numerous government security and emergency management personnel for various activities within the National Capital Region.

In FY15, ARL will:

- Continue to improve and adapt the HYSPLIT model to address threats imposed from radioactive material, volcanic ash, dust, and wildfire smoke, and other hazardous materials.

- Provide reach-back support to NWS and other organizations that use our models.

- Continue to maintain mesonets in Nevada, Idaho, eastern Tennessee, and the National Capital Region.

- Continue to support the Department of Energy at the INL and NNSS.

Great Lakes Environmental Research Laboratory (GLERL)

The Great Lakes Environmental Research Laboratory (GLERL) has developed and continues to upgrade the Great Lakes Coastal Forecasting System (GLCFS). This system produces forecasts of waves, currents, temperature, ice, and other properties of the five Great Lakes and their connecting channels that are used by commercial ship captains, recreational boaters, fishermen, beach managers, oil spill responders, and search and rescue personnel. The GLCFS is currently being upgraded from a system based on the Princeton Ocean Model (POM), to a system based on the Finite Volume Coastal Ocean Model (FVCOM), which allows the model to be executed with variable spatial resolution, among other capabilities. The FVCOM versions of the GLCFS will become operational in FY2014, starting with Lake Erie.

The NOAA Great Lakes Regional Collaboration Team has also recently completed a technical memorandum that describes the particular roles and responsibilities for different NOAA offices and personnel in the Great Lakes in the event of a natural or human-caused emergency such as a severe winter storm, large wildfire, or oil spill. The team will be assisting other NOAA regions in FY2014 to adapt this approach for use in their areas, with consideration of the particular emergencies that might be anticipated in other regions such as hurricanes or earthquakes.

Department of Energy

The Department of Energy (DOE) Meteorological Coordinating Council (DMCC) was established in 1994 to coordinate meteorological activities among the field offices to enhance cost effectiveness and productivity and to leverage synergistic opportunities. DOE has delegated the operation of its site/facility meteorological programs to NOAA and to non-Federal for-profit management and operating contractors. The DMCC membership is therefore composed of subject-matter experts from within the DOE complex, representing the three components with operational responsibilities for the following programs:

- Department of Commerce (DOC/NOAA) under an interagency agreement

- Management and operating (M&O) contractors

- Private contractors

The DMCC operates as a subcommittee of the DOE Emergency Management Issues Special Interest Group (EMI SIG) and has a web page that can be accessed directly or through the web page of the Subcommittee for Consequence Assessment and Protective Actions (SCAPA). DMCC also issues an annual report as part of its presentation to the EMI SIG Steering Committee.

A current DMCC project is to improve the provision of quality-assured meteorological information and execution of transport and diffusion models that meet software quality assurance requirements. Products of the DMCC include evaluations of meteorological requirements contained in DOE orders and guidance documents, site meteorological program peer reviews (i.e., meteorological program assist visits), and, as needed, customized technical assistance. The DMCC developed tools to enable DOE/NNSA sites to perform self-assessments of their individual meteorological monitoring programs and the meteorological aspects of consequence assessment.

Nuclear Regulatory Commission

The Office of Nuclear Regulatory Research (RES) plans, recommends, and implements a program of nuclear regulatory research for nuclear power plants and other facilities regulated by the Nuclear Regulatory Commission (NRC). RES provides technical support, technical tools, and information to identify and resolve safety issues for current and new designs and technologies through testing, data development, analysis, and national and international collaboration. For example, RES is currently evaluating toxic gas dispersion modeling capabilities of several computer codes as licensing tools for assessment of control room habitability specific to dense gas release simulations. RES also develops regulatory guidance and participates in the development of criteria and consensus standards related to the protection of the public health and safety and the environment.

HYDROMETEOROLOGY AND WATER RESOURCES SERVICES

For purposes of this *Federal Plan*, Hydrometeorology and Water Resources Services are those specialized meteorological services and facilities that combine atmospheric science, hydrology, and water resources in order to meet the requirements of Federal, state, and local agencies for information on the effects of precipitation events on infrastructure, water supplies, and waterways. These products and services also meet the needs of the general public in the conduct of everyday activities and for the protection of lives and property.

OPERATIONAL PROGRAMS INCLUDING PRODUCTS AND SERVICES

Interagency Collaborative Programs and Products

Integrated Water Resources Science and Services (IWRSS)

The Integrated Water Resources Science and Services (IWRSS) is a new business model for interagency collaboration. IWRSS brings a consortium of United States federal agencies with complementary water resources missions together to share resources to help solve the nation's water resources issues. Initiated through an Interagency Memorandum of Understanding (MOU), IWRSS's overarching objective is to enable and demonstrate a broad, interactive national water resources information system to serve as a reliable and authoritative means for adaptive water related planning, preparedness and response activities. The goals are to:
- integrate information delivery and simplify access to this data,
- increase accuracy and timeliness of water information, and
- provide summit-to-the-sea high resolution water resources information and forecasts

Currently the collaboration is with three United States federal agencies: USGS, NOAA, and USACE. Other federal agencies are expected to join the consortium in the near future. Two initial charters have been written to support and help define the IWRSS effort. The first charter is the National Flood Inundation Mapping (NFIM); with the second charter being the System Interoperability and Data Synchronization (SIDSRT). Interagency teams have been identified and charged with addressing the tasks described in the charters.

The implementation of the charters represents initial activities that address the IWRSS goals and the objective of the Building Strong Collaborative Relationships for a Sustainable Water Resources Future initiative to build a Federal Support Toolbox for Integrated Water Resources Management (IWRM). Additional information is included in the NOAA/National Weather Service narrative below and at http://www.nws.noaa.gov/oh/nwc/.

NOAA/National Weather Service

National Oceanic and Atmospheric Administration's (NOAA) National Weather Service (NWS) has the primary responsibility among Federal agencies to provide advanced alerts of hydrologic conditions via flood warnings and river forecasts in the United States. The provision of these water prediction and warning services are significantly influenced by anthropogenic activities associated with water management and water use/regulation (e.g. reservoirs, diversions, withdrawals). In order to provide these important services, collaboration with other federal agencies is crucial as 24 Federal agencies have a role in water. The end-to-end forecast process employed by the NWS to provide hydrologic and water resources forecast and warning services leverages data and information from Federal partners and involves the following:

- Meteorological and hydrological observations and anthropogenic data;
- Data collection, management and processing;
- Science and technology infusion;
- River and water quantity forecasting;
- Inundation mapping;
- Dissemination systems;
- Performance measurement; and
- Outreach and education.

NWS relies on routine access to water data and observations from other Federal agencies to produce and provide water forecasts and warnings. These Federal agencies and their respective roles include the:

- U.S. Geological Survey (USGS) - real-time and historical streamflow observations, stream/river cross-sections, rating curves (relating streamflow to river stage), topographic data (for inundation mapping)
- U.S. Army Corps of Engineers (USACE) – water use/regulation data, real-time streamflow observations, and hydraulic models
- Bureau of Reclamation (BOR) – water use/regulation data
- U.S. Department of Agriculture (USDA) – snow observations
- Department of Energy (DOE) – water use/regulation data

NWS hydrology and water resources program also leverages and expands NOAA science and service partnerships for the atmosphere, watersheds, estuaries, and oceans as well as partnerships with universities, the private sector, and the international community to improve and integrate modeling capabilities.

An NWS field infrastructure which includes 122 Weather Forecast Offices (WFO), 13 River Forecast Centers (RFC), the National Operational Hydrologic Remote Sensing Center (NOHRSC), and the National Centers for Environmental Prediction's (NCEP) Weather Prediction Center (WPC) and Climate Prediction Center (CPC) work as a team to provide hydrologic forecast and warning services from floods to droughts to minimize the loss of life and property and to meet the growing water resources service needs of our Nation. NWS hydrologic products and services support decision makers from a spectrum of service sectors including emergency management, agriculture, hydropower, reservoir management, and watershed management, river commerce, municipal and industrial water supply, and recreation.

A new element in the NWS hydrology program infrastructure, the National Water Center (NWC), was completed in early 2014 on the campus of the University of Alabama in Tuscaloosa, AL. The NWC will serve as a cornerstone for collaboration among Federal water agencies and provide a central hub to integrate and advance regional field operations and services. It is currently being staffed to achieve an Initial Operating Capability by May 2015.

WFOs assess and monitor the threat of flash and river flooding 24 hours a day 7 days a week to provide timely and accurate life-saving flood watches and warnings. Toward this end, WFOs integrate a spectrum of RFC, NCEP and NOHRSC guidance, Doppler weather radar (NEXRAD)-based precipitation estimates, and real-time telemetered precipitation and stream gauge observations, to provide routine river forecast services and critical, event-based decision support services. In addition, WFOs work with dam operators to provide timely warnings for floods that result from infrastructure failure such as dam break and levee breaches. Moreover, WFOs routinely conduct local outreach and education to heighten public and partner awareness of flood risks and NWS hydrologic services.

RFCs routinely generate short range (deterministic) through extended range (probabilistic) river forecasts and (deterministic) flash flood guidance. Information from the RFCs serves as the basis for local flood and flash flood warnings, watches, and advisories issued by the WFOs. These RFC products typically incorporate guidance from WPC, CPC, NOHRSC and the WFOs, and emphasize flooding impacts from meteorological events based on geographic area, land use, time of the year, and other factors. In order to provide objective simulations of future river flows, RFCs calibrate, operate, and verify sophisticated hydrologic models based on rainfall, soil characteristics, quantitative precipitation forecasts (QPFs), reservoir regulations, and several other variables. Some RFCs, especially those in mountainous regions, also provide water-supply volume and peak-flow forecasts based on snow pack in high elevations. These water supply forecasts are used by a wide range of decision makers, including those in agriculture, hydroelectric dam operation and electricity generation, and water resources management. RFCs routinely coordinate with their associated WFOs, Federal water partners, stakeholders, and the WPC.

NOHRSC provides comprehensive snow observations, analyses, data sets and map products for the Nation. NOHRSC products and services are used by RFCs and WFOs to develop a variety of hydrologic products such as spring flood outlooks, water supply outlooks, river and flood forecasts, and reservoir inflow forecasts. Additionally, the NOHRSC provides and supports geographic information system (GIS) data sets and applications used by the RFCs in generating automated hydrologic forecast basin boundaries. NOHRSC products and services also support a wide variety of government and private-sector applications in water resource management, disaster emergency preparedness, weather and flood forecasting, agriculture, transportation and commerce.

Personnel at the NCEP's WPC, located in College Park, Maryland, routinely prepare a spectrum of forecast products used by the WFOs to develop local rainfall, snow, and ice forecasts and by the RFCs to develop local river and flood forecasts. WPC coordinates with RFCs, WFOs and other Federal agencies, such as the Federal Emergency Management Agency (FEMA), during major flood events. The WPC also provides an array of surface analyses and short-range forecast products used by NWS field offices and the weather enterprise.

NCEP's CPC monitors and forecasts short-term climate fluctuations and generates guidance to communicate the effects climate patterns can have on the Nation. CPC develops and produces a suite of climate predictions and monitoring products, assessments, and discussions. These forecast products include 6-10 day, 8-14 day, one-month, and three month outlooks which depict the probability the temperature and precipitation will be above or below normal. CPC also routinely produces a U.S. Hazards Outlook, a U.S. Monthly Drought Outlook, a U.S. Seasonal Drought Outlook, and partners with the USDA and the National Drought Mitigation Center to produce the U.S. Drought Monitor. These products are used by WFOs and RFCs to enhance decision support services and serve as an input to RFC-generated, ensemble-based, extended range (probabilistic) river forecasts.

The Community Hydrologic Prediction System (CHPS) is an operational framework that allows for broad systems interoperability at the 13 NWS RFCs, and supports new water resources-related forecasts. It reinforces NOAA's national water information strategy, allowing NOAA's research and development enterprise and operational service delivery infrastructure to be integrated and leveraged with other Federal water agency activities and the private sector. Through CHPS, and under the auspices of the Integrated Water Resources Science and Services (IWRSS) Consortium, NOAA plans to deliver a new suite of high-resolution forecasts (including estimates of uncertainty) for streamflow, soil moisture, soil temperature, and many other variables directly related to watershed conditions via collaboration and sharing of data and models with other Federal, university, and private-sector experts. Furthermore, these activities will enable NOAA to deliver a national database of hydrologic analyses and predictions and generate user-friendly geographic information system (GIS) products for monitoring floods and drought. This activity contributes to the National Integrated Drought Information System (NIDIS).

Advanced Hydrologic Prediction Service (AHPS).

The overarching goals of AHPS are to provide: a) better forecast accuracy by incorporating advanced hydrologic science into NWS models; b) more specific and timely information on fast-rising floods by using tools which make it easier to rapidly identify small basins affected by heavy rainfall and excessive runoff, and predict the extent and timing of the resulting inundation; c) new types of forecast information by incorporating enhanced techniques for quantifying forecast certainty and conveying this information in products which specify the probability of reaching various water levels; d) longer forecast horizons; e) easier-to-use products; and f) increased, more timely, and consistent access to hydrologic information. AHPS includes a suite of web-based products and information designed to support more informed decisions through timely and accurate hydrologic forecasts and warnings.

The NWS continues to implement AHPS which builds on the existing NWS infrastructure, including the Advanced Weather Information Processing System (AWIPS), NEXRAD, and CHPS. AHPS also provides Ensemble Streamflow Prediction (ESP) - a capability that allows the NWS to generate extended range (probabilistic) river forecasts which quantify the forecast certainty. This information enables decision makers to apply risk-based analyses as they prepare for, and respond to, flooding and better balance competing demands on water supply, especially during periods of drought. The ESP capability is being expanded to include new techniques to facilitate the comprehensive generation of probabilistic hydrologic information seamlessly from short- to long- forecast horizons within CHPS. This enhanced capability is referred to as the Hydrologic Ensemble Forecast Service (HEFS) and is being implemented at select river forecast locations across the country.

Another AHPS capability, known as Flash Flood Monitoring and Prediction (FFMP), combines high-resolution radar rainfall observations with GIS technology to provide more accurate and precise flash flood detection. Flash floods, typically caused by intense, small-scale convective systems, are the leading cause of flood fatalities. The added precision provided by FFMP greatly reduces the area warned in flash flood warnings, making them more credible and leading to more effective and efficient public response, which ultimately saves lives. AHPS also provides opportunities to improve NOAA's analysis and forecast capabilities related to coastal water conditions, through joint efforts with NOAA's National Ocean Service (NOS) and Office of Oceanic and Atmospheric Research (OAR).

Integrated Water Resources Science and Services (IWRSS).

Our Nation faces a spectrum of growing water resources challenges which necessitate the provision of new and enhanced services. These challenges include: a) population growth and economic development are stressing water supplies and increasing vulnerability; b) a changing climate is impacting water availability and quality and increasing uncertainty; and, c) an aging water infrastructure is forcing critical, expensive decisions.

In this era of growing challenges and reduced budgets, the collective capabilities of Federal water agencies need to be leveraged to inform and guide increasingly important water decisions. The provision of life-saving NWS forecasts and warnings and other important water resources information is dependent upon routine collaboration between Federal water agencies including NOAA's NWS, the U.S. Army Corps of Engineers (USACE), and the U.S. Geological Survey (USGS). To strengthen this relationship and enhance the quality of Federal water resources services, in May of 2011, NOAA's NWS, the USACE, and the USGS entered into a memorandum of understanding to design, develop, and implement the Integrated Water Resources Science and Services (IWRSS) program. Consisting of a consortium of federal agencies with complementary missions in water science, observation, management and prediction, IWRSS is a new business model for interagency collaboration in the information age.

The IWRSS Consortium envisions an integrated national water modeling framework and information services framework. Together these frameworks will establish a common operating picture for water, improve water modeling and synthesis, and support the production of a comprehensive, seamless and consistent suite of high-resolution tree-top-to-bedrock, summit-to-sea water resources products and information services. Toward this end, IWRSS applies a cross-cutting, multi-disciplinary systems approach to address complex water problems collaboratively. The overarching objective of IWRSS is to enable and demonstrate a broad, integrative national water resources information system to serve as a reliable and authoritative means for adaptive water-related planning, preparedness and response activities. IWRSS builds on progress made under AHPS, CHPS and other NOAA water forecasting services.

NWS Partnerships for Hydrometeorological Products and Services.

Partnerships with a variety of Federal, state, and local agencies are critical to the NWS Hydrologic Services Program. For example, the NWS works very closely on water-related issues with many federal water agencies including: the USGS, the Bureau of Reclamation, and the Bureau of Land Management in the Department of the Interior; with the USACE in the Department of Defense; the Department of Agriculture's Natural Resources Conservation Service (NRCS); and FEMA in the Department of Homeland Security (DHS). Among these partnering activities are stream gaging, flood inundation mapping, river and water supply forecasting, and water management. For example, river stage and flow observations and stage discharge relationships provided by the USGS and reservoir operation information provided by the USACE, are critical to NWS warning and forecast operations for the Nation's rivers.

NOAA/Office of Marine and Aviation Operations (OMAO)

Within the NOAA/Office of Marine and Aviation Operations (OMAO) aircraft fleet, a NOAA AC-695A Commander 1000 (N45RF) and a DeHavilland DHC-6 Twin Otter (N46RF or N48RF) are used annually to conduct important snow-pack surveys in the northern and western continental U.S., Alaska, and southern Canada. During these survey flights, the gamma radiation sensors aboard these aircraft measure the naturally occurring terrestrial radiation emitted from the ground to obtain snow water–equivalent estimates. The data are transmitted to the National Operational Hydrologic Remote Sensing Center (NOHRSC) up to three times a day, and, after further processing, the data are distributed to NWS field offices within five minutes of receipt from the aircraft. These data are used by the NWS to forecast river levels and potential flood events, resulting from snowmelt water runoff. Hydroelectric power interests and other water supply managers also use the data to regulate water storage and delivery.

U.S. Department of Agriculture

NRCS Hydrometeorological Observations

Snowmelt provides the majority of the annual water supply in the Western United States; therefore, having information on snowpack is critical for water supply forecasts and management. The Natural Resources Conservation Service (NRCS) measures snowpack and collects hydrometeorological data in 13 Western States, including Alaska. NRCS' Snow Survey and Water Supply Forecasting (SSWSF) Program conducts snow surveys at high elevations in the mountainous West. The data collection system includes approximately 1,160 active manual snow courses in the United States and Canada (Canada operates their own data collection, but shares data with US), and more than 885 automated Snow Telemetry (SNOTEL) monitoring stations. The NRCS collects data at the manual snow courses in cooperation with a number of different Federal, State, local, and private partners, as well as Canadian agencies. The SNOTEL and manual snow course data, along with data from stream-gages, major reservoirs, and climatological observation stations managed by other agencies, are merged into a hydroclimatic database that is used to produce real-time watershed analyses and water supply forecasts. The purpose of water supply forecasts are to: (1) help irrigators make the most effective use of limited water supplies for agricultural production needs; (2) assist the Federal government in administering international water treaties with Canada and Mexico; (3) assist state governments in managing intrastate streams and interstate water compacts; (4) assist municipalities in planning the early management of anticipated water supplies and drought mitigation; (5) operate reservoirs to satisfy multiple use demands, including hydropower generation; (6) mitigate flood damages in levied areas and downstream from reservoirs; and (7) support fish and wildlife management activities associated with species protection legislation.

The automated SNOTEL network transmits data from remote sites and is served on the web by the national Water and Climate Center in Portland, Oregon. The SNOTEL network provides near-real-time remote hydrometeorological data that significantly improve flood stage forecasts and the monitoring of other life-threatening snow-related events. The primary use of the snow survey data is the production of water supply forecasts for more than 632 Western basins. All SNOTEL data are sent hourly to the National Weather Service (NWS) to assist in forecasting flood events. SNOTEL information enables emergency management agencies to effectively mitigate drought and flood damages and to monitor and assess wildfire potential.

Water supply forecasts are produced bi-monthly each year from January through June, as well as mid-month forecasts and daily guidance for some basins beginning in December. The NRCS furnishes snow measurements that are combined with advanced snow modeling and analysis provided by the National Oceanic & Atmospheric Administration (NOAA). NOAA's National Operational Hydrologic Remote Sensing Center continues to support joint NWS-NRCS efforts. The NRCS typically develops nearly 7,000 seasonal water supply forecasts for 632 streamflow forecast locations in 13 Western States. In addition, the Program provides daily water supply guidance forecasts for 198 of the 632 Western basins. These products provide information for water managers to adapt to weather changes as they occur. The web link for this information is http://www.wcc.nrcs.usda.gov/wsf/daily_forecasts.html.

Historical snow survey data are valuable to climate change researchers to aid in developing reliable projections of climate change. It has been projected that changes to the hydrologic cycle in the Western States, resulting from changes in snowpack conditions and timing of snowmelt, will increase the water supply challenges the States face. Monitoring data provides assistance to water managers at all levels to adapt to the impacts of climate change. Snowpack measurements extend back decades, and in some cases over 100 years, highlighting the value of this high-altitude network for understanding and assessing changes in climate.

The SSWSF Program provides a variety of climate and water supply products that are used to assess drought in the West. These include SNOTEL snowpack and precipitation analyses in the mountains, water supply forecasts, and State Surface Water Supply Indexes (SWSI). These products are critical to the weekly production of the interagency Drought Monitor, a web-based report. Separate from the SSWSF Program and SNOTEL network, the NRCS also manages a cooperative nationwide network of 203 Soil Climate Analysis Network (SCAN) sites in 40 States, as well as Puerto Rico and the U.S. Virgin Islands. These SCAN sites monitor soil temperature and soil moisture, which support national drought monitoring, agriculture production, and climate change research. Presently, more than 400 SNOTEL stations also have soil moisture and soil temperature sensors.

U.S Army Corps of Engineers

In its civil operational activities, the Corps of Engineers (COE) uses a network of about 10,850 land-based gages. About 55 percent of these sites collect meteorological data, 35 percent collect a combination of hydrologic and meteorological data, and 10 percent collect hydrologic or water quality data. The meteorological gages commonly measure precipitation and temperature. All data are used in the regulation of COE dams and other water projects, for flood control, navigation, hydroelectric power, irrigation, water supply, water quality, and recreation. The COE funds or partially funds nearly half of all the gages it uses.

The COE funds NOAA/National Weather Service (NWS) to collect and maintain precipitation information from 876 meteorological sites. The COE funds the NWS for hydro-meteorological studies and funds the U.S. Geological Survey (USGS) for maintaining hydro-meteorological data collection services for 2479 sites. The rest of the sites are maintained by the COE. Services performed by USGS vary by site and by year, and can include site visits, maintenance of equipment, replacement of damaged equipment, field measurements for verification of data and continuous monitoring of data results. About 90 percent of all COE sites provide real-time data via satellite, microwaves, meterbursts, landlines, or radio. Data from COE gage sites are available to NWS, and to other federal, state and local agencies.

U.S. Geological Survey

Hydrometeorological Data Collection and Distribution

The USGS's Water Mission Area (WMA) collects streamflow, precipitation, water quality, ground-water level, and other water resources and climatological data as part of a national network and for a number of projects concerning rainfall-runoff, water quality, and hydrologic processes. Currently, the USGS collects continuous hydrologic and meteorological data at about 9,000 surface water sites, 2,700 ground water-level sites, and 1,500 water quality sites. Periodic records are collected at approximately 1,500 additional surface water sites, 20,000 ground water sites, and 10,300 water quality sites. Precipitation records are collected at about 800 sites.

The USGS has developed new rapidly deployable, mobile streamgages to provide short-term water-level data to critical areas lacking permanent streamgages. Image provided by USGS Office of Surface Water.

Near-real-time streamflow data and ancillary information are provided to NWS RFCs for river forecast locations and to general public for all USGS streamgages via USGS NWIS web site (http://waterdata.usgs.gov/nwis/). Additional historical and real-time water resources data also are available. During floods, these data are supplemented by additional flood flow measurements.

The USGS also collects precipitation samples at a number of sites to determine the atmospheric contribution of chemical constituent loads to runoff, and for defining the effect of atmospheric deposition on water quality and the aquatic environment.

Nuclear Regulatory Commission

Following the accident at the Fukushima Dai-ichi nuclear power plant resulting from the March 11, 2011, Great Tōhoku Earthquake and subsequent tsunami, the Nuclear Regulatory Commission (NRC) issued letters to all its operating nuclear power plant licensees requesting that they reevaluate the flooding hazards at their sites using updated information and present-day regulatory guidance and methodologies. This includes considering the effects from large precipitation events, snow pack, ice effects, and other meteorological events that drive flood events, such as storm surge. Licensees were directed to provide their responses on a set response date between March 2013 and March 2015 based on a set of criteria used to prioritize the nuclear power plant sites. The NRC staff, along with support from its consultants, has begun reviewing the flood hazard reevaluations that have been submitted to date to determine whether additional regulatory actions are necessary to protect against the updated hazards.

SUPPORTING RESEARCH PROGRAMS AND PROJECTS

NOAA/Office of Oceanic and Atmospheric Research (OAR)

Hydrometeorological Testbed (HMT).

The HMT conducts research on precipitation and weather conditions that can lead to both flooding and drought, and fosters transition of scientific advances and new tools into forecasting operations. HMT's outputs support efforts to balance water resource demands and flood control in a changing climate. The HMT is a national program aimed at accelerating the infusion of new technologies, models, and scientific results from the research community into daily forecasting operations of the NWS at both the national (NCEP Weather Prediction Center) and regional (regional forecast centers and local forecast offices) levels. Research at the HMT, which operates as an end-to-end demonstration project with forecasters and researchers joining forces in the operational setting, has focused on improving regional forecasts of extreme precipitation, both too much and too little. This is achieved by a combination of: a) unique observations designed to monitor precipitation, kinematic, and surface processes; b) research aimed at improved understanding of physical processes leading to extreme precipitation; and c) operational forecast experiments to test new tools, methods, and models. HMT partners with organizations across NOAA, including NESDIS, OAR, NWS, NMFS, and NOS as well as regional academic and State authorities.

In FY 15, ESRL/PSD will continue to lead the implementation of the Hydrometeorology Testbed (HMT) – Southeast Pilot Study (SEPS), and associated analysis efforts, aimed at improving NOAA's ability to monitor and predict extreme precipitation in the southeast U.S. ESRL/PSD and HMT are also leading several projects in the California Russian River watershed as part of the NOAA Habitat Blueprint, aimed at sustaining healthy ecosystems.

Great Lakes Environmental Research Laboratory (GLERL)

GLERL conducts research and modeling to advance understanding of the hydrometeorological and hydrologic system of the Great Lakes basin. The Great Lakes is unique among hydrologic catchments worldwide in having about one-third of its area covered by fresh water, thus affording water a stronger second chance to evaporate while within a water body in addition to evapotranspiration from the land areas. Economic sectors such as fishing, shipping, and recreation are linked to the lake levels and water quality of the Great Lakes. A major product from GLERL is the Advanced Hydrologic Prediction System (AHPS, different from the NWS product of the same name) that provides probabilistic outlooks on a seasonal time scale, based on the Climate Prediction Center's meteorological outlooks (and corresponding Canadian products), for lake levels, precipitation, river flow, and other variables. GLERL also has the Great Lakes Hydrometeorological Dashboard, a web-based system for displaying a wide variety of observational and model-generated data for lake levels, precipitation, ice cover, evaporation, and other variables. GLERL's hydrologic activities are coordinated with cooperating groups in both the operational and research realm through the Coordinating Committee on Great Lakes Basic Hydraulic and Hydrologic Data (CCGLBHHD, or "The Coordinating Committee"), including other US Government entities (US Geological Survey, US Army Corps of Engineers, EPA, and NOAA National Ocean Service) and the Canadian Government (Environment Canada).

National Severe Storms Laboratory (NSSL)

Storm-scale Hydrometeorology Research.

Routine water level simulations that capture the complex interaction between rainfall, river flows, waves, tides and storm surge have not been possible until now. NSSL has developed a prototype real-time system that combines observations, weather and water models, and decision support tools to better forecast and prepare for inland and coastal floods. The NSSL Coastal and Inland Flooding Observation and Warning (CI-FLOW) project uses the NSSL multi-sensor rainfall estimates to drive an NWS-distributed hydrologic model that predicts streamflow to help the NWS improve flash flood warnings. CI-FLOW is coordinating with NOAA's Integrated Water Resources Science and Services – an innovative partnership of federal agencies with complementary operational missions in water science, observation, prediction and management. In addition to streamflow prediction, streamflow data from predictive models are used to drive a storm surge model from the University of North Carolina. We believe this system of coupled models, tested during the 2010 hurricane season and in Hurricane Irene in 2011, can be used not only for inundation studies of landfalling tropical systems, but also for land-use studies, algal bloom studies, and water quality assessment studies.

NSSL is working with the NWS to bring enhanced weather information and capabilities to operations. After more than a decade of development, testing and experimental evaluation, NSSL, in collaboration with the National Center for Environmental Prediction (NCEP), is bringing the Multi-Radar/Multi-Sensor (MRMS) system to NWS operations. MRMS consolidates and "cleans up" data from a wide variety of sources including radars, satellites, and surface observations to provide forecasters with the best-possible real-time analysis of the atmosphere at any location. This system has shown considerable improvement in radar-based rainfall estimates and provides forecasters with an ability to "mine" the data set to extract new types of aviation, severe and hydrologic weather information. By placing high-quality, remotely-sensed data in a three-dimensional framework accessible to NWS forecasters, MRMS will be opening the door to exciting new possibilities in improving flash flood and severe weather warnings, as well as forecasts.

NSSL continues to develop, with substantial support from the FAA and Taiwan's Central Weather Bureau, a seamless CONUS radar mosaic system to quantitatively estimate precipitation (QPE) every 5 minutes (called next generation QPE or Q2). Q2 incorporates information from operational and research weather surveillance radars (e.g., NOAA/NWS WSR-88Ds, FAA Terminal Doppler Weather Radars, deployable "gap-filling radars, Canadian operational network radars, etc), GOES satellite data, lightning flash detections, rain gauge accumulations, surface and upper air observations, and more recently, numerical model output to produce an ensemble of many gridded precipitation estimates. These precipitation products, with their different levels of complexity, each have strengths and weaknesses that vary with season, geography, and precipitation type. The Q2 algorithm is unique in the way that it performs quality control on the reflectivity fields, identifies the bright band using radar data, segregates convective and stratiform precipitation and applies differential Z-R relationships, delineates the rain-snow line, and adaptively calibrates infrared satellite data using radar precipitation rates. Research from the limited area deployment of Q2 demonstrated that calibrated satellite-based precipitation outperforms radar-only algorithms in complex terrain during the cool season. The Q2 multi-sensor and radar products are both adjusted by rain gauges using a mean field bias technique as well as a local bias adjustment.

The Flooded Locations And Simulated Hydrographs (FLASH) project also takes advantage of NSSL's high resolution, multisensor precipitation estimates. The rainfall estimates are used to force the Coupled Routing and Excess Storage (CREST) distributed hydrologic model running at a resolution commensurate with the rainfall forcing and flash floods. The FLASH system produces forecasts of flash flooding out to 6 hours in the future at every grid point in the conterminous US. FLASH is presently running in real-time as a demonstration and will be transferred to NWS operations along with MRMS. Weather Prediction Center forecasters evaluated FLASH model outputs during the Flash Flooding and Intense Rainfall (FFaIR) testbed experiment during the summer of 2013. They stated in their final report that FLASH provided better overall guidance in the forecast of flash floods than the operational tools used by forecasters in the NWS. The system offers to the potential for a consistent, nation-wide solution for flash flood forecasting by the National Weather Service.

Department of Agriculture

Natural Resources Conservation Service

Within the Snow Survey and Water Supply Forecasting Program, there are a number of agency initiatives that support research and development in regards to system operations. These initiatives include model development, quality assurance tools, radio and telemetry development, SCAN support, and radio spectrum management.

Nuclear Regulatory Commission

The Nuclear Regulatory Commission (NRC) conducts meteorological research to support licensing activities. Current research activities include updating the methods used to estimate the effects of extreme precipitation events, evaluating the use of paleoflood information to extend the historical record, and applying probabilistic techniques to riverine systems. The extreme precipitation work has been extended to consider the influence of orographic features. In addition, work has been initiated to start looking at flooding, including site specific probable maximum precipitation, from a probabilistic perspective. Preliminary planning is underway to develop an integrated probabilistic approach for flood hazard assessment for nuclear power plants that will include advanced approaches for modeling meteorological and hydrological phenomena and systems. This work is prioritized for those areas of the United States where new nuclear power plants are proposed and will provide the design basis for flood protection systems, or where existing nuclear power plants have submitted analyses for the NRC to review. The work will be done in cooperation with the Department of the Interior's Bureau of Reclamation and the U.S. Army Corps of Engineers. The focus of NRC research in these areas is to accurately assess the potential hazard to safe operation of nuclear facilities from extremely rare hydrometeorological events.

MILITARY SERVICES

For purposes of this *Federal Plan*, Military Services are those meteorological operations, services, and capabilities established to meet the unique requirements of military user commands and their component elements. Programs and services that are not uniquely military in nature are reported under another service category (e.g., Basic Services, Aviation Services [civilian], Surface Transportation Services, or Emergency Response and Homeland Security Services).

OPERATIONAL PROGRAMS INCLUDING PRODUCTS AND SERVICES

For each of the military services with meteorological operational programs (U.S. Air Force, Navy, Army, Marine Corps, and Coast Guard), the discussion below first describes that service's operational organizations, followed by a description of the principal meteorological products and services provided by these organizations.

U.S. Air Force

Operational Organizations

Air Force weather organizations enable military decision-makers to anticipate and exploit the weather for air, ground, space, cyberspace, and intelligence operations. AF weather personnel provide mission-tailored terrestrial and space environment observations, forecasts, and services to the U.S. Air Force (AF), U.S. Army (USA), and a variety of U.S. Government departments and agencies.

The AF Director of Weather (AF/A3W) is aligned under the AF Directorate of Operations (AF/A3), Headquarters Air Force. The Director of Weather oversees Air Force-wide training, organizing, and equipping of AF weather personnel and capabilities, to include the following functions:

- Development of doctrine, policies, requirements, and standards for weather support for worldwide Air Force and Army operations and training

- Evaluation of weather support effectiveness for worldwide operations and training

- Management of weather officer, enlisted, and civilian career fields

- Development and implementation of mid- to long-range plans for the organization, equipment, manpower, and technology necessary to meet future Air Force, Army, and other DoD Agency weather requirements

- Advising Air Staff and subordinate headquarters weather functional managers, regarding manpower, career field management, personnel utilization, training, operations policy and procedures, and technology acquisition

- Advocating and fielding standardized weather equipment to support worldwide training and combat operations

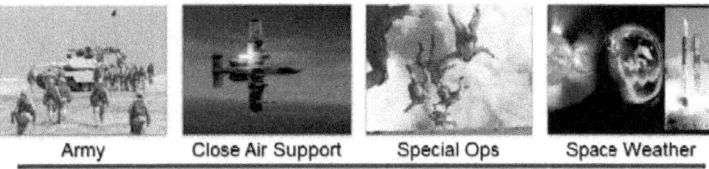

Air Force Weather mission statement

AF weather operations provide a Total Force capability, employing more than 4,000 Active Duty (AD) and Reserve Component military and civilian personnel supporting Air Force and Army conventional and special operations forces (SOF) worldwide. The majority of AF weather personnel are focused on two distinct, yet related functions: characterizing the past, current, and future state of the air and space environment, and exploiting environmental data and information to provide actionable environmental impacts information directly to decision makers.

AF weather personnel act as "eyes forward" to collect, analyze, tailor, integrate and disseminate weather environmental information, including forecasts of future conditions, in support of military operations. Weather personnel must understand warfighter tactics, techniques, and procedures, and help decision makers mitigate weather impacts and take advantage of weather conditions. AF weather personnel support Air Force, Army, Joint, and DoD conventional and special operations at various garrison and deployed locations worldwide.

AF weather personnel aligned with Army units directly support the G-2 intelligence centers, Aviation and Army Fire Support operations. Weather is a vital part of the intelligence estimate and is an essential element that supports the military rapid-response planning process. Weather personnel assigned to these commands provide expertise, products, and services that directly support the Joint Intelligence Preparation of the Operational Environment (JIPOE) process by helping intelligence analysts to effectively evaluate, integrate, and synchronize weather effects for military courses of action.

The AF weather support infrastructure is designed to rapidly deploy and operate in austere expeditionary environments. It is capable of providing sustained, comprehensive, and relevant weather support to all elements of an Air Expeditionary Force, as well as forward deployed air bases and stations of the establishment supporting that force. AF weather forces are organized in a 3-tier structure to maximize capabilities that can be accomplished in the rear area via "reach-back" technology. This minimizes forward expeditionary presence, making a "light and lean" force consistent with the overall AF vision for contingency operations in the 21st century.

Air Force Weather Agency (AFWA). AFWA is a Field Operating Agency, reporting to the AF Director of Weather. AFWA delivers weather training tools and fields all standard weather systems and equipment. In conjunction with acquisition organizations, AFWA builds and maintains the world's most comprehensive weather database of observation, forecast, climatological, and space weather products available on the World Wide Web. AFWA presently oversees two weather groups with distinct operational roles.

- The 1st Weather Group (1 WXG) aligns stateside weather operations by overseeing regionalized operational weather squadrons (OWSs). The 1 WXG has three subordinate OWSs whose areas of responsibility include the continental United States, Canada, parts of Mexico, and surrounding areas: the 15th, 25th, and 26th OWSs. They form the backbone of regionally focused, "reachback" weather operations for much of North America, providing a variety of weather forecast products and support to units assigned to and/or deployed in their respective areas of responsibility.

- The 2nd Weather Group (2 WXG) delivers timely, relevant, and specialized terrestrial, space, and climatological global environmental intelligence to joint warfighters, U.S. DoD decision makers, national agencies, and allied nations for the planning and execution of missions across the complete spectrum of military operations through the operation, sustainment, and maintenance of the AF's $288 million strategic center computer complex, production network, and applications. The 2 WXG is the weather production center in the first tier of the AF weather organizational structure. It delivers worldwide weather products to Air Force and Army decision-makers, unified commands, National Programs, and the National Command Authorities. Within the AF weather 3-tier structure to support forward operations via reach-back technology, the 2 WXG's 2nd Weather Squadron (2 WS) provides global coverage of forecaster-in-the-loop products to exploit environmental information necessary to effectively plan and conduct all levels of military operations, including providing dedicated support to the intelligence community, as well as backup for two national weather centers.

Operational Weather Squadrons. Around the world, the OWSs are the second tier of the AF weather support structure and provide continuous, complete environmental situational awareness. They are responsible for producing and disseminating mission-planning and mission-execution weather analyses, forecasts, and briefings for Air Force, Army, National Guard, and Reserve forces operating anywhere around the world.

- The 15th OWS's area of responsibility includes 149 installations/sites in a 22-state region of the northeastern United States.

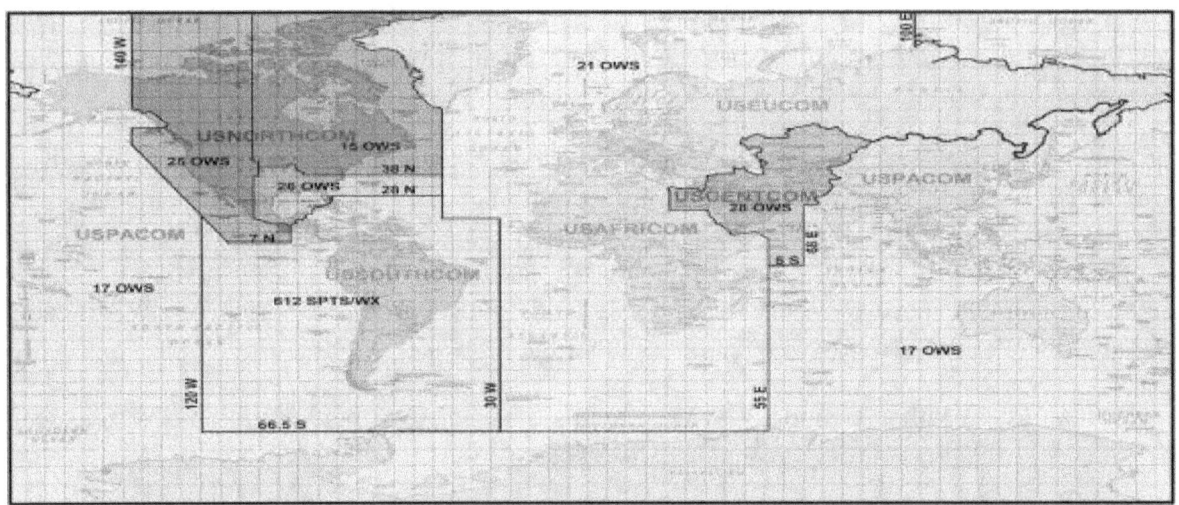

Air Force Operational Weather Squadron (OWS) areas of responsibility (AORs) overlaid on geographic combatant commander AORs.

- The 17th OWS's area of responsibility covers more than 95 million square miles of the Pacific region including Alaska and Hawaii, Australia, Korea, and Japan.

- The 21st OWS's area of responsibility includes sites within the U.S. European Command and U.S. Africa Command areas of responsibility, including Europe, Greenland, Russia, the Caucasus region, Turkey, Israel, and most of Africa.

- The 25th OWS's area of responsibility includes 79 installations/sites in an 11-state region of the western United States.

- The 26th OWS's area of responsibility includes 133 installations/sites in a seven-state region of the south central United States.

- The 28th OWS's area of responsibility includes sites throughout the U.S. Central Command area of responsibility, including southwest Asia and Egypt.

Weather Flights. Deployed in the field and focused on operational weather, AF Weather Flights constitute the third tier of AF weather support, acting as the prime interface with an AF installation's flying and ground operations. Weather Flights are located at military installations around the world and are the "eyes forward" for the responsible OWS.

Special Operations Weather. AF special operations units provide environmental reconnaissance in non-permissive and/or hostile areas and transmit them to a Joint Special Operations Task Force or next-echelon weather element on an as-required basis. Their tailored weather information and knowledge enable planning, command decisions, and execution of SOF operations. AF combat weather technicians assigned to SOF units are expected to know and keep current on the entire environment in the isolated locations to which their unit deploys.

Air Force-Army Weather Support. Weather Airmen aligned with Army units directly support the Army G-2 intelligence centers and Army fire-support operations. AF weather personnel predict the impact weather will have on Army and joint operations, giving leadership at all levels the ability to adjust operational and tactical strategies helping to further mission success. AF weather technicians and meteorologists assigned to support Army units are expected to forecast the weather anywhere their Army unit deploys. Army-trained weather personnel may parachute behind enemy lines and travel with a small platoon of soldiers, providing on-the-scene weather information for a variety of missions.

Weather Specialty Teams. AF weather experts are assigned to weather specialty teams in air and space operations centers. This crosscutting team integrates environmental information at key decision points of air and space operations planning, execution, and assessment. Armed with this information, decision makers can balance operational risks against mission need to optimize timing, tactics, target and weapons selection, and other factors affecting air and space operations.

Air Force Operational Climatology. The 2 WXG's 14th Weather Squadron (14 WS), at Asheville, North Carolina, is co-located with the National Climatic Data Center, one of the environmental data centers under the National Oceanic and Atmospheric Administration's (NOAA's) National Environmental Satellite, Data, and Information Service (NESDIS). The mission of 14 WS is to support DoD climatological requirements for engineering, planning, and operations to maximize effectiveness and efficiency of military operations. For more information on the 14th Weather Squadron and the National Climatic Data Center, see the Climate Services section of this Federal Plan.

Space Weather Operations. AF weather personnel within 2 WXG provide space weather analyses, forecasts, and alert notification for all DoD agencies and U.S. Government systems. With observatories in Australia, Italy, Massachusetts, New Mexico, and Hawaii, AF weather technicians maintain a continuous observational watch on the sun, which emits electromagnetic energy and electrically charged particles capable of causing disturbances in the near-Earth environment and disrupting satellite operations and satellite-based communications. The mission of the AF solar observatories is to monitor solar flares, noise storms, and other releases of energy from the sun and, when necessary, notify military and civilian organizations concerned with space weather, power, and communications in countries throughout the world. For further discussion of the complementary roles of AF space weather operations and the National Weather Service's (NWS) Space Weather Prediction Center, see the section on Space Weather Services.

USAF Reserve Component. The Reserve Component includes weather Airmen in both the Air Force Reserve Command (AFRC) and the Air National Guard (ANG). The AF continues to integrate these forces to more closely align with active duty force operations. AFRC weather personnel augment the active duty force at all three tiers. In some cases, the AFRC provides very unique weather-related services not duplicated in the active duty force, such as AFRC's 53rd Weather Reconnaissance Squadron (53 WRS) (see "AFRC Hurricane Hunters," below).

To augment OWS operations, AFRC organized two operational weather flights, each staffed by AFRC weather personnel, capable of augmenting an OWS either in the CONUS or overseas. Additional AFR weather personnel serve as individual mobilization augmentees assigned to various active AF organizations at all echelons, typically in staff, forecasting, or scientific roles.

Weather traditional reservists at one location work with an AFRC Remotely Piloted Aircraft (RPA) unit. There are also AFR weather personnel in Air Reserve Technician positions, i.e., combined full-time Civil Service/AFR military positions, employed by HQ AFRC as a staff weather officer and by the 53rd WRS as aerial reconnaissance weather officers. Lastly, AFRC civil service and contract weather personnel provide weather services at AFRC-operated bases in the CONUS.

The ANG traditional program consists of 23 numbered weather flights, ranging in size from 13 to 25 personnel, who meet monthly to train for their military mission. These flights provide weather support to ANG and Army National Guard units. Air Combat Command (ACC)-gained ANG wings also have up to four traditional weather positions to provide weather operations for each wing's flying mission. In addition, there are traditional weather positions in two ANG Special Tactics Squadrons (AF Special Operations Command), and four ANG RPA units (e.g., Predator). The ANG also has seven contract and four civil service locations where it is responsible for providing peacetime weather support to airfield operations.

Air Force Products and Services

Satellite Services. The AF operates a satellite data processing center, ingesting and storing worldwide meteorological satellite (METSAT) data. The Defense Meteorological Satellite Program (DMSP), which provides cloud, upper air, and space environmental data, is a vital source of global weather data used to support military operations and has been collecting weather data for U.S. military operations for more than 5 decades. Onboard sensors provide AF weather organizations with visible, infrared, and microwave imagery, plus temperature and moisture sounding data. The DMSP also supplies direct, real-time readouts of regional imagery and mission-sensor data to DoD land-based and shipboard terminals located worldwide.

There are two primary operational DMSP satellites in polar orbit at about 450 nautical miles (nominal) at all times. The primary weather sensor on DMSP is the Operational Linescan System, which provides continuous visual and infrared imagery of cloud cover over an area 1,600 nautical miles wide. Worldwide coverage of weather features is accomplished every 12 hours providing essential data over data-sparse or data-denied areas. Additional satellite sensors measure atmospheric vertical profiles of moisture and temperature. Military weather personnel can detect developing patterns of weather and track existing weather systems over remote areas, including the presence of severe thunderstorms, hurricanes, and typhoons.

In addition to DMSP polar-orbiting data, the AF receives stored data from NOAA's Polar-orbiting Operational Environmental Satellite (POES) constellation and real-time high-resolution data from NOAA's Geostationary Operational Environmental Satellite (GOES) East and West; European METSAT (EUMETSAT) -7, -8, and -9 geostationary satellites, and Meteorological Operational Polar (METOP) orbiters; as well as the Japanese Multifunctional Transport Satellite (MTSAT). The AF also currently receives data from the National Aeronautics and Space Administration's (NASA's) Moderate Resolution Imaging Spectroradiometer (MODIS), Tropical Rainfall Measuring Mission (TRMM), and Aqua Advanced Microwave Scanning Radiometer-E (AMSR-E); NOAA's Washington Volcanic Ash Advisory Center (VAAC); and NOAA's Space Weather Prediction Center (SWPC).

Space Launch Support. AF meteorological support for space launches is discussed in the Other Specialized Services section of this plan.

Air and Space Natural Environment Modeling and Simulation. The Air Force Director of Weather carries out the DoD Air and Space Natural Environment Modeling and Simulation Executive Agent (ASNE MSEA) responsibilities of managing, coordinating, and implementing all aspects of modeling and simulation, relating to the Air and Space Natural Environment domain to include, but not limited to, planning, programming, monitoring, and reporting across all DoD Components in accordance with the Under Secretary of Defense for Acquisition, Technology, and Logistics memorandum to the Secretary of the Air Force designating the Department of the Air Force as the MSEA for air and space natural environment representations. The DoD ASNE MSEA ensures DoD communities that use simulations for their training, acquisition, testing, planning, experimentation, and analysis have the right tools, infrastructure, and databases necessary to represent the air and space natural environment and its effects. More information is available in the Climate Services section of this Federal Plan.

AFRC Hurricane Hunters. The AFRC's 53 WRS, also known as the "Hurricane Hunters," provides another means of collecting vital meteorological data, especially in and around tropical cyclones. Their specially equipped WC-130J aircraft collect temperature, moisture, wind, pressure, and visually observed information at the aircraft location as well as vertical profiles of the atmosphere collected by dropsondes. Hurricane Hunter aircraft penetrate the eyes of tropical cyclones to provide the National Hurricane Center very accurate center fix locations as well as other meteorological parameters, including sea level pressure. In addition to tropical cyclone reconnaissance, the 53 WRS collects meteorological information to improve wintertime West Coast forecasts and to support scientific field programs when possible.

The sun tries to break through the thick clouds surrounding the WC-130J aircraft as it penetrates Tropical Storm Lee on Sept. 2, 2011. The 53rd Weather Reconnaissance Squadron "Hurricane Hunters," were gathering atmospheric data to relay to the National Hurricane Center for their forecast models. (U.S. Air Force photo by Staff Sgt. Valerie Smock)

Air Force Aviation Weather Support. The three 1st Weather Group (1 WXG) CONUS OWSs and the US Air Forces in Europe, Pacific AFs and AF Central Command aligned OWSs all perform "reachback" weather operations, providing a variety of weather forecast products and support to units assigned to and/or deployed into their respective areas of responsibility (AORs). Operational support to the AF, Army, Navy, Marine, Guard, Reserve and regional Combatant Commanders includes graphical analyses, aviation terminal aerodrome forecasts, severe weather watches/warnings and advisories, and mission execution forecasts, such as aviation hazards and en route and target forecasts. Additionally, the OWSs provide flight weather briefings to aircrews, operating within their AOR without home station support or as requested by base or post-level weather forces.

The 15th OWS, located at Scott AFB, IL, provides short-term backup services for the National Weather Services' Storm Prediction Center (SPC) and Aviation Weather Center (AWC). In the event of an extended or catastrophic outage at either SPC or AWC, essential staff would relocate to the AFWA facility at Offutt AFB, NE, where system and communications infrastructure exists to support relocation backup of these critical national missions.

At AF bases and Army posts, AF weather personnel focus on their supported unit's mission requirements. These units provide and disseminate observations and develop tailored mission execution forecasts based on centrally-produced guidance. For AF operations, these weather professionals are normally assigned to a flight under an operations support squadron of an AF flying wing; however, individuals from the weather flight are integrated into flying squadron mission planning and execution processes.

For Army operations, the AF weather personnel are normally aligned with Army units down to the brigade level. The weather Airmen are integrated into all aspects of Army operations. In this capacity, weather forces supporting AF and Army aviation operations infuse critical weather information at key points in the decision cycle to help aircrews maximize wartime capabilities, enhance flight safety, and optimize training effectiveness.

Volcanic Ash Surveillance and Analysis. One of the roles of the 2nd Weather Group is to provide volcanic ash surveillance and analysis for DoD aviation operations worldwide. Analysts continuously monitor all active volcanoes, generating more than 3,500 bulletins per year. Tailored satellite imagery, graphical ash plume forecasts, and text bulletins provide vital information needed to mitigate airborne volcanic ash as a threat to flight safety. The 2nd Weather Group also provides critical backup for NOAA's Washington Volcanic Ash Advisory Center.

Education and Training. AF officers newly accessed into the weather career field are university graduates with degrees in meteorology/atmospheric science. Some AF weather officers currently serving hold undergraduate degrees in other disciplines, but possess extensive technical training and operational weather experience. Further, the AF sends approximately ten officers per year to obtain advanced degrees, both M.S. and Ph.D., through the AF Institute of Technology. These advanced degrees include specialties in meteorology as well as solar and space science. New enlisted weather personnel attend the rigorous nine-month Weather Initial Skills Course at Keeler AFB, MS. At their first weather duty assignment, they gain a combination of operational experience and further skill upgrade training in accordance with the enlisted weather career field education and training plan. Continuation training programs for both officer and enlisted personnel ensure continuous learning appropriate to the supported mission and location.

Air Force Aviation Weather Products and Services

NextGen Development. The AF weather function continues active collaboration with the Next Generation Air Transportation System (NextGen) program, which is described more fully in the Aviation Weather Services section of this Federal Plan. Experiences gained through implementation of DoD's Joint METOC Data Base, AF Weather Web Services (AFW-WEBS), and machine-to-machine data services used by the Air Force's primary automated mission planning systems are providing valuable lessons learned for NextGen's development. Air Force Weather Ensemble Prediction System (AFWEPS) is also providing valuable path-finding insight into the utility and delivery of probabilistic aviation impacts that is a requirement for NextGen.

Weather System Upgrades. In FY 2015, the AF will continue to upgrade weather systems and processes that support AF and Army aviation. Continued refinements of the Joint Environmental Toolkit along with upgraded surface weather sensors will produce more accurate and timely weather observation and forecast products.

U.S. Army

USA Weather Support Structure

Weather support within the Army is conducted by Air Force personnel and equipment in accordance with a United States Army (USA)–United States Air Force (USAF) agreement: Army Regulation [AR] 115-10/Air Force Instruction [AFI] 15-157 (IP), Weather Support for the U.S. Army, 6 January 2010. This inter-Service regulation describes the responsibilities of USAF and USA for providing weather support. Under this agreement, the USAF provides the Army with the necessary labor and unique tactical and fixed weather equipment to meet Army tactical, installation, and airfield support requirements for both Active and Reserve Components. The USAF assigns weather personnel to provide direct and indirect weather support to the Army Commands, Army Service Component Commands, and installations. The Army provides assigned Air Force personnel the equipment necessary to perform their Army support mission in the tactical environment. The Army also provides facilities and host services to Air Force personnel assigned to Army installations. Each Service provides this support on a non-reimbursable basis.

Air Force Weather personnel provide installation, garrison, and tactical weather support on a daily basis to U.S. Army Forces Command, U.S. Army Europe, U.S. Army Pacific, U.S. Army Special Operations Command, Eighth U.S. Army, and the U.S. Army Training and Doctrine Command (TRADOC). The Army provides operational weather support to Army research, development, test, and evaluation (RDT&E) ranges, centers, and other research facilities using the Army Test and Evaluation Command's meteorological teams. U.S. Army Space and Missile Defense Command/Army Forces Strategic Command (SMDC/ARSTRAT) provides weather support to the Ronald Reagan Ballistic Missile Defense Test Site at Kwajalein Atoll through a Meteorological Environmental Test Support contractor.

Headquarters, Department of the Army, Army Commands, and Army Service Component Commands

Headquarters, Department of the Army (HQDA) The Office of the Deputy Chief of Staff, G-2, employs two civilian meteorologists whose primary duties are to establish weather policy within the Army, coordinate on AF weather policy issues with the Air Force Deputy Chief of Staff for Operations and Requirements (HQ USAF/A3), submit validated Army weather requirements and priorities to the HQ USAF/A3, coordinate with the AF on Army-AF and Joint Service weather operational concepts and doctrine, serve as the Army staff lead for meteorological satellite capabilities and issues, and review and coordinate Army-related support issues with the Office of the Secretary of Defense, the Joint Staff, the Department of the AF, other Services, HQDA staffs, Army Commands, Direct Reporting Units, Army Service Component Commands, and other Federal Agencies. FY 2015 activities will focus on meteorological satellite activities and collaboration of meteorological activities within the Army.

The Office of the Deputy Chief of Staff, G-3/5/7 (DCS, G-3/5/7), validates and prioritizes weather support requirements and programs to meet Army requirements, sets priorities for weather support for Army training and contingencies, coordinates with the HQ USAF on Army weather program and resource issues, Army Guard and Reserve weather issues, and Army installation weather support requirements.

Project Manager Distributed Common Ground System – Army (PM DCGS-A) is the materiel developer of weather capability in the Army and reports to the Program Executive Office Intelligence, Electronic Warfare and Sensors (PEO IEWS) and Assistant Secretary of the Army for Acquisition, Logistics and Technology (ASA(ALT)). In addition to developing, integrating, testing, and fielding capability, PM DCGS-A also leverages Air Force weather capability in support of Army operations.

Training and Doctrine Command. Headquarters, TRADOC is responsible for leading the USA in development of and validating requirements for USA-USAF inter-Service weather operations, services concepts, and doctrine required in support of Army operations. TRADOC develops and manages USA weather training programs, documents standard USA equipment for use by Air Force weather personnel in the Table of Organization and Equipment (TOE), and recommends modifications to the TOE and Common Table of Allowances to DCS, G-3/5/7, for approval as required. TRADOC's Army Weather Proponent Office (AWPO) processes tactical Army weather support requirements, represents the Army's warfighting functions by determining needed weather capabilities and processing weather requirements found in Joint and USA conceptual documents and originating from TRADOC centers and schools.

Key mission areas for the next few years: (1) assist the USAF with development and implementation of new weather support concepts to meet the needs of the USA's evolving force structure; (2) update weather support doctrine, policy, organization, concepts, along with tactics, techniques, and procedures; (3) ensure weather lessons learned and after action concerns with AF weather support to USA operations are documented and communicated to Army and Air Force leadership; and (4) ensure USA weather support processes and procedures are trained across the TRADOC schools and centers. These mission areas are accomplished in coordination with the USAF Staff Weather Officers and USA and USAF civilians assigned within TRADOC.

U.S. Army Intelligence Center of Excellence. The U.S. Army Intelligence Center of Excellence (USAICoE) is the functional proponent for USA tactical weather support. As such, TRADOC's AWPO represents the USA Warfighter by processing weather support requirements and developing solutions to satisfy those requirements when they are the responsibility of the USA. The AWPO collects and processes weather requirements from TRADOC schools/centers, USA Medical Command, and USA Corps of Engineers (USACE). It collaborates with HQDA and Headquarters, USAF, to recommend solutions to satisfy those requirements by processing tactical USA weather support requirements through the Joint Capabilities Integration and Development System (JCIDS) process. The AWPO employs one Department of the Army Civilian (DAC) to lead these weather requirement efforts. This DAC oversees the AWPO efforts with the JCIDS process, and in all doctrine, organization, training, materiel, leadership, education, personnel, and facilities (DOTMLPF) work. This JCIDS and DOTMLPF work occurs within USAICoE and in conjunction with other USA Centers of Excellence and Army research and experimentation organizations.

U.S. Army Aviation Center of Excellence (USAACE). USAACE provides 10 courses in the aviation training curriculum that contain weather training for aviators and unmanned aircraft systems operators. USAACE is in the process of acquiring excess automated surface observing systems from the Air Force to enhance flight safety at several training airfields.

Artillery Meteorological Education and Training. The U.S. Army Field Artillery School, Fort Sill, Oklahoma, is the proponent for upper air meteorological support to the Army and home of the Field Artillery Meteorology Course. The AN/TMQ-52A/B MMS-P is a suite of meteorological sensors and associated software/models which will provide the field artillery with current and/or expected weather conditions at a point where the weapon munitions is expected to engage a target (Target Area Met). U.S. Army Field Artillery School: The Military Occupational Specialty (MOS) 13T Field Artillery Survey/Meteorological Crewmember Course is no longer being instructed by the 13T. Field Artillery Meteorological training has been relegated into 13D MOS, which is Field Artillery Automated Tactical Data System Specialist. The 13D have been fielded Computer Meteorological Data- Profiler (CMD-P) that replaced the AN/TMQ-52 Meteorological Measuring Set – Profiler (MMS-P) system.

The CMD-P (AN/GMK-2) is a weather measurement system developed to provide meteorological (MET) data to support artillery and target acquisition units. The AN/TSR-8 Global Broadcast Service (GBS) Satellite Receiver Suite receives polar orbiter MET data & communications satellite data units. Operational Global Atmospheric Prediction System (NOGAPS) initialization data is acquired via a GBS satellite link. CMD-P provides MET data messages covering up to 500 km. The profiler's technology migration from the MMS-P to the CMD-P platform reduces the logistics footprint to a portable computer configuration that is located in the Tactical Operations Center (TOC). The CMD-P interfaces with the Advanced Field Artillery Tactical Data System (AFATDS) via a Local Area Network (LAN) connection and is operated by the AFATDS Operator 13D. The system will no longer require the Global Broadcast Service (GBS) receiver suite as part of the profiler system but will rely on the GBS connection from the TOC LAN (Meteorological data is downloaded on Compact Disc via internet from the Air Force Weather Agency). The system software will be capable of providing Field Artillery Computer MET and Gridded MET messages on demand with or without an operator in-the-loop.

U. S. Army North (USARNORTH). USARNORTH employs one civilian meteorologist as an advisor to the Commander, USARNORTH, on all issues involving meteorology, oceanography, and space weather. This individual applies meteorological and oceanographic policies and objectives to cover all USARNORTH mission requirements, and coordinates on USARNORTH exercise and contingency plans.

U.S. Army Test and Evaluation Command (ATEC). ATEC is responsible for providing operational meteorological support to USA RDT&E. Under responsibilities established in AR 115-10/ AFI 15-157 (IP), ATEC meteorological units provide meteorological data collection and analysis, consultation, and weather forecast and warning services to support USA and other DoD RDT&E activities at eight USA installations. The Meteorology Division at Dugway Proving Ground's West Desert Test Center provides meteorological support to USA RDT&E. Specialized services provided by the division include: (1) technical assistance to the ATEC operational meteorological teams/branches; (2) system administration support to the 4DWX system components and network connections at each ATEC test center; (3) atmospheric model verification and validation, including algorithm evaluation and the generation of validation data sets; and (4) technical assistance to the DoD chemical, biological, radiological, nuclear, and explosive (CBRNE) defense modeling community in the development of new CBRNE hazard assessment models. Division employees also serve on various national and international committees, addressing issues related to meteorological measurements, atmospheric dispersion modeling, CBRNE hazard assessment, and air quality.

U.S. Army Space and Missile Defense Command/Army Forces Strategic Command Support to the Ronald Reagan Ballistic Missile Defense Test Site. A subcommand of SMDC/ARSTRAT provides operational support to the Ronald Reagan Ballistic Missile Defense Test Site, including support for range activities (local and remote missile launches), missile weapons readiness testing, aviation and marine operations, and emergency operations. For further description of this support service, see the Other Specialized Services section of this federal plan.

U.S. Army Products and Services

U.S. Army Artillery. The Profiler Block III provides highly accurate meteorological data to the Advanced Field Artillery Tactical Data Systems (AFATDS) to adjust artillery fire. The system supports battalion and brigade Field Artillery units. The Profiler Block III is connected to the Tactical Operations Center local area network and provides the Advanced Field Artillery Tactical Data Systems (AFATDS) with the meteorological requirement for achieving first round hits and fires for effect. The Profiler Block III uses weather data obtained from the Air Force Weather Agency (AFWA) combined with weather modeling software. It provides meteorological data on demand with or without an operator in-the-loop to the AFATDS. A virtual module product improvement is currently under development and will be available in FY15 with an upgraded weather model and the software is changing to be Common Operating Environment compliant.

Distributed Common Ground System–Army (DCGS-A). Weather Services. DCGS-A weather services enables the Staff Weather Officer and intelligence operators to integrate weather effects in Intelligence Preparation of The Battlefield, Military Decision Making Processes, and Mission Command. DCGS-A exploits AF weather capabilities, employs Ozone Widget technology, and Data Distribution Service to distribute weather information to users throughout the enterprise.

U.S. Navy

Operational Organizations

Oceanographer of the Navy. Naval Oceanography encompasses Meteorology and Oceanography (METOC), Geospatial Information andServices (GI&S), and Precise Time and Astrometry (PTA) under the sponsorship of the Chief of Naval Operations. The Oceanographer of the Navy (OPNAV N2/N6E) is the Chief of Naval Operations' (CNO) principal advisor and the Oceanographer of the Navy (N2/N6E) implements the CNO responsibilities for Meteorology and Oceanography (METOC), Maritime Domain Awareness, Navy Space, Positioning, Navigation, and Timing (PNT), and performs functions relating to external interfaces with national and international operational and research and development oceanographic organizations and activities.

Rear Admiral Jonathan W. White (OPNAV N2N6E) Oceanographer and Navigator of the Navy; *Official U.S. Navy Photograph*

As the Oceanographer and Navigator of the Navy, N2N6E serves as the focal point for matters related to the Naval Oceanography enterprise (NOe) and other marine science fields. The Oceanographer and his staff work closely with U.S. Fleet Forces Command, Pacific Command (USPACOM), and the Office of Naval Research to ensure the proper resources are available to meet mission requirements. As the functional manager for GI&S, he works within the National System for Geospatial-Intelligence (NSG) and liaises with the National Geospatial-Intelligence Agency (NGA), National Reconnaissance Office (NRO), and the Defense Threat Reduction Agency (DTRA). He acts as Naval Deputy to the NOAA Administrator and represents the Naval Oceanography Program in interagency and international forums, including the North Atlantic Treaty Organization (NATO) and the World Meteorological Organization (WMO).

The Navigator of the Navy provides for the standardization of METOC, maritime geospatial information, astrometric and precise-time models, databases, and environmental predictive techniques. He coordinates NOe, Navigation Operations for Policy, architectures with Navy and DoD science and technology and RDT&E, along with related efforts in civilian agencies and develops means for transition from research to operational applications. This integrated approach provides the Fleet with critical information about the physical environment and its impacts on platforms, sensors, weapons, and personnel. Naval Meteorology and Oceanography support is integral to Navy, Marine Corps, and Joint Force capabilities to deter or win regional conflicts or major wars and conduct peacetime operations, including Humanitarian Assistance and Disaster Relief (HA/DR).

This year CNO approved the establishment of the Navy Information Dominance Forces

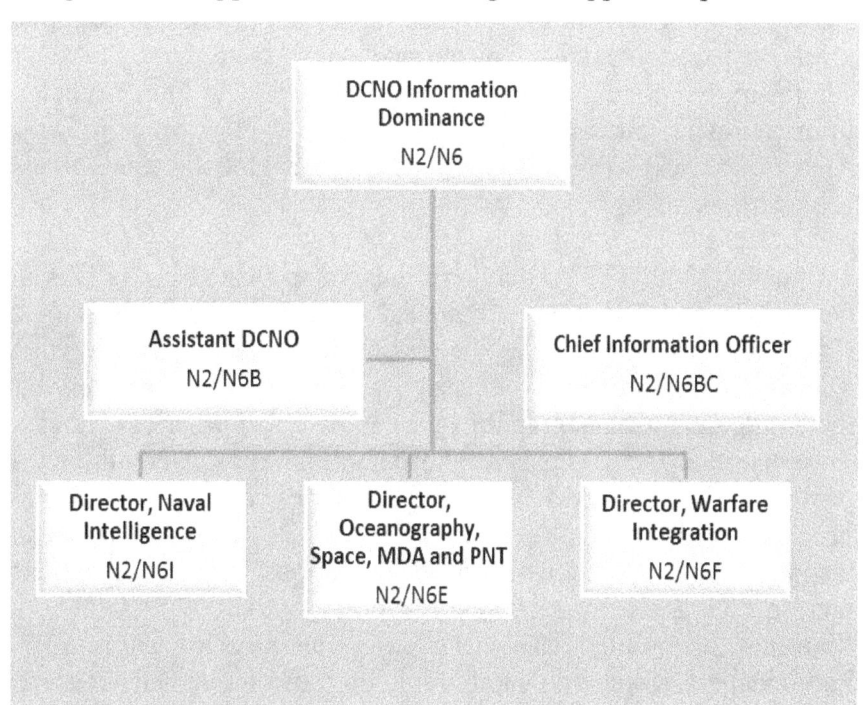

Organization of the Deputy, Chief of Naval Operations (DCNO) for Information Dominance (OPNAV N2N6) showing placement of the Oceanographer and Navigator of the Navy, as the Director, Oceanography, Space, and Maritime Awareness, and PNT.

Command (NAVIDFOR), the Navy's newest TYCOM. NAVIDFOR will incorporate the current Navy Cyber Forces Command and integrate the TYCOM-like functions performed by OPNAV N2/N6, FCC/C10F, the Office of Naval Intelligence (ONI) and the Navy Meteorology and Oceanographic Command (NMOC). With this integrated man, train and equip (MT&E) portfolio, NAVIDFOR will coordinate closely with the platform TYCOMs to deliver enhanced Information Dominance capability and readiness to the fleet.

Task Force Climate Change. The Oceanographer of the Navy serves as Director of TFCC and develops comprehensive approaches regarding the Arctic and global climate change to guide future Navy public, policy, and strategy discussions. A new National Strategy for the Arctic Region was signed 10 May 2013, reflecting growing recognition that the Arctic is opening up for human enterprise and will increasingly become a strategic priority for the United States.

The U.S. Navy Arctic Roadmap was updated in February 2014 prepares the U.S. Navy to respond effectively to future contingencies, delineates the U.S. Navy's Arctic Region leadership role within the Defense Department, and articulates the Navy's support to national priorities. The Arctic and climate change Roadmaps provide all encompassing, chronological, science-based guidance for future Navy action from 2014 through 2030.

Naval Meteorology and Oceanography Command. The Naval Meteorology and Oceanography Command (COMNAVMETOCCOM) is responsible for the Command and management of the Naval Oceanography Program (NOP) utilizing meteorology and oceanography, GI&S, and precise time and astrometry, to leverage the environment to enable successful strategic, tactical and operational battle space utilization across the continuum of campaigning and at all levels of war: strategic, operational and tactical. COMNAVMETOCCOM will retain administrative control (ADCON) over the METOC fourth echelon commands, reporting to the Information Dominance Typer Command (IDFOR), and serving as the operational arm of the NOe. COMNAVMETOCCOM's assets are globally distributed at shore facilities in fleet concentration areas.

Two central METOC Production Centers are located in the U.S. and operate through the Naval Oceanography Operations Command (NAVOCEANOPSCOM) at the Stennis Space Center (SSC) in Mississippi. As the Navy's physical science team, Naval METOC personnel (Navy and Marine Corps) are highly educated sailors, marines, and civilians that measure and collect meteorological, oceanographic, hydrographic and other data relevant to the physical battlespace. They use the data to analyze and determine current conditions, forecast the future state, and provide mission-focused impacts that enable commanders to make well-informed operational decisions across the entire spectrum of warfare. Operational oceanography operates along eight distinct lines of operations: maritime operations, aviation operations, fleet operations, navigation, precise time and astrometry, expeditionary warfare, anti-submarine warfare, and mine warfare. Maritime operations enable the safe operation of ships and submarines at sea through individualized forecasts, monitoring of their movement relative to hazardous weather, and alternate route advisories as warranted. Aviation operations are similar, but focus on the safe navigation of aircraft. Fleet operations represent our oceanography forces deployed on aircraft carriers, amphibious assault ships, independent deployers and task forces. METOC and Navigation services enable safety of surface and subsurface operations and support the Global

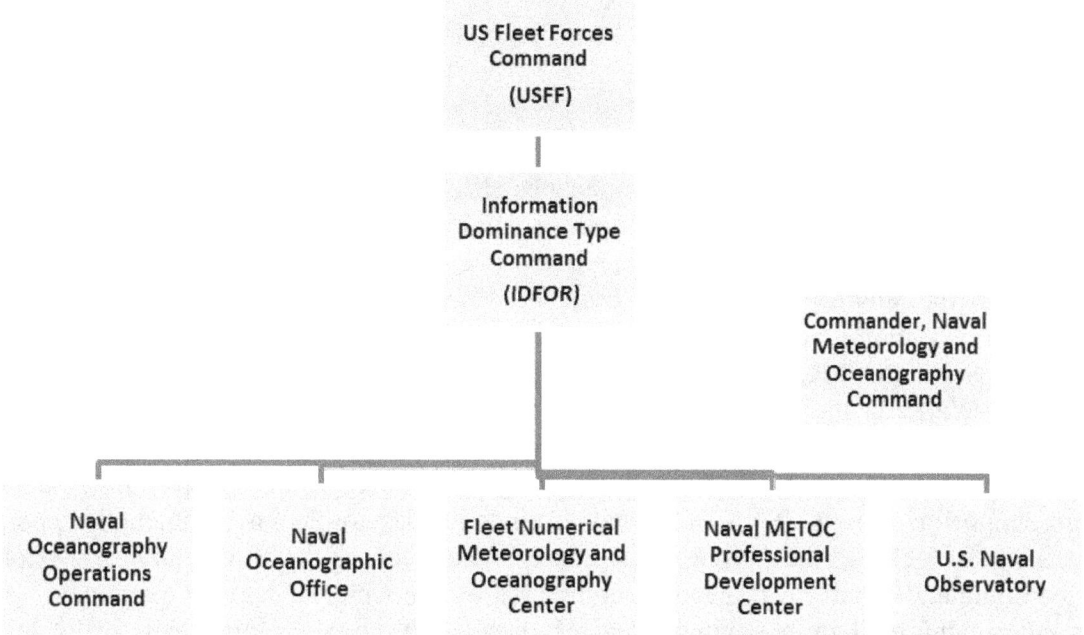

Organization of Naval Meteorology and Oceanography Command within the U.S. Fleet Forces Command structure

Positioning System, networks, communications, and space-based systems. Expeditionary warfare (amphibious warfare, riverine warfare and special operations), antisubmarine warfare, and mine warfare (offensive mining and defensive countermeasures and clearance), as their names imply, support their respective warfare areas via a mixture of forward-deployed and reach-back capabilities required to provide intelligence preparation of the operational environment (IPOE). COMNAVMETOCCOM is focused on providing critical environmental knowledge to the war fighting disciplines of Anti-Submarine Warfare; Naval Special Warfare; Mine Warfare; Intelligence, Surveillance and Reconnaissance; and Fleet Operations (Strike and Expeditionary), as well as to the support areas of Maritime Operations, Aviation Operations, Navigation, Precise Time, and Astrometry. Major activities and additional subordinates within the command include the following:

- NAVOCEANOPSCOM, Stennis Space Center, MS
- Fleet Weather Center Norfolk and Fleet Weather Center San Diego with subordinate units and detachments
- Naval Special Warfare Oceanography Center in San Diego, CA, (with components and detachments in Stuttgart, Germany, Norfolk, and Pearl Harbor)
- Naval Oceanography Anti-Submarine Warfare Centers in Yokosuka, Japan, and Stennis Space Center, MS (with subordinate detachments)
- Naval Oceanographic Office (NAVOCEANO), Stennis Space Center, MS
- Fleet Survey Team, Stennis Space Center, MS
- Naval Ice Center, Suitland, MD
- Fleet Numerical Meteorology and Oceanography Center (FNMOC), Monterey, CA
- USNO, Washington, DC

- Naval Meteorology and Oceanography Professional Development Center, Gulfport, MS, with detachments in Norfolk, VA; San Diego, CA; Pearl Harbor, HI; and Yokosuka, Japan

Naval Oceanography Operations Command. NAVOCEANOPSCOM is an echelon four METOC command that coordinates and manages efforts among field activities under the Operational Oceanography enterprise to optimize warfighting resources, support safe operations, and enhance decision superiority within the battlespace through superior understanding and exploitation of the environment. The Command encompasses warfighting and enabling directorates for: Aviation Operations, Maritime Operations, Fleet Operations, Precise Time and Astronomy, Navigation, ISR (Intelligence, Surveillance, and Reconnaissance), Mine Warfare, Naval Special Warfare, and Anti-Submarine Warfare. Each directorate determines how that directorate's services are delivered globally. Each directorate reports to a single Navy Captain who functions as Naval Oceanography's Director of Operational Oceanography (DOO). The operations support portion of USN/USMC budget funds the day-to-day environmental support to the Department of Defense, the Active and Reserve Components of the Navy and Marine Corps, ten unified commands, and other agencies as directed by the Chief of Naval Operations. Over 1,228 military and civilian personnel conduct these activities at more than 22 locations worldwide.

Fleet Numerical Meteorology and Oceanography Center. FNMOC, is an echelon four activity reporting to CNMOC, is the NOe production center for meteorology. This center plays a significant role in the national capability for operational weather and ocean prediction by implementing, operating, maintaining, and improving Numerical Weather Production (NWP) systems, including global and regional METOC models that extend from the top of the atmosphere to the bottom of the ocean. Since the end of the Navy's Geosat Follow-On (GFO) satellite, the military has been reliant on the Ocean Surface Topography Mission for radar altimetry, a unique source for atmospheric, oceanographic, and geodetic data. Navy is working with NOAA to find partnership opportunities through the JASON radar altimetry program. Through close collaboration with NAVOCEANO, FNMOC is also a key component in the Navy's operational weather and ocean prediction program. Naval operational oceanography is the key contributor to providing the predictive, physical maritime battlespace awareness capability. In the future, vital information from both intelligence and oceanography professionals will be fused into the same common operational picture afloat, providing a view of the battlespace we can only imagine today, ensuring better and faster warfighting decisions.

Naval Oceanographic Office. NAVOCEANO is the NOe's production center for oceanography. Since atmospheric conditions are inherently coupled to oceanographic conditions, the Navy's program in meteorology is closely linked with oceanography, which is the focus of the NAVOCEANO, Stennis Space Center, Mississippi. NAVOCEANO's primary responsibilities include the collection, processing, and distribution of oceanographic, hydrographic, and other geophysical data and products. NAVOCEANO is responsible for the administration of a fleet of ocean-class hydrographic survey vessels and data from both aviation assets and spacecraft NAVOCEANO is headquartered at the Stennis Space Center, which is home to the Navy Department of Defense Supercomputing Resource Center that Navy, Army, and Air Force scientists and researchers use to design tools and weapons systems that support the Department of Defense's global mission. The Navy DSRC is a premier provider of high performance computing services and support to DoD scientists and engineers. It is one of five supercomputing centers established under the DoD High Performance Computing Modernization Program (HPCMP). CNMOC maintains oversight of the Navy DSRC systems that have been operational since 1997.The Navy DSRC provides a high performance computing capability with primary emphasis on support of the largest, most computationally-intensive HPC applications. Our center leads the way for numerous HPCMP-wide initiatives and provides our users with in-depth computational expertise and support. While the HPCMP is primarily focused on DoD research and development programs, the Navy DSRC is unique in that approximately 15 percent of its total capability is apportioned for operational use by naval operational oceanography. Today, the Navy DSRC enables, on a daily basis, operational, global, regional, and port scale ocean circulation, as well as wave and sea ice forecast numerical models supporting worldwide Navy and DoD operations. Navy DSRC's total computational capacity affords us computational space to not only improve our ocean models, but to bring online world-class atmospheric models within the DSRC and begin to more tightly couple ocean and atmospheric physics and energy exchange to provide a more accurate, longer range future state of the atmosphere.

United States Naval Observatory. The U.S. Naval Observatory in Washington, DC maintains precise time and makes it available to DoD users. Coordinated Universal Time (UTC) is the DoD standard and the primary time reference for GPS and other military applications. Naval Observatory's (USNO) Master Clock (MC) is a reference for the internet's Network Time Protocol (an Internet standard that facilitates the transfer of digital data). USNO's mission includes determining the positions and motions of the earth, sun, moon, planets, stars, and other celestial objects; providing astronomical data; determining precise time; measuring the earth's rotation; and maintaining the Master Clock for the United States. USNO astronomers make celestial observations, construct the Celestial Reference Frame (CRF), distribute star catalogs, formulate theories, and conduct relevant research. NOAA Geostationary Operational Environmental Satellite (GOES) and Polar-orbiting Operational Environmental Satellite (POES) constellations of weather satellites, utilize the precise positions of stars to calibrate optical effects in the focal plane of their satellite imaging systems, applying corrections to subsequent images of the earth's weather systems.

Left to right: Dr. Chris Ekstrom, Chief, USNO Clock Development Division, Dr. Steven Peil, USNO Research Physicist, Dr. Bobby Junker, ONR, and Dr. Bill Phillips, NIST/UMD (winner of Nobel Prize in physics in 1997 for laser trapping and cooling of atoms). Official U.S. Navy photograph

Navy Products and Services

Maritime Weather Operations. The Fleet Weather Center in Norfolk, VA supports operations for Commander Second Fleet (North Atlantic); Commander, Fourth Fleet (Caribbean Sea and South Atlantic); Commander, Sixth Fleet (Mediterranean Sea), and Arctic Region operations.

The Fleet Weather Center in San Diego, CA supports operations for the Commander, Third Fleet (East Pacific); Commander, Fifth Fleet (Arabian Sea and Suez Canal); Commander, Seventh Fleet (West and Central Pacific) areas of responsibilities.

Navy meteorologists and forecasters provide Optimum Track Ship Routing (OTSR) and weather forecasts to support transoceanic voyages and coastal operations. OTSR services include:

- Hazardous ocean and weather advisories and divert recommendations to ship Commanding Officers and Masters at sea

- Sortie recommendations for potentially damaging weather conditions in port

- Preliminary climatologic outlooks for transit and mission planning

- Routine ship weather forecasts and aviation weather forecasts for ship-based helicopters, to include high wind and seas warnings and local area warnings for Fleet Concentration Areas

Joint Typhoon Warning Center (JTWC). The JTWC, established by the U.S. Pacific Command, is jointly manned with U.S. Navy and U.S. Air Force personnel. JTWC services include tropical cyclone forecasts, warnings, and other products for DoD Warfighters operating in the Pacific and Indian Oceans. JTWC, located in Pearl Harbor, Hawaii, is an internationally recognized tropical cyclone forecasting center.

FNMOC Numerical Weather Prediction Systems. FNMOC satisfies the military's requirement for an operational global Numerical Weather Prediction (NWP) capability. This requirement is driven by the importance of weather and ocean conditions on modern military operations, the need to use classified weather observations to guarantee the very best weather and ocean predictions in theaters of conflict, and the imperative to produce and disseminate weather and ocean products to military decision makers without fear of interruption or compromise as a result of cyber terrorists or cyber warfare. FNMOC employs four primary models—the Navy Global Environmental Model (NAVGEM), the Coupled Ocean/Atmosphere Mesoscale Prediction System (COAMPS), the Geophysical Fluid Dynamics Navy (GFDN) model, and the Wave Watch III model (WW3)—along with a number of specialized models and related applications.

- NAVGEM is an improvement to the former Navy Operational Global Atmospheric System (NOGAPS) that drives nearly all other FNMOC models and applications in some fashion, and forms the basis for the FNMOC global Ensemble Forecast System.

- COAMPS is a high-resolution, non-hydrostatic regional model, multiply nested within NAVGEM. It has proven to be particularly valuable for forecasting weather and ocean conditions in highly complex coastal areas.

- GFDN is a moving-nest tropical cyclone model, nested within NAVGEM. It is used to forecast tropical cyclone tracks globally.

- WW3 is a spectral ocean wave model that is employed both globally (driven by NAVGEM) and regionally (driven by COAMPS) in support of a wide variety of naval operations.

FNMOC Products and Services. FNMOC's complex and robust operational prediction capability is designed to deliver detailed forecasts of wind stresses and heat fluxes to drive very high-resolution ocean models that provide ocean thermal structure and currents in support of anti-submarine and mine warfare operations, or near-shore wind, sea, and surf forecasts that directly support Fleet Operations through ship-to-objective maneuver. FNMOC models feed directly into applications models, tactical decision aids, and other products that provide direct support to various weather-sensitive activities associated with the optimum path aircraft routing, optimum track ship routing, issuance of high-winds and high-seas warnings, hurricane/typhoon sortie decisions, covert ingress/egress of Special Operations Forces, ballistic missile targeting, cruise missile launch and targeting, radar, EO, and FLIR system performance prediction in support of ship self-defense, naval gunfire operations, understanding the threats posed by airborne nuclear/biological/chemical agents, search-and-rescue at sea, and many other activities.

Many of FNMOC's products are distributed to users over the Web via the PC-based METCAST system, and subsequently displayed and manipulated on the user's PC with the Joint METOC Viewer (JMV) software. This includes all standard METOC fields, synoptic observations, and satellite imagery. FNMOC provides a Web-based capability called WxMap (i.e., "Weather Map"). WxMap, requiring only a Web browser for access, allows the user to select and quickly display predicted METOC fields for any user-defined geographical area. FNMOC benefits greatly from collocation with its supporting R&D activity, the Marine Meteorology Division of the Naval Research Laboratory (NRL/MRY). NRL/MRY is a world-class research organization, with focus on weather-related support to warfighting. FNMOC and NRL/MRY share space, data, software and computer systems, and together with the nearby Naval Postgraduate School represent one of the largest concentrations of weather-related intellectual capital in the Nation. Collocation and close cooperation between research and operations, such as exists between NRL/MRY and FNMOC, is the optimum arrangement for transitioning R&D quickly and cost effectively into new and improved operational weather prediction capabilities.

NAVOCEANO Products and Services. NAVOCEANO is the Navy's center for operational oceanographic support and provides daily analyses and forecasts of the ocean state with a series of global, regional, and coastal ocean circulation and wave models. The core of the system is the dynamic Navy Coastal Ocean Model (NCOM) which predicts three-dimensional ocean properties to 96 hours. The 1/8 degree (14km/7.5nm) resolution Global NCOM covers the world from pole to pole and is coupled with the Arctic Polar Ice Prediction System, which forecasts ice properties for the National Ice Center. Twelve regional NCOMs are on line with 24 planned in 2014. Nested 1/36 degree (3km/1.7nm) regional NCOM domains of order 20 by 20 degree sizes provide high-resolution ocean forecasts in areas of Navy and national interest. Global and regional NCOM products and data fields are shared with our NOAA partners.

NAVOCEANO runs a series of coastal, estuarine, and river domains with resolutions as fine as 1/360 degree (300 m, 1000 ft) or less in the support of mine warfare and homeland security efforts. When appropriate, coastal NCOMs are supplemented by other models including HYDROMAP, DELFT3D, and PCTIDES. The NCOMs are initiated through the assimilation of ocean data from satellites (sea surface temperature and altimetry) and various surface and subsurface observing systems, including ship data, ARGO profiling floats, and gliders. Global NCOM will be replaced by the 1/12-degree resolution Global HYbrid Coordinate Ocean Model, which was developed under the National Ocean Partnership Program by a consortium of government and academic scientists, led by NRL Stennis and including NOAA's National Centers for Environmental Prediction (NCEP).

NAVOCEANO is the Navy's primary processing facility for a number of polar-orbiting and geostationary satellite collection systems, and is nationally recognized for satellite-derived sea-surface temperature and satellite altimeter-derived sea surface topography and wave height observations. These products are shared with NOAA partners and are critically important to successfully running both the NAVOCEANO ocean models and FNMOC's NAVGEM and COAMPS atmospheric models.

U.S. Marine Corps

Operational Organizations

CMC, DC/AVN, APX-9. The senior USMC METOC officer is dual-hatted as the METOC Services Officer and the METOC Occupational Field Manager. The METOC Services Officer is the principal advisor and subject matter expert to the Deputy Commandant, Aviation. The Aviation Department mission is to assist and advise the Commandant of the Marine Corps on all matters relating to Marine Aviation consistent with Marine Corps requirements. Specific duties include, but are not limited to: advise the Commandant of the Marine Corps on aviation safety, aviation policies, and Joint matters relating to aviation; provide principal aviation staff interface with the Chief of Naval Operations; Joint Capabilities Board; USMC liaison to the National Oceanic and Atmospheric Association and the Office of the Federal Coordinator for Meteorology.

U.S. Marine Corps Forces (MARFOR). The MARFOR METOC officer serves as a Special Staff for the Commanding General of the MARFORs and resides within the G-2. The primary mission of the MARFOR METOC Officer is to advise and assist the MARFOR Commanding General in the development of METOC policies and the execution and management of METOC resources by planning, coordinating, and validating the collection, evaluation, interpretation, and dissemination of METOC data. MARFOR billets include assignment to Marine Forces Pacific (MARFORPAC) Camp Smith, HI and Marine Forces Command (MARFORCOM) Norfolk, VA.

Marine Expeditionary Force (MEF). The primary mission of the MEF METOC Officer is to advise and assist the MEF Commanding General in the development of METOC policies and the execution and management of METOC resources by planning, coordinating, and validating the collection, evaluation, interpretation, and dissemination of METOC data. Other duties include: planning for the employment and use of organic METOC assets, equipment, and capabilities; preparing METOC annex (Annex H) for OPLANs or OPORDs and providing METOC's input into the CIS annex (Annex K), the intelligence annex (Annex B), and other annexes, as necessary regarding METOC issues.

Intelligence Battalion (Intel Bn). The Intel Bn provides the largest preponderance of METOC personnel outside of the Aviation Combat Element (ACE) and provides direct support to the MEF and general support to all other elements of the MAGTF. The secondary mission is to provide METOC Support Teams (MSTs) to the Command Element (CE), Ground Combat Element (GCE), Logistics Combat Element (LCE), and Marine Expeditionary Units (MEUs); which are task-organized to support unique mission requirements.

Marine Division (MARDIV). The MARDIV METOC Chief serves as a Special Staff for the Commanding General of the Division and resides within the G-2. The primary mission of the Division METOC Chief is to integrate into operational planning. Other duties include: prepare and present METOC briefs in support of mission planning and execution; forecast and identify environmental factors expected to impact operations; and provide METOC expertise in the areas of Doctrine, Organization, Training, Material, Leadership, Personnel, and Facilities on behalf of the Commanding General of the Division.

Marine Aircraft Wing (MAW). The primary mission of the MAW METOC Staff is to integrate into operational planning and advise the Commanding General on the capabilities and limitations of the METOC personnel and equipment with the MAW. TACC METOC personnel; assist subordinate commands in knowledge development for all METOC personnel in the Wing; and provide METOC expertise in the areas of Doctrine, Organization, Training, Material, Leadership, Personnel, and Facilities on behalf of the Commanding General of the MAW.

Marine Air Control Group (MACG). The primary mission of the MACG METOC Chief is to integrate into operational planning and to advise the Commanding Officer on the capabilities and limitations of the METOC personnel and equipment within the MACG. Other duties include planning and execution of MAGTF operations; analyze and interpret centrally prepared products, alphanumeric data, satellite images, and Doppler radar coverage in order to provide forecast meteorological conditions, space weather, climatological and astronomical products, and impacts assessments.

Marine Corps Installations Command (MCICOM)

Marine Corps Installations East and West (MCI East and MCI West)

Regional METOC Centers (RMC). The Regional METOC Centers (RMC) are operational supporting establishments that are under the cognizance of MCICOM. There are two RMCs; one on each coast. RMC West (RMC-W) is located at MCAS Miramar and has a primary mission of coordinating METOC support for all MCI West stations and bases thus ensuring adequate aviation METOC support for all MCAS, 3rd MAW, and transient aircraft in garrison. RMC East (RMC-E) is located at MCAS Cherry Point and has the primary mission of coordinating METOC support for MCI East stations and bases thus ensuring adequate aviation METOC support for all MCAS, 2nd MAW, and transient aircraft in garrison.

Other CONUS Marine Corps Bases and Air Stations. Marine Corps Bases and Air Stations require continuous METOC support for force protection and flight operations. Although CONUS Marine METOC services have been regionalized with the establishment of the RMCs, each Station still maintains a small contingency of METOC personnel to provide on-site local area environmental expertise.

CMC, DC/CD&I, IID. The Combat Development Directorate (CDD) METOC officer serves to integrate, across battlespace functional areas, METOC requirements with regards to ISR near, mid, and far-term requirements for the support of MAGTF, Joint, and Combined operations. Other duties include: ensure that Doctrine, Organization, Training, Material, Leadership, Personnel, and Facilities solutions represent a thorough analysis of capabilities needed to provide timely, relevant, and tailored METOC combat information and intelligence to commanders and staff; and coordinate with other integration divisions in areas of mutual interest and represents the USMC METOC community in USMC, other service, and joint forums in which METOC concepts and requirements pertain.

CMC, MCSC. The MARCORSYSCOM METOC Project Officer leads the research and technology development, acquisition, engineering and production, maintenance, life-cycle sustainment and disposal of METOC systems as required by the Marine warfighter and the MAGTF.

Marine Liaison, Oceanographer of the Navy (N2/N6). The United States Marine Corps Liaison and Requirements Officer responsible for the funding of all programs associated with METOC support to the MAGTF resides within OPNAV N2/N6E1. The N2/N6E1 USMC Requirements Officer represents the Deputy Commandant for Aviation in all aspects of METOC policy and procedures, requirements, acquisitions, and finance and must maintain familiarization with defense acquisition, technology, and logistics life cycle management framework. This officer also monitors Marine Corps METOC programs to ensure requirements are being met and advises the Program Sponsor of program status and milestone accomplishments.

Marine Liaison, Commander Naval Meteorology and Oceanography Command. The United States Marine Corps Liaison Officer, Commander Naval Meteorology and Oceanography Command is the principal advisor and subject matter expert for all MAGTF METOC support capabilities. Duties include: aligning the METOC support requirements for littoral and expeditionary warfare operations to include the proper employment and optimal utilization of MAGTF METOC capabilities.

Marine Aviation Weapons and Tactics Squadron One (MAWTS-1). The MAWTS-1 METOC Officer serves as the METOC subject matter expert for MAWTS-1 and advises the MAWTS-1 Commanding Officer on OccFld 68XX and DoD initiatives that impacts the management and employment of Aviation METOC resources. Other duties include: plan, coordinate, and provide METOC support and curriculum to the Weapons and Tactics Instructor (WTI) Course; manage the MCWP 3-35.7, MAGTF METOC Support as the doctrinal proponent; maintain NAVMC 3500.38 68XX, METOC T&R manual as syllabus sponsor; create courseware that facilitates the tactical application of METOC support to the MAGTF.

Marine Corps Detachment (MARDET) Keesler AFB. The mission of MARDET, Keesler AFB is to train entry and career-level Marines for service with the Marine Corps Operating Forces, sustain their transformation process while they obtain the technical skills of their future MOS. Marine Corps METOC personnel must attend the Meteorological Oceanographic Analyst Forecaster (MOAF) Course. This training includes meteorology, oceanography, computers, satellite, meteorological reports, chart analysis, air mass soundings analysis, space environment, and climatology.

Chemical Biological Incident Response Force (CBIRF). The CBIRF METOC Analyst serves to provide essential meteorological and hazardous prediction information in support of CBIRF`s real world and training operations coupled with providing daily and weekly weather forecasts and climatological briefs to the corresponding staff to support planning operations and typical battalion operations. Other duties include: research forecasting techniques or tools as well as new plume modeling software or tools; responsible for coordinating and maintaining liaison with local, state, and federal agencies such as the Defense Threat Reduction Agency (DTRA) or Joint Task Force Civil Support (JTF-CS).

Operational Supporting Establishments

Regional METOC Center (RMC). The Marine Corps' RMCs provide 24 hour direct (in person) and indirect (web, e-mail, and phone) regionalized METOC information, products, and services in support of Marine Corps operations and other military operations as may be directed from a garrison environment. RMCs have the secondary mission to coordinate and supervise the training of all METOC forecasters in the MCICOM chain of command. As part of the regionalization concept of the RMCs, they are responsible for 24/7 METOC support for the station they are aboard and after field closure for all other stations within their area of responsibility (AOR).

Other CONUS Marine Corps Bases and Air Stations. Station METOC support responsibilities include: METOC support to base training, tenant units, and transient aircrews; and liaison with the RMC to provide a continuous meteorological watch (METWATCH). The RMCs have final issuing authority for all terminal aerodrome forecasts (TAF) and weather WWAs for the bases and air stations within their respective area of responsibility. RMCs have final issuing authority for all terminal aerodrome forecasts (TAF) and WWAs for the stations and bases within their respective region.

Marine Corps Aviation Support. The Marine Aircraft Wing (MAW) conducts the complete range of air operations in support of the MEF, to include anti-air warfare, offensive air support, assault support, aerial reconnaissance, electronic warfare, and control of aircraft and missiles. The MAW serves as the principal headquarters for the ACE. Most of the MAGTF's METOC support assets reside within the MAW, specifically at the MACG and its subordinate Marine Air Traffic Control Detachments. These assets are organized, structured, and capable of supporting a variety of MAGTF and ACE-specific operations as defined by the size, scope, and mission requirements. Dedicated METOC support is available for all MAGTF elements from within the MAW/ACE.

Marine Corps Products and Services

METOC Support Capabilities—Meteorological Mobile Facility (Replacement) Next Generation [METMF(R)NEXGEN]. The highest level of METOC capability within the MAGTF is the METMF(R) NEXGEN. The METMF(R) NEXGEN provides the MAGTF with a lightweight, highly mobile, fully integrated FORCENet compliant meteorological system capable of sustaining METOC operations in direct or general support of all elements of the MAGTF. It provides a METOC capability similar to that found in garrison METOC facilities.

An Automated Weather Observing System (AWOS) collects weather data at the Mountain Warfare Training Center in Bridgeport, CA., Mar 26, 2014. This Meteorology and Oceanography site allowed members of the Marine Air Traffic Control Mobile Team, Marine Air Control Squadron 2, to relay vital weather information to both air and ground units during Mountain Exercise 2-14. Providing accurate weather information to participating Army CH-47 and AH-64 pilots was paramount for their air support and allowed 3rd Battalion 3rd Marines to better plan and execute their mission. (Photo by SSgt Kevin M. Brady)

The METMF(R) NEXGEN provisions for all functions of environmental sensing and data ingest, for the efficient collation and integration of collected data, and for user-friendly graphic user interfaces (GUIs) and software tools necessary for accurate interpretation and value-added production. The METMF(R) NEXGEN enables the Marine METOC personnel to effectively turn relevant METOC data into actionable environmental information which in turn can facilitate timely operational decision-making. The smallest level of METOC capability is the Naval Integrated Tactical Environmental System - Variant IV (NITES IV). The NITES IV is designed as a scalable, flexible, and mobile system for a first in and last out capability. The primary differences between the METMF(R) NEXGEN and NITES IV are size, logistics, scalability, mobility, and that the METMF(R) NEXGEN has organic sensing capabilities while the NITES IV sensing capabilities are limited. Thus, the NITES IV relies heavily on reach-back communications and MFC databases for data and products. The NITES IV provides METOC personnel access to METOC data and products which are then analyzed and tailored for a specific mission.

Naval Integrated Tactical Environmental System (NITES IV). The NITES IV is a modular system, used to provide limited METOC support in a stand-alone mode with increasing capabilities realized with the addition of SIPRNET/NIPRNET connectivity. The NITES-IV suite consists of three laptops. Each laptop is designed to perform a different function, but all three are loaded with the same software and can perform the tasks of the others. Mission requirements, network availability, and embarkation space dictate how best to employ the NITES IV.

Vaisala TacMet Meteorological Observation System. Organic to the METMF(R) NEXGEN and NITES IV, the TacMet is a field deployable, compact weather station for various field operations. Offering broad sensor capability, the local MAWS201M network system is used to provide weather data to support air operations on a temporary or semi-permanent basis. The local system enhanced configuration offers full aviation support, interfacing with intelligent sensors such as ceilometer, present weather sensor, and electric field mill. These sensors are capable of being deployed up to 150 feet away from the systems. A standalone version of this sensor is also available to the METMF(R) NEXGEN in the form of two (2) Remote Sensor Sub-systems (RSS). The RSS includes meteorological sensors from pre-positioned, remote locations. This portable, scalable system comprises a sensor set with wind speed and direction, barometric pressure, air temperature, relative humidity, and precipitation sensors.

METMF(R) NEXGEN provides critical enabling capabilities to collect, process, disseminate and integrate essential elements of information regarding the time-sensitive characterization of the physical environment. Official U.S.M.C. photograph.

U.S. Coast Guard

The U.S. Coast Guard (USCG) is a military, multi-mission maritime service and one of our Nation's five armed forces. The USCG protects the public, the environment, and the economic interests in the Nation's ports and waterways, along the coast, on international waters, and in other maritime regions, as required to support national security. The USCG has a long history of environmental observations and science support. Support for meteorological operations and supporting research is detailed in other sections of this plan.

SUPPORTING RESEARCH PROGRAMS AND PROJECTS

Interagency Supporting Research Programs and Projects

National Unified Operational Prediction Capability Research Partnering Initiative

The Navy, Air Force/Air Force Weather (AFW), and NOAA/NWS partner in this tri-agency management organization. The National Unified Operational Prediction Capability (NUOPC) vision is a national numerical weather prediction (NWP) system with interoperable components built on common standards and a common framework (the Earth System Modeling Framework), with managed operational ensemble diversity and a national global NWP research agenda to accelerate science and technology infusion. NUOPC focuses on the next-generation systems for global NWP with full implementation by 2020, allowing for possible future expansion into other areas of numerical environmental prediction. NUOPC established its Initial Operational Capability in January 2011, which provides a National multi-model global ensemble exchanged between NOAA, the Navy and the Air Force on a one degree grid for 72 variables out to 16 days. Upgrades to ensemble systems will continue and the resolution of the exchanged ensemble members is scheduled to be doubled by the end of 2013 to one half degree, and doubled again to one quarter degree in the next several years as computing and communications capabilities grow to support this requirement. The partners continue to use and develop ensemble-based products such as the new Wave Watch 3 Ensemble, based on this multi-model system that brings substantial skill improvements over any single model solution.

The NUOPC effort partners with the Earth System Predictions Capability (ESPC) to improve and extend forecast skill through coupled systems. This effort is research focused to stimulate inter-agency interactions and collaborations in developing systems to improve forecasts at time scales from near-term weather to inter-seasonal/inter-annual. It also is working to couple domains such as atmosphere, ocean, ice, land and eventually into ecosystems. Improvements in predictions in these broader domains and timescales will result in better severe weather warnings (hurricanes, tornadoes, snowstorms), better cost avoidance for weather sensitive industries (agriculture, transportation, utilities, defense), and better informed decision making for industry, defense, and the general public. Finally, NUOPC efforts will work to operationalize findings from ESPC for the next generation environmental prediction system. More information is available at the NUOPC website at: www.weather.gov/nuopc/ and the ESPC website at: www.espc.oar.noaa.gov/.

U.S. Air Force Supporting Research Programs and Projects

Technology Transition Initiatives. The overarching objective of the AF meteorological and space environmental technology transition program is to give capability designers, operational weather personnel, and weather information users the technology and tools to gain and maintain the advantage over a potential adversary. AF capability needs in the atmospheric and space environment sciences are articulated in the Initial Capabilities Document for the METOC Environment, Capability Review assessments, AF Strategic Plans, the AF Weather Operations Functional Concept and Enabling Concepts (Characterize the Environment, Exploit Environmental Information, and Net-Centric Operations), and supporting concept and implementation plans. The AF also uses cooperative development and testing agreements with other governmental agencies and laboratories, as well as with for-profit companies. The Air Force Institute of Technology offers AF graduate students in the atmospheric and space environmental sciences opportunities to research topics of immediate operational interest to U.S. Military Services.

AFW-WEBS is a centralized Web service capability providing access to environmental information appropriate to all levels of operation and command. The program is designed to leverage net-centric capabilities and geospatial display services via AFW-WEBS to provide the operational warfighting community a single point of access to the total AF authoritative environmental content from sources across the AF enterprise. By CY 2015, AFW-WEBS will have evolved into the single Web site optimized for accessing authoritative AF meteorological information and services. All appropriate meteorological information will be serviced in geospatially enabled formats for direct integration into warfighter systems and decision cycles.

Consistent environmental characterizations of key mission-impacting weather parameters improved by the FITL process will be used as a common source for both visualized web content and for direct M2M services accessed by warfighter systems. AF weather forces will employ AFW-WEBS capabilities to improve the quality, accuracy, and effectiveness of all-weather support processes. Finally, C2, ISR, and mission planning systems will employ AFW-WEBS products and services to help more decision-makers maintain better situational awareness and use the knowledge gained from this process to plan and execute more effective missions.

Cloud Forecasting. In applied meteorological R&D, the AF is improving cloud forecasting techniques by increasing the resolution, using a new cloud interpretation/typing scheme, integrating available satellite (to include non-traditional METSAT) into the cloud analysis, incorporating cloud optical properties, and blending numerical weather prediction with forecast cloud advection techniques. The AF has transitioned key advances in tactical decision aids into operations, permitting improved forecasting of electro-optical system performance and generation of cloud and target scene visualizations for training, system development, and mission rehearsal.

Weather Forecast Modeling for Air Force and Army Operations. The Weather Research and Forecasting (WRF) model is a community model and is another area of AF participation in research and development in collaboration with National Center for Atmospheric Research (NCAR). The AF initially implemented WRF operationally in 2006 and continues development at NCAR, along with test and evaluation of real-time runs of the WRF model runs as well as the WRF-Chem (which takes into account chemical and aerosol constituents). The Land Information System analyzes the current state of the land surface to provide information to DoD and civilian agencies, and, through coupling with WRF, will improve forecasting performance near the surface and in the low levels of the atmosphere. This enables AF weather forces to provide better environmental characterization for missions such as (but not limited to) low-level aircraft operations, the dispersion of aerosol contaminants, and the employment of precision-guided munitions. It also allows for assessment of trafficability for ground forces.

In early 2012, the AF implemented the Air Force Weather Ensemble Prediction Suite (AFWEPS). AFWEPS output, at both the mesoscale and global scale, provides better meteorological intelligence for military operations by objectively quantifying the forecast certainty of mission-impacting meteorological parameters to optimize operational risk management for all echelons of decision making. It provides probabilistic algorithms for high-impact variables and quantifies biases, allowing concise, focused products.

Tactical Decision Aids. The AF collaborates in the development of several tactical decision aids, including the Target Acquisition Weapons Software (TAWS), the Infrared Target Scene Simulator (IRTSS), and Integrated Weather Effects Decision Aid (IWEDA). TAWS provides a joint mission-planning tool for combining platform, weapon, target, background, and weather factors to depict three-dimensional target acquisition and lock-on range and recognition range versus time.

- TAWS can be used to predict environmental impacts on night vision goggles and low light-level systems used by air, naval, and ground forces to execute nighttime operations.

- IRTSS uses detailed terrain information and multi-spectral imagery with TAWS weather inputs to generate forecast target scene images for mission rehearsal.

- IWEDA uses environmental data with force, mission, and/or individual weapons rules of engagement or performance parameters to automatically generate mission-impact forecasts for large-scale planning efforts such as air tasking order preparation. It aids in selecting platforms, systems, or sensors, based on system rules with critical values and a forecast of weather conditions. Results are displayed on a red/yellow/green weather effects matrix and overlaid on a background map.

TAWS, IRTSS, and IWEDA integrate environmental impacts into the mission execution forecasts for operations, command and control, and for mission planning systems throughout the military planning and execution cycle. AFRL, the Navy's Space and Naval Warfare Systems Command, NRL, and the U.S. Army Research Laboratory (ARL) are developing modular programs as part of the IWEDA initiative. The Tactical Decision Aids program continues adding weapons systems and targets to the inputs to these decision aids at the request of users from the Services.

U.S. Army Supporting Research Programs and Projects

Army Materiel Command (AMC)

AMC is responsible for the RDT&E of equipment to satisfy the USA's requirements for meteorological support. AMC provides meteorological and climatological support to RDT&E projects, involving electro-optical sensors and atmospheric and obscurant effects on systems and their performance. AMC has several major subordinate commands and elements carrying out weather R&D responsibilities, including the Research Development and Engineering Command, which has responsibility for the USA's Research Development and Engineering Centers and ARL.

Battlefield Environment Division, ARL Computational and Information Sciences Directorate. The Battlefield Environment Division of the Computational and Information Sciences Directorate in ARL conducts basic and applied research in atmospheric science and technology to provide actionable environmental intelligence crucial to the success of current and future operations. The research program focuses on: (1) measurements, modeling, and theoretical investigations of aerosols, acoustics, electro-optics, and adaptive optics to support advances in detection and identification within the battlefield environment; (2) research to measure, model, predict, and understand the dynamics of the boundary layer atmosphere and its effects on Army systems and operations; and (3) research necessary to meet the Army's requirements for detailed atmospheric analyses and very-short-range predictions (nowcasts) over mission-execution battle space domains in complex terrain areas and within urban environments. This mission effort will be carried out through the: (1) development of high-resolution physics-based and semi-empirical models for urban and complex-terrain meteorology; (2) development of Decision Support Tools (DSTs) that describe the impacts expected, and the resulting performance degradation due to, adverse environmental conditions, to include weather and terrain, for both friendly and threat systems; and (3) development of improved propagation and scattering models to be employed in novel techniques that can enhance the detection, identification, and localization of biological aerosols and acoustic and optical sources, and to enable evaluations that will improve the performance of acoustic and electro-optic sensors. ARL provides a liaison to the Joint Polar Satellite System (JPSS) program office located at the NASA Goddard Space Flight Center to coordinate on Army satellite data, information requirements, and applications development. The following three research programs are planned within ARL-BED for FY 2015.

Atmospheric Sensing. The objectives of this research program include: (1) experimental and theoretical research enabling high resolution characterization of atmospheric and other environmental conditions that may affect the performance of acoustic and electro-optic systems, and the conception of methods to mitigate these effects; (2) development of aerosol, acoustic, and electro-optic DSTs that utilize observed or modeled atmospheric parameters to improve warfighter performance within the battlefield environment; (3) development of theoretical and experimental technologies for the characterization of atmospheric constituents focused on the detection of threat agent aerosols; and (4) development and advancement of environmental remote sensing science and technologies to increase knowledge of the battlefield environment and to support Army operations. This program includes the Intelligent Optics Laboratory (IOL) that explores and develops new advanced Army battle space tactical and long-range atmospheric laser communication and imaging systems that will achieve high bandwidth communication, improve battlefield visualization, and allow for utilization of advanced command and control techniques.

Atmospheric Modeling. The objectives of this program are to: (1) develop high-resolution, physics-based prognostic models and semi-empirical diagnostic models for ultra-fine-scale meteorology in urban, mountain, forest, and jungle terrain; (2) develop, modify, or adapt data assimilation tools that can ingest on-scene, traditional, and non-traditional weather observations, and fuse this information with forecasts to provide analyses and mission-execution Nowcast products for operations and short-term mission planning; (3) design and develop the means to process these models and transfer meteorological data over a distributed network; (4) develop capabilities to use ensemble model probabilistic forecast output in weather DSTs, and develop new aids to exploit such new databases; and (5) support the Artillery Meteorology community with expertise in meteorological modeling and atmospheric effects on artillery fires and target area meteorology.

Accurate wind forecasts are essential to deliver airdropped supplies to the precise location. (U.S. Army photo)

Atmospheric Dynamics. The objectives of this effort are to: (1) develop applications to monitor and assess airborne hazards in complex terrain; (2) perform experiments to characterize the fine-scale effects of local terrain on winds, turbulence, and vertical fluxes; (3) develop DSTs, associated databases, user tools that describe the impacts expected, and the resulting performance degradation due to adverse weather conditions for both Friendly and Threat systems, (4) assess the accuracy and value-added to the Warfighter of Nowcasts and DSTs; and (5) develop atmospheric propagation models and improved theory for characterizing the atmosphere, and the effects of turbulence on emerging sensing applications in previously under-exploited spectral regions including Terahertz for communications and imaging, Short-wave Infrared active and passive imaging, and Near-IR for directed energy systems.

ARL Army Research Office. The Army Research Office, Research Triangle Park, North Carolina, manages the Army's extramural basic research program in the atmospheric sciences. These programs are concerned with understanding the dynamical and physical properties, processes, and constituents of the atmospheric boundary layer through measurements, simulations, and theoretical considerations. The basic research program is conducted through the peer-reviewed, individual investigator program and occasional special initiatives. The focus of the research is on the atmospheric boundary layer over land, where the Army operates. Objectives of the research are to develop, from first principles, the physical basis for understanding the boundary layer, thereby leading to better understanding, modeling, and quantifying of atmospheric effects on soldiers, materials, and weapon systems. The research examines quantification, classification, and dispersion of battlefield materials; the effects of heterogeneous terrain features on airflow; and the development of natural obscurations throughout the diurnal cycle. The Atmospheric Science Program is especially focused on supporting high-risk, high-impact research that has the potential to change the way we view our research philosophies.

U.S. Army Corps of Engineers R&D

The Corps of Engineers (COE) is responsible for reviewing emerging Army systems for environmental effects, as stated in AR 70-1 and AR 115-10/AFI 15-157 (IP). The COE Engineer Research and Development Center (ERDC) develops Tactical Decision Aids (TDAs) and geospatial analysis tools to interpret and help the Warfighter understand the impact of weather on terrain and provide actionable information of terrain, atmospheric and weather effects on units, systems, platforms and soldiers in support of Mission Command and Intelligence, Surveillance and Reconnaissance planning. TDAs will transition to the Distributed Common Ground System–Army (DCGS-A), the Joint Battle Command-Platform and the Commercial Joint Mapping Tool Kit.

ERDC supports Army weapon systems RDT&E with all-season solutions for mitigating adverse environmental effects on Army operations. Basic and applied research is conducted on energy and mass transfer processes at and near the terrain surface. ERDC develops databases and models for predicting the state of the terrain including surface temperature, soil moisture, tactical decision aids, and geospatial tools supporting mobility analysis and sensor performance. These products transition to research and engineering programs including advanced technology demonstrations and specific programs of record. FY 2015 planned efforts include (1) development of downscaling methodologies enabling weather products to feed Army scale applications to enable terrain and weather influences and model predictions for austere entry, military maneuver, observations and fields of fire, cover and concealment, obstacles, key terrain, and avenues of approach analysis; (2) weather and terrain influences and predictions for seismic, acoustic, radar and radio frequency modalities into the sensor mission planning tool, Environmental Awareness for Sensor and Emitter Employment (EASEE); and (3) research into the impacts weather and climate influences have on regional socio-cultural and socio-economic stability.

Army Test and Evaluation Command

ATEC is responsible for providing operational meteorological support to Army RDT&E. Under responsibilities established in AR 115-10/AFJI 15-157, ATEC meteorological units provide meteorological data collection and analysis, consultation, and weather forecast and warning services to support Army and other DOD RDT&E activities at eight Army installations.

Enhancements to ATEC Four-Dimensional Weather System. The Army RDT&E Meteorology Program is continuing to collaborate with the National Center for Atmospheric Research (NCAR) on enhancements to the ATEC Four-Dimensional Weather (4DWX) system, which is the backbone of the meteorological support infrastructure at the Army test ranges. ATEC 4DWX modeling capabilities include Weather Research and Forecasting (WRF) model based on real-time four-dimensional data assimilation at eight Army test ranges, and Global Meteorology on Demand, a globally relocatable mesoscale modeling system to support Army RDT&E (including ATEC distributed and virtual testing) at locations other than the Army ranges. Output from the 4DWX mesoscale model forecasts and analyses is used as meteorological input to atmospheric dispersion, noise propagation, ballistic trajectory, and other range applications models to simulate many tests and their associated impacts. The 4DWX system contributes to improved test planning and conduct, selection of more representative locations for test sensors, inclusion of realistic atmospheric effects in virtual testing, and forensic analyses of meteorological effects on test results.

Major 4DWX system components include a central data archival/retrieval system for all range and external meteorological and model data, the WRF high-resolution mesoscale meteorological model, an innovative real-time data assimilation system, and a variety of user-configurable displays.

The DoD High Performance Computing Modernization Office (HPCMO) in 2006 provided the ATEC program with a high-performance computer (HPC) that since 2007 has enabled operational 30-member ensemble 4DWX forecasts to support DPG test operations. With this HPC approaching the end of its expected life cycle, in 2012 DPG purchased a replacement system. The original HPC is still operational; therefore the replacement system is in use for model evaluation studies until the original system fails. The HPCMO recently awarded ATEC a 2014 Designated Support Partition, a large allocation of computer resources at the Navy's DoD Supercomputing Research Center at Stennis Space Center, for extension of the Ensemble 4DWX system currently in operations at Dugway Proving Ground, to all the ATEC test ranges.

System enhancements during FY 2015 will include continued WRF and data assimilation development focused on forecasting improvements at each range, in addition to advances which apply generally to all WRF applications, including continued work on hybrid data assimilation in particular radar data assimilation; improved lightning potential prediction; development of range climatologies; continued development of Very Large Eddy Simulation capabilities; continued development of an on-line automated real-time quality control for observations; and continued enhancements to the AutoNowcaster implementations at Redstone Test Center and at White

Sands Missile Range. In addition, preparations are underway to install a weather surveillance radar at the Cold Regions Test Center. Also, ATEC will continue collaboration with the Army Research Laboratory to evaluate the utility of 4DWX data as input to the ARL MyWIDA weather impacts decision aid, and plans to evaluate an upgraded sound propagation model under development at the Naval Surface Warfare Center, for potential use at Aberdeen Proving Ground and Ft A.P. Hill, Virginia.

U.S. Navy Supporting Research Programs and Projects

Earth System Prediction Capabilities. The National Unified Operational Prediction Capability (NUOPC) initiative exemplifies how naval operational oceanography interagency partnerships contribute to enhanced capabilities. Navy is partnering with NOAA and other agencies on ESPC (Earth System Prediction Capabilities), on a zero-hour to 30-year model of the atmosphere, the ocean, and ice. The ESPC focus is on next-generation systems for Global Numerical Weather Prediction with full implementation by 2020. The ESPC primary deliverable is a multi-model global ensemble forecast system, coupling the domains of land, sea, atmosphere, ice, and space. Improvements in predictive capability are expected to result in better severe weather warnings (hurricanes, tornadoes, snowstorms), better cost avoidance for weather sensitive industries (agriculture, transportation, utilities, defense), and better informed decision making for industry, defense, and the general public. Additional information is included in the Basic Services section of this plan.

Electromagnetic Maneuver Warfare (EMW) Under the mission pillar of Assured Command and Control, in direct support to EMW, the Navy is engaged in improving Electro-Magnetic (EM)/Electro-Optic (EO)/Electronic Warfare (EW) Spectrum Management, especially in Anti-Access/Area-Denial (A2/AD) environments by improving exploitation opportunities across current and emerging fleet sensor systems. The NOe under COMNAVOCEANCOM direction, has partnered with Program Manager Ships (PMS 405) in coordinating atmospheric and oceanographic efforts for Solid State Laser (SSL) Technology Maturation (TM) and Quick Reaction Capability (QRC). An Integrated Product Team (IPT) is charged with providing 24/7 atmospheric characterization and performance prediction for SSL operations at any maritime location of Navy interest. Current focus is on modeling and measurements, needed in developing vertical temperature/moisture and turbulence profiles. CNMOC is the Navy authoritative source for what atmospheric models to tap into and FNMOC is the Navy's lead atmospheric modeling center. Other participants include ONR 34, NRL, and COMNAVSPAWARSYSCOM. Tech Solutions is a SPAWAR-led R&D effort for a weather sensing system onboard naval surface ships called Automated Shipboard Weather Observation System (ASO). The design of a variety of laser systems, for example, the Modular Universal Laser Equipment (MULE), has used the R&D turbulence models to optimize systems for operation in the atmospheric environment expected in different operational theaters.

SPACE WEATHER SERVICES

For purposes of this *Federal Plan*, Space Weather Services are those specialized meteorological services and facilities established to meet the needs of users for information on space weather conditions and space weather storms that can affect terrestrial systems, space systems, Earth's atmosphere, and the space environment. Space weather services include monitoring and alerting of space weather storms and their effects on technological infrastructure and human safety. Early warning of an approaching space weather storm, so that timely protective response is possible, is an important part of space weather services.

OPERATIONAL PROGRAMS INCLUDING PRODUCTS AND SERVICES

NOAA/National Weather Service

Space Weather Prediction Center

The National Centers for Environmental Prediction's (NCEP) Space Weather Prediction Center (SWPC), within the National Oceanic and Atmospheric Administration's (NOAA) National Weather Service (NWS), is the Nation's official source of space weather alerts, watches, and warnings for conditions in the space environment that impact systems and technologies that are vulnerable to space weather. The SWPC provides real-time monitoring and forecasting of space weather events, conducts research in solar-terrestrial physics, and develops techniques for forecasting space weather storms. These services are provided to promote public safety and mitigate economic loss that could result from disruption of critical systems such as satellite operations, communications and navigation systems, and electric power distribution grids. The SWPC operates the national civilian Space Weather Forecast Office, coordinating with Air Force Weather Agency (AFWA) personnel at Offutt Air Force base in the production of joint products, ensuring a consistent message on the space weather forecast across both civilian and Department of Defense (DOD) customer bases. The SWPC provides services to customers on a 24 hour-per-day, seven day-per-week basis. SWPC products and services include observations and forecasts of solar, interplanetary, geospace, ionosphere, and thermosphere conditions that impact technologies on Earth and in space. The SWPC also develops and evaluates new models and products and

transitions them into operations. The SWPC takes a leading role in advocating and specifying new space-environment sensors for operational use.

The SWPC provides services to a broad user community of government agencies including DOD, NASA, DHS, DOE, and DOT. The SWPC provides support to industries, public institutions, and private individuals including the electric power industry, the airline industry, the satellite industry, oil exploration, agriculture, and many users of high frequency (HF) communication and satellite navigation. The SWPC also serves as the primary international World Warning Agency for the International Space Environment Service (ISES). It exchanges international data (solar wind, X-ray, sunspot, corona, magnetic, and ionospheric measurements) in real-time and issues a consensus set of daily forecasts for international use.

U.S. Air Force

The mission of AF weather personnel is to enable AF and Army decision-makers to anticipate and exploit the weather for air, ground, space, cyberspace, and intelligence operations. As this applies to the 'Space Weather Services' category, AF weather personnel provide space environmental information, products, and services required to support Department of Defense (DoD) operations as required, providing actionable environmental impacts information directly to decision makers.

The AF's 2nd Weather Squadron (2 WS), Offutt Air Force Base, Nebraska, is the Department of Defense's reach-back center for space environmental services operations. These personnel apply a detailed understanding of the space environment to translate raw data into useful military intelligence information, which can be integrated into the Common Operating Picture.

Personnel assigned to 2 WS monitor the sun's emissions and provide mission-tailored analyses, forecasts, and warnings. Their products are used for mission planning and environmental situational awareness by national agencies, DoD operators, Warfighters, and decision makers. Solar emission of highly-energetic particles, X-rays, and radio bursts can produce the following effects on DoD operations:

- Electrical anomalies and degrading of components to satellites and other equipment in orbit above the protective levels of the atmosphere.

- Impacts on electromagnetic signals, influencing High Frequency (HF) communication, Ultra High Frequency (UHF) communication, and Global Positioning System (GPS) satellite navigation signals.

- Increased drag on satellites in low-earth orbit.

- Increased interference or false returns to sunward or poleward looking radars.

- Potential health impact of radiation exposure to high-altitude aviators and those flying over polar regions.

The 2 WS space weather technicians, both at Offutt Air Force Base (AFB), Nebraska, and at solar observatories around the globe, are always focused on the sun. They provide timely, relevant, and accurate space weather information to DoD personnel, issuing approximately 35,000 forecaster-in-the-loop and automated textual and graphical products warning of significant solar activity daily.

Space environmental information is obtained through a combination of ground- and space-based systems. For the near-earth environment, i.e., ionosphere, ground-based and space-based systems complement each other to provide highly accurate point source verification and specification as well as enable global coverage and theater-wide situational awareness. For solar data, ground-based systems provide reliable observations of the sun in optical and radio frequencies, and space-based observations provide information (or data) unobtainable from the ground. Space-based systems provide in situ measurements of the space environment; i.e., solar wind and magnetosphere.

The 2 WS operates the Solar Electro-optical Observing Network (SEON), a network of five ground-based observing sites located around the globe providing 24-hour coverage of solar phenomena at optical and/or radio wavelengths. The network sites are:

- Detachment 1, Learmonth, Australia

- Detachment 2, Sagamore Hill, Massachusetts

- Detachment 4, Holloman AFB, New Mexico

- Detachment 5, Kaena Point, Hawaii

- San Vito, Italy (contract site)

Solar optical and radio telescopes at Learmonth, Australia. Source: U.S. Air Force.

The SEON network sites utilize the Radio Solar Telescope Network (RSTN) and/or the Solar Observing Optical Network (SOON). The RSTN is composed of the Radio Interference Measuring Set (RIMS) and the Solar Radio Spectrograph (SRS) and is used to monitor solar radio bursts at eight specific frequencies as well as a spectral band. The SOON is used to monitor solar flare activity, which are sometimes the precursor to coronal mass ejections that may interact with the earth's magnetic field to create geomagnetic storms. The SOON images the sun in the hydrogen-alpha wavelength, which reveals the complex solar activity in the lower atmosphere or chromosphere, as well as imaging the sun in the continuum (white light), which shows sunspots on the sun's surface, or photosphere. The SOON also enables the production of magnetograms by analyzing right-hand and left-hand circularly polarized light to image the line-of-sight component (Doppler shifting) of the magnetic field in the photosphere. When solar emissions are observed over threshold levels, solar analysts transmit activity messages that are used to prepare mission-tailored analyses, forecasts, and warnings used for mission planning, mission execution, and environmental situational awareness.

The AF employs a worldwide network of ground-based ionosondes and other sensors to provide environmental data in the ionosphere. It manages the NEXt-generation IONosonde (NEXION) fielding that started in summer of 2009, which will culminate in 20 NEXION sites worldwide. The AF has a network of 10 Scintillation Network Decision Aid (SCINDA) sites, which became operational in 2014. These sites characterize interference to SATCOM and GPS radio signals in the global equatorial regions. NASA's Jet Propulsion Laboratory operates a complementary global network of more than 125 sensors, deriving ionospheric line-of-sight total electron content from GPS signals, and provides these data to the AF. Additional data are provided by the U.S. Geological Survey (USGS), which operates a network of ground-based magnetometers, primarily in North America, that provide the AF with critical measurements of the earth's geomagnetic field and its variances.

From space, the Defense Meteorological Satellite Program (DMSP) Special Sensor Precipitating Electron and Ion Spectrometer measures low-energy precipitating electrons that cause the aurora and other high latitude ionospheric phenomena. The DMSP Special Sensors-Ions, Electrons, and Scintillation sensor provides top-side, in-situ measurements of the ionospheric environment, complementing ground-based sensors. These data are utilized to assess the impact of ionospheric conditions on ballistic-missile early warning radar systems and long-range communications. Additionally, the data are used to monitor global auroral activity and to predict the effects of the space environment on satellite operations. The Solar X-Ray Imager aboard NOAA's GOES-15 satellite monitors solar X-ray emissions and provides near real-time display at AFWA and the SWPC in Boulder, Colorado. The AF also leverages space-based data from NASA and other agencies.

AF weather organizations will continue to lead the Department of Defense in space weather operations in FY 2015 and beyond. In FY 2015, the AF will continue to upgrade its solar equipment and processes, along with providing new or upgraded facilities for some solar locations.

U.S. Geological Survey

The Geomagnetism Program (http://geomag.usgs.gov) of the USGS Geologic Hazards Science Center provides real-time, ground-based measurements of the Earth's magnetic field, which are an important contribution to the diagnosis of conditions in the near-Earth space environment of the Sun, the solar wind, the magnetosphere, the ionosphere, and the thermosphere. During geomagnetic storms, brought about by the complex interaction of the Earth's magnetic field with that of the Sun's, both high- and low-frequency radio communications can be difficult or impossible, global positioning systems (GPS) can be degraded, satellite electronics can be damaged, satellite drag can be increased, and astronauts and high-altitude pilots can be subjected to enhanced levels of radiation.

Ground-based geomagnetic observatory data are complementary to those collected by space-based satellites; indeed, most of the hazardous effects on technological systems brought about by magnetic storms occur at or near the Earth's surface. The Geomagnetism Program monitors ground-level magnetic activity by operating 14 magnetic observatories in the United States and its territories. Data from these observatories are transmitted to the Program's headquarters in Golden, CO in near-real time. USGS and foreign observatory data streams are used to calculate storm-time disturbance indices, such as the magnetospheric ring-current index Dst, which serves as a standard proxy measure of magnetic storm intensity and is an important input to numerous operational physics-based models of the magnetospheric-ionospheric system. Data and storm-time disturbance indices are transmitted to SWPC, U.S. Air Force, and NASA in real time. In addition, USGS data are utilized by private industry, foreign space weather agencies, and academia.

USGS observatories are operated in cooperation with INTERMAGNET (http://www.intermagnet.org), an international consortium that coordinates the operation of over 120 geomagnetic observatories in 40 different countries; the role of INTERMAGNET was recently summarized in an article in the *EOS* newsletter, which is published by the American Geophysical Union. The U.S. Geological Survey recently began a novel public-private cooperative relationship with the oil and gas drilling industry to operate a magnetic observatory in Alaska. The USGS Geomagnetism Program is an integral part of the National Space Weather Program (NSWP).

Federal Emergency Management Agency

The Federal Emergency Management Agency (FEMA) has initiated an Interagency Planning effort to develop a Concept of Operations Plan (CONPLAN) that will identify roles and responsibilities of each agency during a Space Weather Event. FEMA is working with their interagency partners to define the scope of this CONPLAN and to create a workable scenario on which to base the plan. Currently, this effort is in its infancy stage and participants in this effort are still being identified.

SUPPORTING RESEARCH PROGRAMS AND PROJECTS

NOAA/National Weather Service

Space Weather Prediction Testbed

The SWPC operates the Space Weather Prediction Testbed (SWPT) to provide the operational Space Weather Forecast Office with new models, products, and forecast techniques. The SWPT performs applied research and evaluates scientific developments from other agencies to develop and identify new capabilities to improve the prediction skill of space weather forecasts. The SWPT also validates and verifies new research results and works to transition research developments into operational products and services. Through these activities the SWPT achieves its principal objective of infusing the benefits of new research and technology into operational space weather products and services, in order to improve the utility and capabilities of the SWPC alerts, watches, warnings, and forecasts for its customers.

SWPT Activities include:

- Maintaining awareness of scientific advances and new techniques being developed to identify improved data-analysis techniques, forecast models, and observational systems that have potential for significantly improving the forecast guidance provided by space weather forecasters;

- Conducting, supporting, and managing focused research on data-analysis techniques/ algorithms, forecast models, and observational systems that have the potential to significantly improve the forecast guidance provided to space weather customers;

- Developing, testing, validating, and verifying promising numerical codes and forecast techniques, emerging from the research community to determine their potential benefits for possible use in operations;

- Communicating priorities and operating procedures to maintain fair and open interactions with all stakeholders (operational, research, academic, international, and commercial) and to stimulate improvements in space weather analysis and forecasting applications;

Current projects at the SWPT include the following:

- Improvements in the one-to-four day forecasts of space weather storms by improving the definition and parameterization of coronal mass ejections (CME) for input into operational solar-heliospheric models that forecast the propagation of solar disturbances through interplanetary space from the Sun to the Earth.

- Improving services to the electric power industry by implementing a geospace/magnetosphere model so that SWPC can provide customers with regional forecasts and specifications of space weather impacts on Earth.

- Improving estimates of auroral impacts by the development and transition of an auroral forecast model that will provide estimates of where and how intense the aurora will be. These forecasts will be used in other space weather models as well as by the general public interested in observing aurora.

- Improving services to Global Positioning System /Global Navigation Satellite System (GPS/GNSS) customers by the development of the Whole Atmosphere Model which extends the NCEP Global Forecast Systems weather model up to the near-space environment thereby providing specification and forecasts of the impacts of the lower atmosphere weather systems on space weather.

U.S. Geological Survey

Research conducted within the USGS Geomagnetism Program targets space-weather applications and geomagnetic hazards. Recent work has focused on mapping magnetic activity across North America and between magnetic observatory locations, and estimating the storm-time induction of electric fields in the Earth's electrically conducting lithosphere. Other research projects have focused on analysis of individual historical magnetic storms, long-term changes in geomagnetic activity, and long-term changes in solar-terrestrial interaction, all of which are important for understanding the potential hazard posed by magnetic storms that will occur in the future.

The research staff of the USGS Geomagnetism Program provides leadership and technical guidance to INTERMAGNET, an international consortium that is dedicated to promoting the global integration of real-time magnetic-observatory operations and the dissemination of their data. USGS research staff members also represent the USGS in numerous national and international forums. The importance of magnetic observatory data for space and solid earth research was recently highlighted in a feature article in the journal Physics Today, which is published by the American Institute of Physics. The role of the USGS Geomagnetism Program within the larger NSWP was recently summarized in a feature article in the journal *Space Weather*, which is published by the American Geophysical Union.

National Aeronautics and Space Administration

The objective of the Heliophysics Division of the National Aeronautics and Space Administration's (NASA) Science Mission Directorate (SMD) is to understand the Sun and its interactions with Earth and the solar system, including space weather. The three areas of concentration are theory development, data collection and analysis, and modeling of the resulting scientific understanding. To support this effort, the division operates a fleet of 18 missions involving 29 spacecraft. The region of space that must be covered is huge, extending from the Sun through the Earth's near-space environment and outward to the edges of the solar system.

To meet national and societal needs, NASA coordinates its space weather activities with several interagency and international partners, including the National Oceanic and Atmospheric

Administration (NOAA), which is responsible for delivering operational space weather products and services to the nation. Currently, five NASA research missions provide data that have become essential to our nation's space weather protection community. This is done by either direct broadcast from the satellite to a network of NASA and non-NASA receiving antennae, or by data that is processed in near real-time and made accessible via the internet. The Advanced Composition Explorer (ACE) spacecraft provides data on the condition of the solar wind upstream of the Earth's magnetic field. Additional space weather information is provided by the two Solar Terrestrial Relations Observatory (STEREO) spacecraft which broadcast beacon images of the far side of the Sun; as well as coronal mass ejection alerts from the European Space Agency (ESA)/NASA Solar and Heliospheric Observatory (SOHO) and super high-resolution images from the Solar Dynamics Observatory (SDO). Also, the two Van Allen Probes provide near-real-time information on the conditions in the Earth's radiation belts.

NASA also supports the development of space weather technology and predictive capability with its Research Program and the Living with a Star Science Program. With the Research Program's instrument development funding, NASA is developing the next generation of instruments that will be capable of observing extreme space weather conditions. NASA works to continually improve the understanding of space weather that enables improvements to space weather prediction models. Also, within the Research Program, as a quality assurance activity to validate solar and space physics research models and to prepare them for transition into operational activities, NASA operates a Community Coordinated Modeling Center (CCMC), an interagency collaborative activity involving the National Science Foundation (NSF), NOAA, and the DOD.

Department of Energy

The U.S. Department of Energy National Nuclear Security Administration (DOE/NNSA) supports the NSWP through the collection and distribution of operational data, through participation in research missions with space weather applications, and through the development of space weather models such as DREAM and AE/AP-9. One of the most significant contributions is the collection and distribution of space weather data from DOE/NNSA instruments on U.S. government satellites in geosynchronous and GPS orbits. DOE/NNSA geosynchronous observations have been available continuously since 1979 and, since 1989, measurements span energies from a few electron volts (eV) to tens of MeV. Geosynchronous observations are also available in real time from multiple satellites which constitutes an important resource for driving real-time specification and forecast models.

DOE/NNSA space weather resources on GPS satellites include both particle measurements for the space radiation environment and impulsive RF measurements that provide important information on ionospheric structure and density with global coverage (24 satellites). GPS observations cover the time period from 1983 to present so, as with other geosynchronous measurements, they provide an important resource for space weather climate models and for validation and testing of specification and forecast models.

DOE/NNSA, through the national laboratories, has also provided important space weather capabilities through the construction and operation of scientific instruments that also provide important space weather information. These include the plasma spectrometer on ACE (L1 solar wind), particle detectors on RBSP (ring current, radiation belts, solar particles), Forté (ionospheric structure and density), and others. DOE/NNSA has also supported the development of space weather models. These include the next-generation climatology model (AE/AP-9) for spacecraft design and the DREAM model which is a real-time assimilative model for the radiation belts. Both models rely heavily on geosynchronous and GPS observations. A third is RAM-SCB which is a model of the ring current. DOE/NNSA strives to partner with other entities with space weather interests and is exploring new ways in which its data, models, and space weather services can be more fully utilized to support the national space weather enterprise.

National Science Foundation

The National Science Foundation (NSF) supports the NSWP in pursuing the program's objective to perform the research and technology transfer needed to improve the specification and forecasts of space weather events that can cause disruption and failure of space-borne and ground-based technological systems and that can endanger human health. NSF supports space weather through dedicated programs as well as through basic space physics research programs. Space weather relevant research efforts include the development of large-scale space weather forecast models, construction and operation of advanced ground-based instruments and networks for the observation of space weather parameters, and the development and demonstration of innovative and creative small space weather satellites. NSF NSWP support in FY 2014 was estimated at $14 million and is expected to be around the same level in FY 2015. In FY14, NSF created a dedicated Space Weather program to go alongside science programs in Aeronomy, Magnetospheric Physics and Solar-Terrestrial Research.

U.S. Air Force

Air Force Research Laboratory

The Air Force Research Laboratory (AFRL) supports the Air Force space weather mission by executing research conducted by external agencies and by conducting in-house research on space weather. In space weather research, AFRL programs focus on ionospheric impacts to radio frequency systems, charged particle specification and forecasts, solar disturbance prediction, and neutral density effects on low-earth orbiting spacecraft. Working closely with the DMSP System Program Office at the Space and Missile Systems Center, under a memorandum of agreement, AFRL supports the development and upgrading of operational space weather sensors, models, and software products including space environment sensors on the DMSP spacecraft, state-of-the-art ground-based scintillation detectors, total electron content sensors, ionospheric characterization, solar radio and optical emissions observing, and the Operational Space Environment Network Display suite of Web-based products.

SURFACE TRANSPORTATION SERVICES

For purposes of this *Federal Plan*, Surface Transportation Services are those specialized meteorological services and facilities established to meet the weather information needs of the following surface transportation sectors: roadways, long-haul railways, the marine transportation system, rural and urban transit, pipeline systems, and airport ground operations. The roadway sector includes State and Federal highways and all State and local roads and streets. The marine transportation system includes coastal and inland waterways, ports and harbors, and the intermodal terminals serving them. Rural and urban transit includes bus and van service on roadways and rail lines for metropolitan subway and surface "light-rail" systems. Operational and supporting research programs for Aviation Services are often also relevant to airport ground operations, but program budgets counted in Aviation Services are not double-counted here under airport ground operations, and vice versa.

OPERATIONAL PROGRAMS INCLUDING PRODUCTS AND SERVICES

NOAA/National Weather Service (NWS)

National Weather Service's (NWS) Marine and Coastal Weather Services is the lead for the Nation's marine and coastal weather services, encompassing a vast area from -coastal waterways and near-shore bays and inlets to the open oceans spanning much of the northern and western hemispheres. The program is aimed at ensuring safe and efficient transportation, fishing, recreation, and offshore mineral exploration, in support of both commercial and recreational interests, and with consideration of the expanding and weather-sensitive U.S. coastal population. Forty-seven coastal Weather Forecast Offices (WFOs), the National Hurricane Center and the Ocean Prediction Center provide forecasts, analyses, watches, warnings and advisories of maritime conditions as well as coastal and tropical hazards. These services are provided for coastal waters, offshore and high seas waters, and Great Lakes near-shore and open lake waters. Coastal WFOs have responsibility for forecasts and warnings extending up to 60 nautical miles from the shore. The National Centers has are responsible for offshore and high seas waters, meeting U.S. obligations to maritime weather safety under the International Convention for Safety of Life at Sea, to which the United States is a signatory.

Using observational data sources such as buoys and satellites, and numerical model guidance, NWS forecasters monitor weather conditions continuously over their responsible maritime domain. They produce and disseminate routine forecast products and analyses, watches, warnings, and advisories which describe maritime weather, sea ice and oceanographic conditions including tropical storms and coastal storm hazards. NWS coastal marine warning and forecast products describe wind, waves, visibility, icing, storm surge, coastal flooding, severe weather, high surf, and rip currents. NWS tropical storm products describe track and intensity as well as associated coastal hazards such as storm surge, waves, and inland impacts.

The Marine and Coastal Services Program collaborates with a wide range of partners within and outside of NOAA. The program relies on the NWS Office of Operational Systems and NOAA's National Environmental Satellite, Data, and Information Service (NESDIS) for the collection of marine and coastal observations and the delivery of marine and coastal products to users. Through the joint National Ice Center, the Program collaborates with NESDIS, the U.S. Navy, and the U.S. Coast Guard (USCG) to provide ice warning and advisory services. It supports the Navy, the USCG, the U.S. Maritime Administration, and the U.S. Army Corps of Engineers (USACE) to operate the Nation's Marine Transportation System safely. It collaborates with the DOD, FEMA, DHS, and USACE to provide tropical cyclone warning and forecast services; with the USCG, Navy, Air Force, and private entities to disseminate weather information to mariners. It supports NOAA's National Ocean Service (NOS) on the PORTS and TIDES programs; and through the World Meteorological Organization to coordinate maritime weather and ice safety services with national meteorological and hydrological services world-wide for consistent and seamless services that cover world oceans. It also collaborates with NOAA's Office of Response and Restoration, Department of Defense, USCG and Department of Homeland Security to support emergency responses to maritime incidences such as hazardous material spills; maritime domain search, rescue, and recovery operations; and to maintain maritime domain situational awareness.

The NWS and Federal Highway Administration (FHWA) are working together to improve the incorporation of weather information by state Departments of Transportation (DOTs) in their decision making for high impact weather events. Currently, the NWS provides weather information through a variety of dissemination mechanisms used by some state DOTs for the purpose of ensuring motorist safety, and for protecting life and property. This includes coordination on forecasts and wording of variable message signs along roadways, engagement with state DOTs on 2-way NWSChat, and evaluation of road models based on NWS grids. NWS provides direct decision support to some state DOTs in certain non-routine situations, from real-time to seasonal events, which threaten public safety or can cripple road and bridge infrastructure. Support has been provided outside of standard products for fires (particulates and winds), dense fog, dust storms, hazardous material incidents/spills, coastal and inland flooding (flash flood, probabilistic river forecasts, precipitation frequency Atlas and runoff), and dangerous winter conditions (black ice, snow accumulation, melt related flooding) or tropical storms. Direct support includes single-slide IDSS (Impact-based Decision Support Services) graphics prepared for specific DOTs and experimental local graphical travel pages. State DOTs have also captured NWS graphics for redisplay directly on their public websites or internal web pages for each mile marker as well as for content in their 5-1-1 Travel Information Systems. To improve readiness, NWS provides seasonal training workshops and site visits with state DOTs.

NOAA/National Ocean Service

Marine Transportation System Services

The National Ocean Service (NOS) is the primary civil agency within the federal government responsible for the health and safety of our nation's coastal and oceanic environment. Largely through the Center for Operational Oceanographic Products and Services (CO-OPS) program line (http://tidesandcurrents.noaa.gov/), NOS acquires water levels, currents, and other physical oceanographic and meteorological data and distributes these data and circulation predictions as elements of an integrated NOS program. This program provides a comprehensive science-based suite of information required by the marine transportation community to ensure safe and efficient transportation, including the transport of hazardous materials. NOS also provides coastal

Visibility station in Mobile, AL used to determine fog.
Photo credit: NOS

oceanographic and meteorological products required by the NWS to meet its short-term weather and forecasting responsibilities, including tsunami and storm surge warnings. NOS manages several observing systems and programs; however, four in particular are heavily linked to the capability of NOAA to meet the marine transportation needs of the nation:

National Water Level Observation Network (NWLON). NOS manages the NWLON, which officially consists of 210 stations located along the coasts of the United States and the Great Lakes, from which water level data as well as other oceanographic and meteorological data are collected and disseminated. NWLON provides data and supporting information to a number of NOAA and other federal programs, such as the NOS Nautical Charting Program, NOS Shoreline Mapping Program, NWS Tsunami Warning System, NWS storm surge warning/forecast activities, and the Climate Services Program. Approximately 182 of the 210 NWLON stations contain at least one meteorological sensor (an anemometer, a barometer, an air temperature sensor and at some Great Lakes stations a relative humidity sensor), and 150 stations are outfitted with a full suite, which includes dual anemometers, a barometer and an air temperature sensor. Water level and meteorological data are automatically formatted into SHEF bulletin format for inclusion into the NOAA AWIPS system. In FY12, two NWLON stations in Texas were upgraded to Sentinels, which are specially-designed water level stations that are built to withstand a Category 4 hurricane, and these stations include a full suite of meteorological sensors. For more information see http://tidesandcurrents.noaa.gov/nwlon.html.

Physical Oceanographic Real-Time System (PORTS®). PORTS® is a decision support tool which improves the safety and efficiency of maritime commerce and coastal resource management through the integration of real-time environmental observations, forecasts, and other geospatial information. PORTS® measures and disseminates observations and predictions of water levels, currents, salinity, waves, bridge air gap and many meteorological parameters, needed and requested by the mariner to navigate safely. There are 23 existing PORTS® systems that comprise a total of 86 PORTS® water level stations. Currently, 71 of these stations contain at least one meteorological sensor (anemometer, barometer, air temperature sensor or a visibility sensor). In FY15, one new PORTS® is planned in Morgan City, LA. This station will include meteorological sensors including winds.

The PORTS® systems come in a variety of sizes and configurations, each specifically designed to meet local user requirements. PORTS® is a partnership program in which local operating partners fund the installation and operation of the measurement systems. The largest of NOS' existing installations is composed of over 100 separate instruments. The smallest consists of a single water level gauge and associated oceanographic and meteorological instruments. Regardless of its size, each PORTS® installation provides information that allows shippers and port operators to maximize port throughput while maintaining an adequate margin of safety for the increasingly large vessels visiting United States ports. In addition, prevention of maritime accidents is the most cost effective measure that can be taken to protect fragile coastal ecosystems. One major oil spill can cost billions of dollars and destroy sensitive marine habitats critical to supporting coastal marine ecosystems. PORTS® provides information to make navigation safer, thus reducing the likelihood of a maritime accident, and also provides the information necessary to mitigate the damages from a spill, should one occur. An extensible PORTS® can be integrated with other marine transportation technologies such as the Coast Guard's Automated Identification System AIS, Electronic Chart Display Information Systems ECDIS, and Vessel Traffic Systems VTS. Visibility sensors are the most recent sensor type to be integrated into the PORTS® systems, and there are currently two visibility stations installed in Mobile Bay PORTS®, three in the San Francisco PORTS® and three in the new Jacksonville PORTS®. More visibility installations are planned for Narragansett Bay and Chesapeake Bay PORTS® in FY14/15. For more information see http://tidesandcurrents.noaa.gov/ports.html.

Visibility Station in the New Jacksonville PORTS

National Operational Coastal Modeling Program (NOCMP). NOCMP serves a variety of users with oceanographic nowcast and forecast products for ports, estuaries and the Great Lakes. The integration of PORTS® technology and numerical circulation models allows nowcasts and predictions of up to five parameters (water level, current speed and direction, winds, water temperature and salinity) within the boundaries of the twelve models at locations where physical measurements are not available. In FY14 a new model in San Francisco Bay was implemented, along with nested grid models in Mobile, Gulfport and Pascagoula to enhance the Northern Gulf of Mexico model. Ongoing developments will enable the operational forecast systems to incorporate ecological forecast models and integrate the output with circulation measurements to provide information on transports of materials in the ecosystem essential for effective marine resource management and homeland security. Developed in FY13, model information is now automatically formatted into SHEF bulletin format for inclusion into the NOAA AWIPS system. For more information see http://tidesandcurrents.noaa.gov/nocmp.html.

The NOS Continuous Real-Time Monitoring System (CORMS). CORMS was designed to operate on a 24 x 7 basis to ensure the accuracy and working status of oceanographic and meteorological observations acquired via the NWLON and PORTS® programs. CORMS improves the overall data quality assurance of real-time measurements, reduces NOAA's potential liability by not publicly disseminating inadequate data, and makes the observations more useful for all applications. CORMS ingests real-time data from all field sensors and systems, including the operational nowcast/forecast models, determines data quality, and identifies and communicates the presence of invalid or suspect data to real-time users/customers who rely on the data. CORMS is especially vigilant during storm and tsunami events to ensure the full set of products and services is being disseminated in a timely fashion. An advanced version of this system, CORMS 3, provides personnel with alerts as soon as any sensor data are suspect or any communications problems arise. This enables speedier communication to instrument labs and field crews who may fix the station remotely or initiate emergency maintenance, thereby decreasing downtime of a particular station or sensor.

U.S. Coast Guard

Although no Coast Guard cutters or shore units are solely dedicated to meteorology, they collectively perform a variety of functions in support of the national meteorology program. USCG ocean-going cutters and coastal stations provide weather observations to the NWS. Coast Guard communications stations broadcast NWS marine forecasts, weather warnings, and weather facsimile charts. They also collect weather observations from commercial shipping for the NWS.

USCG conducts the International Ice Patrol (IIP) under the provisions of the International Convention for Safety of Life at Sea. The IIP uses sensor-equipped aircraft to patrol the Grand Banks of Newfoundland to locate and track icebergs that pose a hazard to North Atlantic shipping. Direct observations are supplemented and extrapolated using a numerical iceberg drift and deterioration model. IIP determines the geographic limits of the iceberg hazard and, twice daily, broadcasts iceberg warning bulletins and ice facsimile charts which define the limits of the iceberg threat during the iceberg season (spring and summer). IIP annually archives data on all confirmed and suspected icebergs, and forwards these data to the National Snow and Ice Data Center. These data can be accessed via the IIP web page, www.navcen.uscg.gov/?pageName=IIPHome. Archived data contains all iceberg sighting data along with the last model-predicted position of each berg.

The Coast Guard participates with the Navy and NOAA in supporting the National Ice Center, a multi-agency operational center that produces analyses and forecasts of Arctic, Antarctic, Great Lakes, and coastal ice conditions. The Coast Guard also collaborates with NOAA in operating the National Data Buoy Center (NDBC) which deploys and maintains NOAA's automated network of environmental monitoring platforms in the deep ocean and coastal regions. Five Coast Guard personnel fill key technical and logistics support positions within the NDBC. Coast Guard cutters support the heavy lift deployment and retrieval of data buoys and provide periodic maintenance visits to both buoys and coastal stations, expending approximately 180 cutter days annually. Coast Guard aircraft, small boats, and shore facilities also provide direct NDBC support.

Meteorological activities are coordinated by the Office of Marine Transportation Systems at Coast Guard Headquarters. Field management of Coast Guard meteorological support services is performed at the Coast Guard Area and District levels.

SUPPORTING RESEARCH PROGRAMS AND PROJECTS

DOT/Federal Highway Administration

Road Weather Management Program

The Federal Highway Administration (FHWA) coordinates a number of research and development activities aimed at improving safety, mobility, environmental quality, and national security on the nation's highways. These activities include identification and mitigation of weather impacts on the roadway environment. The FHWA does not operate either the highways or their supporting weather systems but seeks to improve operations in partnership with other public agencies (primarily State Departments of Transportation), national laboratories, private firms, and universities across the transportation and meteorological communities. Since 1999, FHWA's weather-related research activities have been centered in the Road Weather Management Program (RWMP) within the Office of Transportation Operations in coordination with the Intelligent Transportation Systems (ITS) Joint Program Office, which is housed in the Research and Innovative Technology Administration (RITA).

The goals of the RWMP are limited but span many areas to include: improving understanding of weather impacts on highway transportation systems, demonstrating a nation-wide system for observed road weather data, research new environmental data sources, enhancing road weather (e.g. pavement temperature) and traffic modeling with weather inputs, enhancing mechanisms for communicating road weather information to users, and developing decision support tools. For FY2014– FY2015, RWMP major research projects include expanded road weather observed data management and decision support applications, weather and road data from vehicles or mobile devices (mobile data), and weather-responsive traffic management.

Road Weather Observing, Data Collection and Management – The Weather Data Environment (WxDE)

Known as Environmental Sensor Stations (ESS), the standard method for observing road weather conditions is with fixed sensors near and/or actually embedded in the road surface that report common atmospheric weather variables plus pavement and subsurface road temperature, road wetness and pavement chemical concentration. Owned and operated by state, provincial or local transportation agencies, nearly 2,500 ESS are deployed across North America and together comprise one of the largest weather observing networks.

From 2006-2013, a U.S. DOT-sponsored experimental system entitled *Clarus* collected, formatted, quality checked, and displayed ESS road weather data from across North America. Upon meeting its goals, the *Clarus* System was decommissioned on June 15, 2013. However, the operational capability of the *Clarus* System is currently being transitioned to the NWS. FHWA and NWS signed a memorandum of understanding (MOU) in November 2010 to establish a framework for cooperation and coordination for projects like the *Clarus* transition to operations. In FY 2014, these agencies have been continuing to collaborate in the transition efforts. An important area of current research for both agencies is the gathering of weather and road data from vehicles.

Recognizing the continued need for road weather observations in order to execute the RWMP research, the Program started a new project in FY 2014 called the Weather Data Environment (WxDE). The primary purpose of the WxDE is to provide a data and interoperability platform for supporting connected vehicle research. In particular, the WxDE will support the intelligent aggregation, normalization, warehousing, processing, and dissemination of road weather data gathered from a variety of fixed and mobile sources. In fall 2013, an interim demonstration instance (see image to right, source: http://wxdedemo.leidoshost.com/dataMap.jsp) of the WxDE was launched to the community and since then, numerous improvements have been added to the system. In fall 2014, the demonstration instance will be transitioned to an operational production instance and released for general access.

USDOT Connected Vehicle Research – Vehicle Weather Observations

Connected Vehicles Research is a multimodal initiative that aims to enable safe, interoperable networked wireless communications among vehicles, the transportation infrastructure, and personal communication devices.[1] This research aims to leverage the potentially transformative capabilities of wireless technology to gather much more system data ultimately making surface transportation safer, smarter, and greener. Far beyond reliance on fixed or passive sensors, this emerging mobile technology has the potential to provide more extensive real-time travel and weather information to both the public sector and private industry.

Because vehicles were not designed as weather stations, direct weather sensor readings from original vehicle equipment are limited to mostly air temperature and pressure, but when

---

[1] See http://www.its.dot.gov/connected_vehicle/connected_vehicle.htm.

combined with other vehicle data they could prove useful. Some of the inferred weather variables from vehicle data are precipitation rate, visibility, and road surface condition. The RWMP has been working on a Vehicle Data Translator (VDT) to process weather-related data from cars and trucks in order to better characterize the driving conditions along standard distance (e.g. one mile) or user-defined road segments.

In partnership with the National Center for Atmospheric Research (NCAR) and the states of Nevada, Michigan, and Minnesota, the FHWA has been carrying out an expanded project in FY 2014 to demonstrate how data already resident on state fleet vehicles could be collected, processed, transmitted and used for providing motorist advisory warnings and enhancing maintenance decision support systems. The project has been helping to determine requirements, standards, and procedures for the collection and processing of weather, road condition, and vehicle status variables from mobile sources. Mobile weather and road condition data will also be integrated into the WxDE. The vision is for both public and private decision-makers to have the benefit of decision support tools that are supported by data from millions of vehicles through the connected vehicle initiative.

Road Weather High-priority Research Areas

The RWMP has identified the following high-priority research areas for the next ITS Strategic Plan for FY 2015-2019:

- Data and Applications – Continue research on road weather sensor design and integration within vehicles; data communications and standards; data management and quality checking; interpretation of observed mobile data and proper assimilation methods for incorporation of mobile data with other observed and modeled data; and, improved methods for information display and delivery including an understanding of human factors associated with user interpretation.
- State and Local Implementation Issues – From the existing RWMP projects, get a better understanding of the barriers and opportunities for public agencies to make investments on Connected Vehicle RWM applications, including the extent of infrastructure investments.
- Multi-Modal Integration – Continue research to build the bridge between ongoing state DOT efforts and connected vehicles and automation, in accordance with updates to the National ITS Architecture.
- Architecture & Standards – Develop a Connected Vehicle Reference Implementation Architecture for RWM data and applications.
- Early Stage Applications – Engage the solutions developer community to research and develop applications that leverage a flexible RWM connected vehicle communications environment.
- Connected Vehicle Pilots – Create real-world environments, urban and rural, in which public- and private-sector cooperation in implementation can educate the deployment community on the opportunities for replicating deployment.
- Data Exchange Facilitation & Interoperability – Coordinate, internationally and domestically, information sharing specifications, architecture, and standards necessary for enhanced road weather data sharing across the public and private sectors within the

transportation and meteorological communities. Develop and promote strategies that will foster interoperability among agencies for ingesting and disseminating weather data more universally.

- Surface Modeling – High spatial and temporal resolution surface modeling (integration of national weather prediction, pavement, traffic, and maintenance operations models)
- Vehicle Automation – Understand the impact of road weather in connected vehicle automation and role of automation-enabled road weather applications in improving safety, mobility and environment.
- Expanded Regional Pilots – Demonstrate data management capabilities to support multi-modal RWM operations and data fusion, including crowd-sourced information.
- Climate Change and Sustainability – Climate change, especially meeting the information needs during extreme weather events, and role of ITS in meeting the needs. Analyzing and anticipating the effects of climate change will ensure sustainable activities and strategies for deployment and operational decisions related to the ITS environment.

Weather Support for Traffic Managers

Unlike the national aviation system which has been a heavy user of weather information for decades, ground traffic management centers have been slow to integrate weather information into their operations. Since 2006, the RWMP focused a series of research projects on Weather-Responsive Traffic Management (WRTM) specifically addressing four areas: data collection and integration, human factors, WRTM strategies, and traffic analysis and modeling.

Initial research has been completed in many areas such as driver behavior in inclement weather; traffic speed and volume adjustment guidelines for various precipitation and visibility conditions; WRTM state of the practice review of management strategies; test and evaluation of those strategies; and message guidance for road weather advisory and control information.

Only until the last few years have traffic models incorporated weather data or the effects of weather. The RWMP completed research on the integration of weather into several traffic models including dynamic traffic prediction and assignment systems. In FY 2014, deployment began on a limited basis in some US cities. Research is also being conducted on the use of mobile weather data for traffic management.

The FHWA will continue participation in several OFCM projects including the WIST Working Group and the Committee on Integrated Observing Systems (CIOS), among others. The FHWA is also participating in NOAA efforts to explore the development of a national mesonet system, and is leading the Department's effort regarding the transportation Societal Benefit Areas of the National Earth Observations assessment. Nearing the end of several research projects on data management and applications, the RWMP has begun to look at other problem areas and update the program's research agenda. For example, research has begun on the impact of weather on trucking, especially in the area of delay costs in truck heavy corridors/regions. The RWMP looks forward to building on past successes and partnering with organizations that share the same passion for reducing the impact of weather on the nation's surface transportation systems.

DOT/Federal Railroad Administration

The Federal Railroad Administration (FRA) has outlined plans to support research on improving the collection, dissemination, and application of weather data to enhance railroad safety through its Intelligent Weather Systems project included in the FRA's 5-year Research and Development Strategic Plan. These programs address safety issues for freight, commuter, intercity passenger, and high-speed passenger railroads. Intelligent weather systems for railroad operations consist of networks of local weather sensors and instrumentation—both wayside and onboard locomotives—combined with national, regional, and local forecast data to alert train control centers, train crews, and maintenance crews of actual or potential hazardous weather conditions.

FRA intends to examine ways that weather data can be collected on railroads and moved to forecasters, and ways that forecasts and current weather information can be moved to railroad control centers and train and maintenance crews to avoid potential accident situations. This is one of the partnership initiatives identified in the National Science and Technology Council's National Transportation Technology Plan.

Operationally, the FRA relies on the meteorological data streams coming for the National Weather Service's Storm Prediction Center when issuance of regulatory waivers to railroads during times of severe weather is necessary.

NOAA/National Ocean Service (NOS)

Marine Transportation Research

Ocean Systems Test and Evaluation Program (OSTEP). OSTEP facilitates the transition of new oceanographic and meteorological sensors and systems to an operational status, in support of the NWLON and PORTS® programs. OSTEP tests instruments to ensure that they meet NOS requirements, develops operational deployment and implementation processes, and establishes quality-control criteria. OSTEP also develops defensible justification for the selection of instruments used for CO-OPS installations, and subsequent validation procedures for the devices traceable to U.S. National Standards or other accepted standards. Ongoing testing will reveal correlations of visibility data to other meteorological data types, and will result in a possible change in the standard sensor configuration of PORTS® visibility stations.

WILDLAND FIRE WEATHER SERVICES

For purposes of this *Federal Plan*, Wildland Fire Weather Services are those specialized meteorological services and facilities established to meet the requirements of the wildfire management community at the Federal, state, tribal, and local levels. The primary areas of service are to support the reduction of wildfire initiation potential and the mitigation of both human and environmental impacts once initiation does occur. Services can include support to first responders and land managers and climate services tailored to wildland fire management.

OPERATIONAL PROGRAMS INCLUDING PRODUCTS AND SERVICES

Interagency Collaborative Programs and Products

Fire Weather Services in the National Coordination Structure for Wildland Fire Management

Just as the service category for aviation weather derives from the need to understand and prepare for the influences of weather and other atmospheric conditions on the activity of flying aircraft, wildland fire weather services are needed to understand and predict the influences of weather and other atmospheric conditions on fire in the environment, particularly with the objective of assisting in the activity of managing and controlling such fires. Wildland fire weather services are therefore an integral part of the larger activity of wildland fire management.

National Wildfire Coordinating Group

The National Wildfire Coordinating Group (NWCG) is a collaborative group of intergovernmental partners with a shared vision and national responsibilities for wildland fire management. The partners focus on firefighter and public safety by improving coordination and integration through sharing talents, information, and resources. The NWCG develops and maintains standards, qualifications, and training for use by its member organizations. The NWCG enables member agency efforts to be consistent and coordinated while working collaboratively toward common goals. The Executive Board of the NWCG includes the five Federal wildland fire management agencies: The Bureau of Land Management (BLM), Bureau of Indian Affairs (BIA), Fish and Wildlife Service (FWS), and National Park Service (NPS) in the U.S. Department of the Interior, and the U.S. Forest Service (USFS) in the Department of Agriculture. The Executive Board also includes representatives from the U.S. Fire Administration within the Federal Emergency Management Agency, Department of Homeland Security, USFS Research, and three entities with responsibility for wildfire management on non-Federal forest lands: the National Association of State Foresters (NASF), the Intertribal Timber Council (ITC), and the International Association of Fire Chiefs. Oversight, policy coordination, and strategic direction for NWCG are provided by a series of federal and non-federal groups from the senior executive to the fire program levels.

The NWCG is organized into four branches, fourteen committees, and numerous subgroups, representing the many business areas of wildland fire (e.g. aviation, equipment, fuels, and qualifications). The NWCG committee most directly and frequently involved with capabilities for informing the wildland fire community about fire weather is the Fire Environment Committee (FENC) in the Equipment and Technology Branch. This Committee has chartered several permanent subcommittees:

- The Fire Weather Subcommittee (FWS) maintains the Interagency Wildland Fire Weather Station Standards and Guidelines, which addresses the network of permanently located Remote Automated Weather Stations (RAWS), portable stations used for incident response and prescribed fire, and manual fire weather stations. The Fire Environment Observation Unit under the FWS consists of the agency RAWS program managers who maintain the network of 2200 stations for their wildland fire agencies.
- The Fire Danger Subcommittee provides interagency direction to the Forest Service for the National Fire Danger Rating System (NFDRS) – a strategic planning tool and its national processor, the Weather Information Management System (WIMS). RAWS and manual station observations are key weather inputs to the NFDRS in WIMS.
- The Fire Behavior Subcommittee provides interagency guidance for the use of fire weather observations for determining fire behavior predictions from a variety of tactical applications. These predictions include the work of Fire Behavior Analysts working closely with Incident Meteorologists on incidents.
- The National Predictive Services Subcommittee oversees and provides guidance to the Predictive Services Program, which provides an important range of fire weather capabilities to the wildland fire community through the Predictive Services Units discussed below.

In 2012, the FENC developed guidance released by the Executive Board on management of the RAWS network: http://www.nwcg.gov/pms/pubs/426/index.htm. The RAWS network is generally located in remote areas. RAWS observations fill a critical spatial gap in the initialization of the National Digital Forecast Database (NDFD).

National Interagency Fire Center

The National Interagency Fire Center (NIFC), located in Boise, Idaho, is the nation's support center for wildland firefighting. Eight different agencies and organizations are part of NIFC: the five wildland management agencies, the National Weather Service (NWS) in the National Oceanic and Atmospheric Administration (NOAA), the National Association of State Foresters, and the

Entrance to the NIFC in Boise, showing the logos of the participating Federal agencies and the National Association of State Foresters.

U.S. Fire Administration. Decisions are made using the interagency cooperation concept because NIFC has no single director or manager.

The National Interagency Coordination Center (NICC), located at the NIFC, coordinates the national mobilization of resources for wildland fire and other incidents throughout the United States. Wildfire suppression is built on a three-tiered system of support: the local area, one of eleven geographic areas, and the national level. When a fire is reported, the local agency and its firefighting partners respond. If the fire continues to grow, the agency can ask for help from its Geographic Area Coordinating Center (GACC). When a geographic area has exhausted all its resources, it can turn to the NICC for help in locating what is needed, from air tankers to radios to firefighting crews to incident management teams.

National Predictive Services Program

Under the Predictive Services Program, meteorologists who specialize in fire weather services team with intelligence specialists and wildland fire analysts at the GACCs and the NICC to form Predictive Services Units. Each GACC and the NICC has a Predictive Services unit staffed with one or two meteorologists and an intelligence specialist. The NICC unit and Pacific Northwest GACC include a wildland fire analyst, and some of the GACC units add a fire behavior specialist during fire season. The Predictive Services units act as centers of expertise to produce integrated planning and decision support tools that enable more proactive, safe, and cost-effective fire management. The Predictive Services Program functions under the guidance of the National Predictive Services Subcommittee of the NWCG.

NOAA/National Weather Service

The National Weather Service (NWS) Fire Weather Services support Federal, state, and local land management, supporting partners such as the Department of Interior and the Department of Agriculture. On the national level, the Office of Climate, Water and Weather Services (OCWWS) directly supports tactical and strategic operations at NIFC. Also nationally, the NWS Storm Prediction Center issues assessments in advance of the development of critical fire weather patterns up to 8 days in advance. NWS Weather Forecast Offices (WFOs) also issue a complete Fire Weather Forecast twice daily, with updates as needed. The forecast contains weather information relevant to fire control and smoke management for the next 36-48 hours. The appropriate dispatch zones and crews use this information to plan staffing levels, equipment placement, prescribed burn conditions, and to assess the daily fire danger. Once per day, NWS meteorologists issue forecasts for specific wildland observation sites for input into the National Fire Danger Rating System (NFDRS). NFDRS determines land use restrictions and informs the public of the daily fire danger via the Smokey Bear awareness campaign. The WFOs also determine if a Fire Weather Watch or a Red Flag Warning needs to be issued. These products alert the public and other agencies that conditions are creating the potential for weather conditions creating extreme fire behavior. Finally, on a request basis, NWS forecasters issue spot forecasts for specific fire incidents, prescribed burn projects or other all-hazard incidents. In addition, the National Center for Environmental Prediction (NCEP) produces high resolution model guidance for winds, relative humidity, convection, precipitation and heat that may impact fire growth and environmental smoke transport over next 3-6 hours.

Upon request, NWS also provides on-scene assistance at large wildfires or other disasters, including HAZMAT incidents, by deploying Incident Meteorologists (IMET) to work with Incident Management Teams. These forecasters come from many different WFOs of all major NWS regions and, in some cases, support incidents more than a thousand miles from their home station. IMETs travel quickly to the incident site and then assemble a mobile weather center capable of providing continuous meteorological support for the duration of the incident. They gather other weather information through a remote connection and provide stand-up and on-the-spot forecasts/analysis to firefighters and agency heads. The IMET program is coordinated and implemented nationally by the National Fire Weather Operations Coordinator, National Science and Dissemination Meteorologist, and the National Fire Weather Program Manager, located at the NIFC.

The NWS has implemented regional digital weather files to complement currently-provided spot forecasts. The weather output enables Fire Behavior Analysts to directly input gridded weather data into fire danger assessments. These improvements are particularly important near zones where planned communities meet the wildland forests (known as the Wildland-Urban Interface or WUI). Recent improvements also include the creation of two fire weather specific gridded weather elements, an improved spot forecast program with access to smoke trajectory modeling. In addition NOAA and the USFS signed a fire weather research MOU that serves to coordinate applied fire weather research among the two agencies. Finally, the NWS renewed the Interagency Agreement for Meteorological Services with land management Agency Federal partners in September of 2012. The Agreement is valid through September 2017. NWS will continue excellent interagency relations with the wildland fire community through implementation of a new Interagency Agreement for Meteorological Services.

USDA/U.S. Forest Service

The U.S. Forest Service (USFS) uses meteorological data and interpretation skills data for decision making regarding wildland fire management. The Forest Service Fire and Aviation Management (FAM) program operates a network of approximately 950 remote automated weather stations (RAWS) in a national network of over 2600 stations. The network provides real-time information which is key in the highly utilized Weather Information Management System (WIMS) used by fire agencies across the country. The data collected is crucial to supporting active wildfire decision-making including use in the Wildland Fire Decision Support System and associated fire modeling tools as well as for decision-making for prescribed fire operations.

The program provides liaison with the Satellite Telemetry Interagency Working Group (STIWG) and its associated Technical Working Group and with the NWS, the wildland fire management agencies in the Department of the Interior (BLM, FWS, BIA, and NPS), State fire protection agencies, and the NWCG on the delivery of fire weather data and forecasting, critical for safety and effectiveness of firefighting and for flash flood warnings. The Forest Service RAWS Program manages the Interagency RAWS Website to support the program. The website address is http://raws.fam.nwcg.gov/. These stations form the basis for the assessment of fire danger, the pre-positioning of firefighting resources, and the conducting of prescribed fire operations. The costs include maintenance support contracts, maintenance training sessions, contracts for the delivery of this information to agency personnel, fire weather forecasters, and state forestry agencies that use the data in real-time for critical decisions.

The agency weather program works with the National Predictive Services Group at the NIFC to provide technical support and oversight to the 11 GACCs. It also works closely with the Forest Service Research and Development staff in the oversight of the five Fire Consortia for Advanced Modeling of Meteorology and Smoke (FCAMMS) locations. This effort, in cooperation with NOAA and the Environmental Protection Agency, provides valuable fire weather, smoke forecasting and air quality information to fire and air quality programs. The FCAMMS and Predictive Services Group provide critical information for both planning of wildland fire activities as well as operational decision-making.

Forest Service FAM has also initiated the Wildland Fire Air Quality Response Program to address the increasingly severe smoke impacts from wildfires. The Program encompasses a national emergency response cache of smoke monitoring and meteorological equipment which display data to internal decision-makers as well as the public through an array of websites including EPA's

Wildland fires in the wildland-urban interface are a continuing threat to lives and property.

AirNow. The program has developed a cadre of interagency personnel called Air Resource Advisors who are available for deployment to wildfire incidents to provide air quality impact information, monitoring, predictive smoke modeling and collaboration with states and those affected by smoke. Operational smoke modeling capability is another element utilizing the FCAMMS tools and recent refined model capabilities with finer spatial resolution and greater temporal scales. The Program also coordinates and collaborates with CDC, EPA and many states in guiding air quality response to wildfire smoke impacts.

The Wildland Fire Decision Support System (WFDSS) integrates emerging science and technology in support of risk-informed decision making. It is a web-based system for documenting decisions, supporting analyses, and completing operational plans applicable to and used for all wildland fires. It promotes access to numerous information analysis tools in the areas of fire behavior modeling, fire weather information, economic principles, air quality and smoke management, and information technology to support effective wildland fire decisions consistent with Land and Resource Management Plans and Fire Management Plans. The WFDSS greatly reduces text input requirements by using spatially oriented and graphically displayed information. The system incorporates a progressive decision documentation and analysis process that can be scaled and adapted to match situational changes. Through WFDSS, information is assembled, consolidated, and processed for decision makers in a way that fosters collaboration and, ultimately, provides better opportunities to improve large wildland fire strategic decision making.

U.S. Geological Survey

The U.S. Geological Survey (USGS), in cooperation with the USFS, routinely provides weekly forecasts of fire danger for the conterminous US and provides these forecasts to the National Interagency Fire Center. The forecasts are derived from an integration of vegetation condition observed from satellite and meteorological forecasts provided from the NWS's National Digital Forecast Database (NDFD). The NDFD forecasts provide meteorological information necessary for the calculation of live and dead fuel moisture, a critical element in determining wildland fire danger.

For active fire, the Basic Fire Behavior and Short-Term Fire Behavior components of the WFDSS use forecasted weather from the NDFD. NDFD incorporates Remote Access Weather Station (RAWS) location to derive forecasted weather data. This forecast information, along with geospatial data provide by the USGS are used to derive live and dead fuel moisture characteristics and wind conditions to aid in the prediction of fire behavior.

Landslides Hazards Program. Debris flows and flash floods that originate from steep watersheds burned by wildfire pose considerable hazards to downstream communities and structures. Fires throughout the western U.S. have impacted hundreds of thousands of acres of public land and made it susceptible to increased runoff and debris-flow activity. Science-based information on post wildfire debris-flow hazards is critically needed by Federal, State, and local agencies to issue warnings and to mitigate the impacts of post-fire hazards on people, their property, and natural resources. A joint NOAA/USGS, flash flood and debris flow warning system for recently burned basins in southern California was established in 2005 by linking the existing NWS Flash Flood Monitoring and Prediction (FFMP) system with rainfall intensity-duration thresholds for burned areas developed by the USGS. Such a system is being used to issue Outlooks, Watches and Warnings that are disseminated to emergency-management personnel and the public through the NWS existing protocol. The USGS has also developed models for characterizing potential post-fire debris flow susceptibility that, when compared with forecast or measured precipitation, have been used to generate maps of potential hazards in real-time, which have been disseminated to the Federal Emergency Management Agency (FEMA) and the public through existing NWS protocol. The USGS has also developed models for characterizing potential post-fire debris flow susceptibility that, when compared with forecast or measured precipitation, have been used to generate maps of potential hazards in real- time, which have been disseminated to FEMA and to State and local agencies. Since its inception, numerous advisories have been given to residents and public officials, resulting in saved lives and reduced property damage.

SUPPORTING RESEARCH PROGRAMS AND PROJECTS

Department of Agriculture

U.S. Forest Service

The research and development (R&D) mission of the USFS is to develop and deliver knowledge and innovative technology to improve the health and use of the Nation's forests and grasslands—both public and private. R&D provides this information to landowners, managers, policymakers, and the American people to help inform their decisions and actions. USFS researchers work independently and with a range of partners to provide land managers with information and technology to make management and land use decisions on issues such as invasive species, healthy watersheds, wildfires, climate change, and traditional and alternative forest products. The USFS R&D workforce includes scientists and technicians in the biological, physical, and social science fields, working in partnership with researchers from other agencies, academia, nonprofit groups, and industry. A few of the key accomplishments:

- Forest Service scientists made improvements to the Water Erosion Prediction Project (WEPP) model for forest conditions. One of the more recent improvements within the WEPP technology is the addition of shallow lateral flow as one of the primary sources of runoff from steep forested watersheds. The resulting predictive model shows the importance of lateral flow in delivering phosphorus from steep forested hill slopes to forest streams.

- Forest Service scientists combined an advanced ecosystem process model with data from FIA and remote sensing to separate the effects of disturbance factors from nondisturbance factors. Results showed that disturbance factors had the strongest effects overall, but with significant regional and temporal differences. This is the first time such separation of causes has been possible at the continental scale, and this new information can be used to support development of policies and approaches to improve sustainable forest management and provide for cleaner air and water.

- Forest Service research has enhanced our understanding of the effects of weather on forest fires and has developed tools to predict fire and smoke behavior that protect public health. Forest Service scientists developed standards for characterizing and measuring effects of airborne pollutants on wild and planted forest lands, and for monitoring how ecosystems may be affected by changes in concentrations of pollutants over time. This research will increase the accuracy of models that land managers use to minimize effects of air pollution on public lands and wilderness areas.

The 2012 Planning Rule requires national forests to consider air quality when developing Plan components and to treat air resources similar to soil and water resources. Forest Service managers directly monitor and use models developed by Forest Service researchers to measure or estimate the amount of atmospheric deposition occurring on National Forests and how this deposition is affecting Forest resources. Long-term air quality and resource monitoring on and near National Forest System (NFS) lands and Class I areas (area classification that receives the highest degree of protection under the Clean Air Act) has helped establish air pollution trends and existing condition of the resources. With this information an Air Quality Portal has been developed as an online decision management tool to assist forest managers in evaluating air pollution impacts on their forests.

National Aeronautics and Space Administration

The National Aeronautics and Space Administration (NASA) supports innovative, near-term demonstrations of its scientific results, technology developments, and satellite observations for societal benefit through its Applied Science Program element of the Earth Science Division (ESD). These projects serve as a bridge between NASA-generated data, derived products, information, knowledge and the decision-making needs of public and private organizations. End-users of NASA's products are able to apply Earth observations and model results to support activities that influence productivity, enhance quality of life, and strengthen the economy.

Wildland Fire Weather Applications: New Opportunities. In FY 2014, the NASA ESD, Applied Sciences Program peer-review-selected nine (9) proposed wildland fire projects to advance to a Phase II (Decision Support Phase, 3-years), where projects are expected to successfully transition their capabilities to operational use by partner organizations. Partner organizations are expected to ramp-up their support and provide operational continuity of the transitioned capabilities. The projects advancing to Phase II focused on fire-related multidisciplinary topics including health and air quality, ecological forecasting, fire weather, water resources, disasters, and climate, and addressed all stages of wildland fire activities (pre-fire, active-fire, and post-fire assessment). In FY15, some of the projects are expected to provide baseline data, models and information, in a quasi-operational environment, allowing early assessment of adaptation by their partners.

Of the nine selected wildland fire applications projects, four had a focused fire-weather forecasting and modeling component. Those four projects are highlighted here:

- *Wildland Fire Behavior and Risk Forecasting* focuses on improving forecasting of wildland fire behavior and risk through inclusion of NASA / NOAA remote sensing observations, integrated within NOAA fire weather forecasting systems. The system is based on a coupled atmosphere-wildland fire model, which combines the Weather Research and Forecasting model (WRF) with a fire spread model (WRF-SFIRE) and a dynamic fuel moisture model. Dynamic data from NASA and NOAA orbital platforms (moisture estimation, burn scar mapping, and fire detection) are assimilated into the simulations. The system is accessed through NOAA FX-Net system or a website, and downloadable KML files for visualization in Google Earth. Expected operational partners to advance the capabilities include NOAA, US Forest Service and BLM.

- *An Integrated Forest and Fire Monitoring and Forecasting System for Improved Forest Management in the Tropics* focuses development of a near-real-time alert system (FireCast) for use in non-traditional partner countries (such as Bolivia, Peru, Madagascar, and Indonesia) to enable improved fire monitoring, forecasting and improving monitoring of illegal forest activities (harvesting, burning, etc.). Firecast will include monitoring by NASA / NOAA orbital observation assets (MODIS, VIIRS, and Landsat) and development of fire risk models based on observed vegetation condition class and fire weather variables. Firecast will make a strong contribution to enhancing conservation and sustainable development goals for each of the target countries.

- *TOPOFIRE: A System for Monitoring Insect and Climate-Induced Impacts on Fire Danger in Complex Terrain* (integrates NASA remote sensing and climate products into a decision support tool that delivers a suite of high spatial resolution real-time information essential to wildland fire management. Data from thousands of temperature and humidity sensors distributed across the US will enable refinement of the North American Land Data Assimilation System (NLDAS) downscaling algorithms, and the generation of high resolution historical and real-time gridded climate datasets. The fire spread simulator FARSITE will be modified to assimilate gridded microclimate data, which will be coupled with high resolution hydrologic outputs to produce on-demand fire behavior simulations that account for topo-climate and disturbance-induced structural change. Crowd-sourced weather and fuels information from incidents will be assimilated into hydro-climatic and fire behavior models, returning terrain and location-corrected fire behavior and incident risk information via smart phone to the firefighter.

- *Development and Application of Spatially Refined Remote Sensing Active Fire Data Sets in Support of Fire Monitoring, Management and Planning* builds on proven science algorithms to produce new spatially refined active fire detection satellite products that yield significantly improved fire information. In addition, these products are used to initialize and validate fire growth predictions in a coupled weather-fire model that accounts for fuel information NASA / NOAA orbital sensors) and weather forecasts (including wind forecasts) to optimize the modeled outputs. The significance is that this approach can now be applied to monitor and predict the growth of a fire or a group of simultaneous wildland fires in a management unit from first detection until containment. The project focuses on near real-time operational delivery of new satellite remote sensing fire data, including customized products tailored for the end user community. Key partner agencies include the USDA Forest Service and the National Weather Service.

Each of these projects will result in an improvement in the capacity of the fire management community to monitor, model, predict and manage fire behavior to reduce impacts on the community and still optimize ecological balances and benefits that fire can provide. The integration of fire weather information and models can and will improve our capacity to understand fire interplay in the ecosystem.

OTHER SPECIALIZED SERVICES

For purposes of this *Federal Plan*, Other Specialized Services include weather and climate information services and facilities established to meet the special needs of user agencies or constituencies not included in basic services or the preceding service categories. This service category includes any efforts to integrate the social sciences into meteorological operations, applications, and services not already described in the preceding sections.

OPERATIONAL PROGRAMS INCLUDING PRODUCTS AND SERVICES

National Aeronautics and Space Administration

The National Aeronautics and Space Administration (NASA) provides operational weather support to spaceflight operations through the Human Exploration and Operations Mission Directorate (HEOMD).

Kennedy Space Center Weather Office

The HEOMD Weather Office at NASA Kennedy Space Center (KSCWO) has oversight responsibility for operation and maintenance of the weather information infrastructure required for NASA's manned spaceflight programs, and Expendable Launch Vehicles (ELV) programs. The infrastructure is a multi-agency partnership between NASA, the Department of Defense (DOD), and the Department of Commerce (DOC), and includes KSCWO, NASA's Marshall Space Flight Center (MSFC) and Johnson Space Center (JSC), the DOD's US Air Force (USAF) 30th and 45th Space Wings, and the DOC's National Oceanic and Atmospheric Administration (NOAA) National Weather Service (NWS) Spaceflight Meteorology Group (SMG). KSCWO also provides daily staff meteorological support to Kennedy Space Center (KSC) and NASA programs operating from KSC or the Eastern Range.

The ELV program operates from many locations, including Cape Canaveral Air Force Station (CCAFS), Vandenberg Air Force Base (VAFB) in California, NASA Wallops Flight Facility (WFF) in Virginia, and the US Army Ronald Reagan Ballistic Missile Defense Test Site on Kwajalein Island. KSCWO ensures that DOD weather support at DOD sites meets NASA requirements through training, technology, and tools. The KSCWO works with non-DOD sites and their weather service providers (such as the NWS or commercial companies) to provide similar assurance at those sites for NASA launches.

KSCWO is the NASA lead for the joint NASA and USAF Lightning Advisory Panel (LAP), which provides independent scientific assessments of changes to the lightning launch commit criteria (LLCC) and technical guidance about lightning-related issues on facilities and ground operations. The Department of Transportation (DOT) Federal Aviation Administration (FAA) utilizes the same criteria for lightning flight commit at commercial spaceports.

In FY 2014, the KSCWO:

- Supported the Tri-Program through infrastructure and requirements concept studies for the Space Launch System (SLS), Multi Purpose Crew Vehicle (MPCV), Ground Systems Development and Operations (GSDO), and Experimental Flight Test 1 (EFT-1).

- Supported NASA ELV launches from WFF and the Eastern and Western Ranges, as well as commercial and DOD launches from the Eastern Range.

- Supported infrastructure and concept of operations development for commercial launch programs and the NASA launch facility at Wallops Island, VA.

- Supported the Morpheus Lander Project by providing forecasts and weather consultations from the USAF 45th Weather Squadron for multiple Morpheus test firings at KSC. Continued procurement of a replacement for the aging 50 MHz Doppler Radar Wind Profiler (DRWP) that supports all launches from KSC and the Eastern Range.

- Installed a new "Lightning Mapping Array" sensor suite.

- Supported the Lightning System Upgrade (LSU) on the Eastern Range for data communication and validation.

- Continued assisting the FAA with the development of Lightning Flight Commit Criteria for the commercial sector.

In FY 2015, the KSCWO will:

- Continue to support the ELV program.

- Continue to support Wallops Island, Kodiak, and other launch facilities in addition to the DOD Ranges.

- Continue to support commercial launch operators in developing weather infrastructure, requirements, and concepts of operation.

- Continue assisting the FAA with the development of Lightning Flight Commit Criteria for the commercial sector.

- Support the Tri-Program, and continue support for planning and design of the test flight programs.

- Continue to work with the Eastern Range and other Ranges to define the requirements and infrastructure for weather support to NASA ELV and manned spacecraft in the post-Shuttle era.

- Support certification testing of the new 50 MHz DRWP.

Spaceflight Meteorology Group

The SMG is located at JSC. In FY 2014, the SMG:

- Operated with significantly reduced staff of 2 meteorologists.

- Provided consultation on the Orion/Multi Purpose Crew Vehicle program (MPCV), specifically Exploration Flight Test 1 (EFT-1) scheduled for December 2014.

- Participated in NASA/JSC Mission Control Center landing simulations for EFT-1

- Provided meteorological consultation and real-time weather forecasting support for the EFT-1 Underway Test off the coast of San Diego, CA

- Provided extensive coordination and collaboration with NASA and DoD to ensure surface and upper air observations occurred during the EFT-1 URT in February 2014. Among other aspects, SMG facilitated U.S. Army meteorologists at Yuma Proving Ground, AZ, dispatched to a U.S. Navy ship for weather balloon support. All weather support objectives were met.

- Coordinated with NCEP to obtain model upper air data archives, to assess (per NASA request) model forecast accuracy for the EFT-1 landing site off Baja, Mexico.

- Supported Orion parachute drop test events at Yuma Proving Ground, AZ, with surface and upper wind forecasts.

- Supported NASA with enroute forecasts for recovery personnel flying from Houston to Kazakhstan and back to Houston - to retrieve U.S. astronauts landing on the Russian Soyuz spacecraft after flying on the International Space Station.

- Supported the JSC Morpheus Lander Project by providing forecasts and weather consultations for multiple Morpheus test firings at JSC in Houston.

- Supported the NASA Langley Scientifically Calibrated in Flight Imagery (SCIFLI) project by providing cloud forecasts for SCIFLI thermal imaging of commercial launch vehicles.

- Provided support for several large outdoor events at JSC to ensure compliance with certain weather monitoring and notification guidelines. Events include Memorial Tree Planting ceremonies, large group picnics and photos, and Chili Cookoffs.

- Provided ongoing lightning and severe weather customized advisories for JSC management, weather sensitive operations, and employees.

- Maintained forecast and verification database for EFT-1 landing site.

- Provided tropical cyclone seasonal outlook briefings to JSC management and various directorates and divisions.

- Continued Excel-based tools to track SMG forecast issuances and decision support emails and briefings.

- Participated in school outreach with local schools and with schools across the U.S. via the JSC Educational Office Skype Studio.

In FY 2015, SMG will:

- Provide consultation on the Orion/Multi-Purpose Crew Vehicle program (MPCV), specifically Exploration Flight Test 1 (EFT-1) scheduled for December 2014.

- Provide support to the EFT-1 Underway Tests.

- Participate in NASA/JSC Mission Control Center landing simulations for EFT-1

- Provide weather decision support to the EFT-1 test scheduled for December 2014.

- Implement AWIPS enhancements focused on EFT-1 support.

- Support Orion parachute drop test events at Yuma Proving Ground, AZ, with surface and upper wind forecasts.

- Support NASA with enroute forecasts for recovery personnel flying from Houston to Kazakhstan and back to Houston - to retrieve U.S. astronauts landing on the Russian Soyuz spacecraft after flying on the International Space Station.

- Provide lightning and severe weather customized advisories for JSC management, weather sensitive operations, and employees.

- Support the NASA Langley SCIFLI project by providing cloud forecasts for thermal imaging of commercial launch vehicles.

- Provide weather support to large outdoor events at JSC.

- Provide lightning, severe weather, and tropical storm/hurricane customized advisories for JSC management, weather sensitive operations, and employees.

The Space Radiation Analysis Group

The Space Radiation Analysis Group (SRAG) is located at JSC and is responsible for ensuring that the radiation exposure received by astronauts remains below established safety limits. SRAG is responsible for daily assessment of the space weather environment for human exploration, and works directly with flight control teams to assess real-time mission impacts of adverse space weather conditions. Some of the key space weather-related activities that SRAG is conducting in FY 2014 and FY 2015 include:

- Daily monitoring of the space weather environment and impact assessment for crewed missions.

- Developing and maintaining protocols for radiation environment contingency response and analysis.

- Operating instruments both internal and external to crewed spacecraft for radiation environment characterization, quantification and impact assessment.

- Building and processing of crew worn passive dosimeters and maintaining all crew radiation exposure records.

- Design and Development of next generation radiation sensors that reduce mass, volume, and power without compromising data integrity.

- Developing collaborations for maintaining necessary space radiation measurements for instruments operating beyond planned lifetimes.

- Leading the development of space weather forecasting models.

- Transitioning maturing space weather forecast models from research to operations.

- Leading development of state-of-the-art radiation transport tools for vehicle design radiation analysis.

- Implementing state-of-the art radiation risk models used to ensure astronauts are within lifetime career radiation risk limits.

- Providing annual, end-of-active service, and ad hoc astronaut radiation risk reports.

Marshall Space Flight Center

The Natural Environments Branch (NEB) develops and implements weather support requirements for the Space Launch System (SLS) and other programs, including development and evaluation of launch constraints.

In FY 2014, the NEB:

- Continued development of terrestrial environment specifications, and provided atmospheric data and models for vehicle design and operation for the Space Launch System (SLS) and Multi-Purpose Crewed Vehicle (MPCV) programs.

- Began developing day-of-launch procedures relating to upper air wind and thermodynamic requirements for the SLS and MPCV programs.

In FY 2015, the NEB will:

- Continue to develop and improve atmospheric wind and thermodynamic climatological data sets and models for vehicle design and operational safety margin analysis for the SLS and MPCV programs.

- Continue developing day-of-launch procedures relating to upper air wind and thermodynamic requirements for the SLS and MPCV programs.

U.S. Air Force Space Launch Support

Air Force weather organizations provide meteorological and space weather products to the Nation's space and missile programs, including a wide range of weather observing services at the Air Force Eastern Range and KSC. AF weather organizations also provide tailored forecasting for NASA's manned and unmanned launches and for commercial launches from KSC. In addition, AF weather personnel provide specialized meteorological information for the Air Force Western Range at Vandenberg Air Force Base (AFB), California; the Pacific Missile Range, which includes Point Mugu and San Nicholas Island, California, and Barking Sands, Hawaii; White Sands Missile Range, New Mexico; Kwajalein Missile Range, Republic of the Marshall Islands; and other Department of Defense (DoD) research and test facilities as directed. The 45th Weather Squadron directly supports the 45th Space Wing of Air Force Space Command at Patrick AFB, Florida, and Cape Canaveral Air Force Station, Florida. The 30th Operations Support Squadron Weather Flight supports the 30th Space Wing across the Western Range at Vandenberg AFB, California.

U.S. Army Space and Missile Defense Command

Support to the Ronald Reagan Ballistic Missile Defense Test Site (RTS)

The RTS meteorological services contractor provides support for range activities, including local and remote missile launches, missile weapons readiness testing, aviation and marine operations, and emergency operations.

A full suite of meteorological surface, upper air, satellite, radar, and lightning observing systems are available. Surface systems include an intra-atoll mesonet and an FAA-approved Automated Weather Observing System (AWOS III-P/T), supporting range and International Civil Aviation Organization Army Airfield operations at Kwajalein. Upper air sounding systems (1680 MHz), utilizing Global Positioning System (GPS) radiosondes, are located on Kwajalein and Roi-Namur. One portable GPS upper air system (403 MHz) is available to provide soundings at remote locations. A dual-polarized Doppler S-band weather radar provides weather surveillance from Kwajalein Island, and a Doppler C-band weather radar is available for operations at Wake Island. Both are volume-scanning radars that support prediction of lightning events. Two Polar-orbiting Operational Environmental Satellite (POES) satellite receivers (one mobile) and one geostationary satellite receiver provides access to satellite imagery, cirrus cloud detection, and cloud height, with data processing and analysis provided through McIDAS management and display systems. A lightning detection network of four sensors is available to the RTS meteorologist at Kwajalein. A thunderstorm sensor that includes a field mill supports lightning prediction and detection at Wake Island. One thunderstorm sensor is attached to the AWOS III-P/T. RTS provides rocketsondes locally and at remote locations where radar tracking can support.

A rocketsonde launch on Kwajalein Atoll. U.S. Army Photo.

In cooperation with NASA Goddard Space Flight Center, RTS Weather continues to support global climate studies through the Tropical Rainfall Measurements Mission and the follow-on program of Global Precipitation Measurement. Solar-Earth radiation fluxes monitoring with a suite of radiation measurements systems have continued since 1989 in support of work at NOAA'S Earth Systems Research Laboratory (ESRL).

National Park Service/Fish and Wildfire Service

NPS Air Quality and Visibility Monitoring

The National Park Service (NPS) monitors air quality and visibility in a number of national parks and monuments. Gaseous pollutant data are collected on continuous and integrated (24-hour to weekly) bases. Surface meteorological data are collected and analyzed for hourly averages. Precipitation chemistry is determined on week-long integrated rainfall samples. Twenty-four-hour-average particle concentrations (mass, elemental analyses, some chemical constituent analyses) are measured every third day. Atmospheric light extinction is measured continuously and relayed to a central location for analyses.

Joint Air Quality Monitoring

The Fish and Wildlife Service (FWS) Air Quality Branch and the NPS Air Resources Division operate under an interagency agreement and are located in Lakewood, Colorado. Expertise from both agencies is pooled to address the air quality issues that are the responsibility of the Assistant Secretary of the Interior for Fish and Wildlife and Parks.

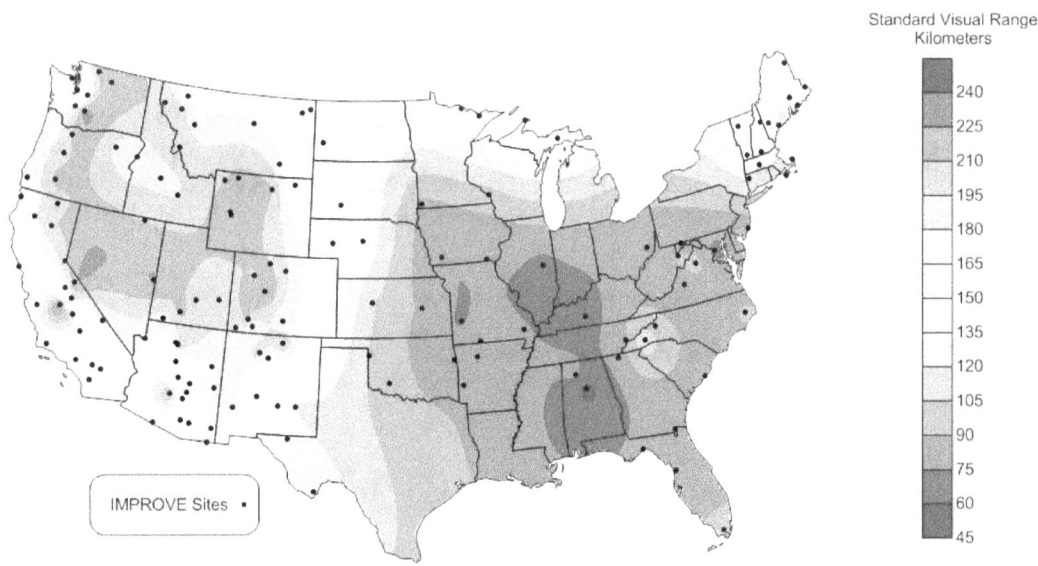

2012 Annual Average Standard Visual Range

Map of one-year average standard visual range (SVR) from 2012, in kilometers, calculated from IMPROVE particle concentrations. Also shown are the locations of most of the IMPROVE and IMPROVE protocol sites in the contiguous U.S.

The NPS oversees the operation of the Interagency Monitoring of Protected Visual Environments (IMPROVE) network and the IMPROVE Protocol network in cooperation with the Environmental Protection Agency (EPA), NOAA, the U.S. Forest Service (USFS), the FWS, the Bureau of Land Management, and various State organizations. Currently, the network has about 170 sites, mostly funded by the EPA in support of its regional haze regulations and through other cooperators. The enhanced network allows a better characterization of visibility and fine particle concentrations throughout rural and remote areas of the country (see figure above).

Nuclear Regulatory Commission

Assessments and evaluations of radiological impacts.

At the present time, the Nuclear Regulatory Commission (NRC) is a user of meteorological information rather than a performer of research in this field. Meteorological data are used to assess radiological impacts of routine airborne releases from facilities and to evaluate the impact of proposed changes in plant design or operation on unplanned releases. The NRC also maintains an interest in the effects of extreme meteorological events on the safe operation of nuclear facilities. The NRC uses current meteorological information and climatological predictions of extreme meteorological events over a range of timeframes, including long-term events greater than 100 years, to evaluate new reactor designs and sites, as well as existing reactor locations. Information of this type is also important for developing scenarios of climatological impacts on the isolation of long-lived nuclear wastes.

Within the NRC, the Offices of Nuclear Reactor Regulation and New Reactors conduct reviews of nuclear power plant siting, design, construction, and operation, while the Offices of Nuclear Material Safety and Safeguards and Federal and State Materials and Environmental Management Programs conduct similar reviews of materials and waste facilities. All of these reviews include consideration of meteorological factors. Employees of these NRC offices also conduct rulemaking activities to establish regulatory requirements, and the NRC Regional Offices assure that NRC licensees comply with the regulatory requirements.

SUPPORTING RESEARCH PROGRAMS AND PROJECTS

U.S. Army

U.S. Army Research Institute of Environmental Medicine

U.S. Army Research Institute of Environmental Medicine (USARIEM) conducts basic and applied research on the effects of heat, cold, high terrestrial altitude and nutritional status on the health and performance of Warfighters and combat crews operating military systems. USARIEM provides subject-matter expertise regarding the physiological impact of exposure to extreme heat, cold, and altitude on military personnel. Products include a series of technical medical bulletins (TB MED), thermal indices, and decision aids.

Applied research in thermal physiology and biophysical modeling is directed towards improving Soldier performance and minimizing health risks in climatic extremes. The research focus is on the sensitivity of Soldiers to local weather conditions (ambient temperature, dew point, wind speed, and solar radiation) and its impact on human performance. The overall goals of USARIEM weather-related research programs are to develop methods to effectively monitor and, where possible, extend the envelope for both training and operations.

USARIEM is investigating methodologies for integrating real-time local environmental data and Warfighter physiological data into predictive modeling processes. The effective combination of these two real-time data streams will enable environmental strain and performance status predictions for Warfighters. USARIEM is collaborating with other organizations, including the Army Research Laboratory-White Sands Missile Range (ARL-WSMR) under a Technology Program Agreement (TPA), to ensure access to soldier-scaled weather information and meaningful guidance concerning the weather effects on Soldiers.

National Park Service

Air Quality Research

The National Park Service (NPS) conducts research to develop and test air quality models to assess long-range transport, chemical transformation, and deposition of air pollutants. These models are used to estimate source contributions to, and to identify source regions responsible for, observed pollutant loadings.

The NPS is conducting research in the area of atmospheric nitrogen loading to high-elevation ecosystems in the Rocky Mountains, which have documented effects from nitrogen deposition. Measurements taken at Rocky Mountain National Park, in Colorado, indicate that routine monitoring networks may underestimate nitrogen deposition on the order of 30 percent by not analyzing for organic nitrogen and not routinely monitoring for ammonia gas. Source apportionment analyses indicate that under high loadings in the spring season, much of the nitrogen deposited at the park originates in the urban and agricultural areas of Colorado to the east of the park. By contrast, nitrogen loadings during the summer months had a significant contribution from Colorado, but higher loadings were noted from source regions out of the state. The NPS is participating in the July 2014 Deriving Information on Surface conditions from Column and Vertically Resolved Observations Relevant to Air Quality (DISCOVER-AQ) and the Front Range Air Pollution and Photochemistry Experiment APPÉ) field campaigns taking place in the Denver, Colorado area extending into the mountains, which includes Rocky Mountain National Park. Measurements of oxidized and reduced nitrogen compounds as well as organic nitrogen compounds will be made to integrate with commensurate airborne and ground measurements in the region. Analysis will continue into 2015.

NPS research in Grand Teton National Park in Wyoming, where effects on aquatic ecosystems from nitrogen deposition have also been documented, continues. Field measurements of atmospheric reactive nitrogen were completed in 2011. Source apportionment analysis is continuing with the availability of updated 2011 emission data.

NPS is also examining the effects of recent oil and gas development in North Dakota on Theodore Roosevelt National Park. Concentrations of oxides of nitrogen, ozone, volatile organic compounds and speciated fine particles were measured during a second field campaign from

November 2013 to March 2014 to assess atmospheric deposition and visibility impairment at the park. Analysis will continue into 2015.

National Aeronautics and Space Administration

Kennedy Space Center Weather Office

The KSCWO is described in the section above on Operational Programs. In FY 2011, the Director of Research for the KSCWO was a co-investigator on a climate research proposal to a FY 2011 NASA/ROSES solicitation. The proposal was selected for award in FY 2012 and work on the research was begun. In FY 2013 and FY 2014, the KSCWO continued work on the project along with the Principal Investigator from Embry Riddle Aeronautical University and co-investigator from the University of Central Florida. The project is entitled "Vulnerability Analysis of the Environment, Facilities, and Personnel of KSC to Extreme Weather Events and Climate Anomalies Resulting from Global Climate Change." The KSCWO ceased participation in the project in January 2014 upon retirement of the KSCWO Director of Research.

Applied Meteorology Unit

The Applied Meteorology Unit (AMU) is a joint venture between KSCWO, USAF 45th Space Wing (45 SW), and NOAA NWS. The AMU is co-located with the 45th Weather Squadron (45 WS) located at CCAFS. The AMU develops, evaluates, and transitions weather technology into operations.

In FY 2014, the AMU:

- Developed a capability for the NASA Launch Services Program (LSP) for the 30th Operational Support Squadron (30 OSS) at VAFB Launch Weather Officers (LWOs) to rapidly assess Global Forecast System (GFS), North American Mesoscale (NAM) and Rapid Refresh (RAP) model forecasts of upper-level winds by calculating the differences between the model data and current upper-level wind speed and direction observations from VAFB Automated Meteorological Profiling System (AMPS) soundings during launch operations. Developed code in Excel to ingest and graphically display real-time model and observational data and delivered the capability as an Excel GUI to the LWOs.

- Developed a capability for the NASA LSP for WFF LWOs to rapidly assess GFS, NAM and RAP model forecasts of upper-level winds by calculating the differences between the model data and current upper-level wind speed and direction observations from WFF soundings during launch operations. Developed code in Excel to ingest and graphically display real-time model and observational data and delivered the capability as an Excel GUI to the LWOs.

- Provided improved accuracy for severe weather notifications and better prepare decision makers at KSC, CCAFS, and Patrick Air Force base (PAFB) to implement appropriate mitigation efforts by building a sounding database from the 1500 UTC CCAFS rawinsondes from 1989-2012, updating severe weather report database using nationally available severe weather records from the National Climatic Data Center and the Storm Prediction Center, and calculating stability parameters based on the 1500 UTC CCAFS sounding database. Delivered a severe weather GUI in the Meteorological Interactive Data Display System based on the findings.

- Acquired dual-Doppler software, the Weather Decision Support System Integrated Information (WDSS-II), and ran test cases using the 45 SW Weather Surveillance Radar (WSR), National Weather Service in Melbourne (NWS MLB) Weather Surveillance Radar – 1988 Doppler (WSR-88D), and Orlando International Airport Terminal Doppler Weather Radar. Using a dual-Doppler capability with these radars could support operations at KSC, CCAFS and the NWS MLB County Warning Area in east-central Florida. Evaluated the dual-Doppler system to assure the output proves to be useful and easily interpreted by the forecasters. Will further collaborate with NWS MLB and the 45 WS to determine the most useful products for display and help determine if a real time implementation of WDSS-II is possible.

- Developed upper-level (UL) wind profile temporal pair databases and conducted a statistical analysis of wind changes at the Eastern Range (ER), Western Range (WR) and WFF for use by NASA LSP space launch vehicle teams in their commit-to-launch decisions. The intent of these databases is to help LSP improve the accuracy of launch commit decisions by applying wind change statistics based on measured historical data, as opposed to modeled data, into UL wind assessments.

- Completed evaluating configurations for a high-resolution model for the ER and WFF to better forecast a variety of unique weather phenomena in support of space launch activities. Ran test cases for the warm and cool seasons using several Weather Research and Forecasting (WRF) model domain configurations. Results comparing the WRF model forecasts against wind tower, accumulated precipitation, and sounding data show that Advanced Research WRF (ARW) outperforms the WRF Non-hydrostatic Mesoscale Model (NMM).

- Established WRF model with data assimilation (DA) for the ER and WFF to better forecast a variety of unique weather phenomena. Provided a recommended local DA and numerical forecast model design optimized for the ER and WFF to support space launch activities. The model will be optimized for local weather challenges at both locations.

- Completed market research of commercial, government, and open source software that might be able to ingest and display the 3-D lightning data from the KSC Lightning Mapping Array, the 45 SW WSR, the NWS WSR-88D, and the vehicle flight path data so that all can be visualized together. The results indicated there was no off-the-shelf software with 3-D capability but several software packages could be developed to provide the capability.

- Supported launch operations for five Atlas V, two Delta IV and four Falcon 9.

In FY 2015, the AMU will:

- Implement a real time modeling capability using the WRF Environmental Modeling System (EMS) on the AMU modeling clusters and display the model output in the Advanced Weather Interactive Processing System (AWIPS II). Run a triple-nested WRF EMS in near real-time in a rapid update mode and conduct a verification of the model output using the National Center for Atmospheric Research Model Evaluation Tools (MET) verification package. The grid spacing, domain sizes, and forecast periods of the model runs will depend on the cluster's capabilities. Most likely, the outer domain will have a 12-km grid spacing and include the east coast of the United States, the middle domain will have a 4-km grid spacing and cover central Florida, and the inner domain will have a 1.33-km grid spacing and cover the KSC/CCAFS area.

- Implement the NASA Natural Environments Branch at MSFC (MSFC NE) splicing algorithm used to merge data from boundary layer (915 MHz) and tropospheric (50 MHz) Doppler radar wind profilers at KSC and CCAFS into the AMU-developed LSP Upper Level Winds tool. The Upper Level Winds tool currently does not use any intelligent splicing algorithm to merge the top level of the boundary layer profiler with the bottom of the tropospheric profiler. Implementing the MSFC NE algorithm will produce a smoother profile when the data from the two sensors are displayed together.

- Continue to fine tune the Gridpoint Statistical Interpolation (GSI)/WRF model configuration based on individual range needs at the ER and WFF. Work with NASA's Short-term Prediction Research and Transition Center (SPoRT) to explore the feasibility of implementing a GSI/WRF rapid update cycle. Explore the possibility of ensemble modeling to improve forecasting of unique weather phenomena at each range.

- Support launch operations for Atlas V, Delta IV and Falcon 9.

Environmental Protection Agency

Air Quality Research

Meteorological support to the Environmental Protection Agency (EPA) Office of Research and Development and Office of Air and Radiation, EPA regional offices, and to State and local agencies includes the following activities:

- Conducting basic and applied research in air quality modeling

- Conducting field studies for air quality model development and air quality model evaluations

- Developing and applying multi-scale and multi-pollutant air quality models for pollution control, direct and indirect exposure assessments, and emission control strategy assessment

- Reviewing of meteorological aspects of environmental impact statements, state implementation plans, and pollution variance requests

- Providing Air Quality Index forecasts to state and local agencies for health advisory warnings

- Understanding the relationships between air quality and human health

- Understanding the atmospheric loading of pollutants to sensitive ecosystems

- Understanding the interactions of global climate change and air quality

- Emergency response planning in support of homeland security

Meteorological expertise and guidance are also provided for developing the national air quality standards, modeling guidelines, and policy development activities of the EPA. In light of the 1990 Amendments to the Clean Air Act and the recent national rules, air quality models and the manner in which they are used are expected to continue to grow over the next few years. In the area of pollutant deposition, the evaluation of nitrogen, oxidant, sulfur, and aerosol chemistries will help to clarify the roles of model formulation, cloud processes, aerosols, radiative transfer, and air/surface exchanges in air quality model predictions, leading to a better understanding of model predictions relative to control strategy assessments. Further development and evaluation of existing air quality models will take place to accommodate the inter-pollutant effects, resulting from the variety of control programs that are now or may be in place, such as the new National Ambient Air Quality Standards for ozone and particulate pollution. These inter-pollutant effects include trade-offs among controls on ozone, sulfur oxides, nitrogen oxides, and volatile organic compounds, as well as developing predictable methods of forecasting the impacts on various measures of air quality.

With respect to the fine particulate model development, air quality models are being enhanced to accurately predict aerosol growth from precursors over local and regional-scale transport distances. As the concentration thresholds for the standards decrease, it will be important to understand intercontinental transport of pollution and how this would affect our ability to meet and maintain standards in the future. To assist in the evaluation of the contribution of various sources to regional air degradation, inert tracer and tagged species numerical models have been developed. These models will introduce separate calculations for inert or reactive chemical species emitted from a particular source or region. The calculations will proceed to simulate transport and transformation to a receptor point, where the contribution of emission sources can be discerned.

Atmospheric research, regarding the effects of climate change on regional air quality, involves both analytical and statistical climatology as well as linking global climate models with regional chemical transport models, and the development of coupled models to better simulate the interactions between meteorology and atmospheric chemistry. Currently research is underway to test the efficacy of these models to accurately simulate the effects of aerosols on radiation.

Research in human exposure modeling includes both micro-environmental monitoring and modeling and the development of exposure assessment tools. This research entails linking air quality models to exposure models to understand the relationships between air quality and human health. Micro-environmental algorithms are being developed based on field data to predict air quality in buildings, attached garages, and street canyons. These improved algorithms are then incorporated into micro-environmental simulation models for conducting human exposure assessments within enclosed spaces in which specific human activities occur.

In addition to the above major areas, dispersion models for inert, reactive, and toxic pollutants are under development and evaluation on all temporal and spatial scales; i.e., indoor, urban, complex terrain, mesoscale, regional, and global. Other efforts include modeling nutrient deposition to the Chesapeake Bay and Gulf Coast, mercury deposition to the Florida Everglades, and the determination of meteorological effects on air quality. Atmospheric flow and dispersion experimental data obtained from wind tunnel and convection tank experiments in the EPA Fluid

Modeling Facility will be used to continue development and evaluation of these models along with providing researchers with insight into the basic physical processes that affect pollutant dispersion around natural and man-made obstacles. For example, the transport and dispersion of airborne agents in the Manhattan, New York, and the Pentagon were simulated in the wind tunnel to help build confidence in the modeling assessment of the source-receptor relationships for horrific events such as the one that occurred on September 11, 2001. The impacts of roadway configuration, noise barriers, and vegetation on air quality near roadways are being assessed, and improvements are being made to the EPA's AERMOD model to better simulate the transport and dispersion of pollutants from roadways.

Over the past 25 years, numerous air quality simulation models have been developed to estimate reductions in ambient air pollutant concentrations, resulting from potential emission control strategies. Separate models were developed, for example, for tropospheric ozone and photochemical smog, for acid deposition, and for fine particles. Distinct models also existed for addressing urban scale problems and the larger regional scale problems. It has been recognized, however, that the various pollutant regimes are closely linked chemically, spatially, and temporally in the atmosphere. The principal purpose of the Community Multi-scale Air Quality (CMAQ) modeling project was to develop a "one-atmosphere," flexible environmental modeling tool that integrates the major atmospheric pollution regimes in a multi-scale, multi-pollutant modeling system. This system will enable high-level computational access to both scientific and air quality management users for socio-economic applications in community health assessments and ecosystem sustainability studies.

The CMAQ model (first released in June 1998) is used by Federal and state agencies, industry, and academia and is updated periodically to reflect the state-of-science. The latest version of CMAQ, which includes science enhancements and computational efficiencies, was released in February 2012. It is also intended to serve as a community framework for continual advancement and for use in conducting environmental assessments. To this end, EPA has established a Community Modeling and Analysis System at the University of North Carolina in Chapel Hill, North Carolina, to provide user support and training to modelers at the state agencies and universities. New versions of the CMAQ modeling system and associated documentation (Installation and Operations Manual, User Manual, Science Document, and tutorial) are publicly available. Additional information is available on the division web site at http://www.epa.gov/amad.

From FY 2005 to FY 2008, the EPA worked closely with the NWS National Centers for Environmental Prediction (NCEP) in the continued development, evaluation, and use of a coupled meteorological-chemical transport model (WRF-CMAQ) for predicting ambient air quality over the continental United States. NWS implemented the CMAQ modeling system, to provide daily forecast guidance for ozone nationwide on an operational basis and fine particulate matter forecast on an experimental basis. State and local air quality management agencies are responsible to forecast local air quality and provide health advisory warnings.

The EPA, through participation in the interagency Information Technology Research and Development (IT R&D) Program, is developing a modeling framework that supports integration of diverse models (e.g., atmospheric, land surface, and watershed). The EPA's IT R&D work also enables increased efficiency in air quality-meteorological modeling through research on parallel implementation of the CMAQ modeling system. The evolving research seeks to improve the environmental management community's ability to evaluate the impact of air quality and watershed management practices, at multiple scales, on stream and estuarine conditions. The following primary objectives are directed toward this goal:

- Developing a prototype multiscale integrated modeling system with predictive meteorological capability for transport and fate of nutrients and chemical stressors

- Enabling the use of remotely sensed meteorological data

- Developing a computer-based problem-solving environment with ready access to data, models, and integrated visualization and analysis tools for water and air quality management, local and regional development planning, and exposure-risk assessments

A variety of research areas are being pursued such as the integration of hydrology and atmospheric models; coupling of meteorology and atmospheric chemistry calculations to account for the influence of radiatively active atmospheric pollutants on atmospheric dynamics and subsequent effects on air pollution; enhanced atmospheric dry deposition models; multi-scale and spatially explicit watershed modeling tools; and model-coupling technology for integrating media and scale-specific models.

The EPA also maintains good working relationships with foreign countries to facilitate the exchange of research meteorologists and research results, pertaining to meteorological aspects of air pollution. For example, agreements are currently in place with Canada, the United Kingdom, Greece, Japan, Korea, China, India, and Mexico, and with several European countries under the NATO Committee for Science for Peace.

APPENDIX A
ACRONYMS

3DWF	3D [three dimensional] Wind Field [model]
4DWX	Four-Dimensional Weather System
AAWU	Alaska Aviation Weather Unit
ACC	[USAF] Air Combat Command
ACCESS	Advancing Collaborative Connections for Earth System Science
ACE	Advance Composition Explorer
AD	Active Duty
ADA	Air Domain Awareness
ADAS	AWOS Data Acquisition System
ADDS	Aviation Digital Data Service
AFB	Air Force Base
AFCENT	Air Force Central Command
AFI	Air Force Instruction
AFRC	Air Force Reserve Command
AFRI	Agriculture and Food Research Initiative
AFRL	Air Force Research Laboratory
AFW	Air Force Weather
AFWA	Air Force Weather Agency
AFWEPS	Air Force Weather Ensemble Prediction Suite
AFW-WEBS	Air Force Weather Web Services
AgriMet	[Bureau of Reclamation] Agricultural Weather
AHPS	Advanced Hydrologic Prediction Service
AIP	Airport Improvement Program [FAA]
AIRMoN	Atmospheric Integrated Research Monitoring Network
AMC	U.S. Army Materiel Command
AMS	American Meteorological Society; [FAA] Acquisition Management System; Autonomous Modular Sensor; Analysis and Modeling Subsystem
AMSR	Advanced Microwave Scanning Radiometer
AMSR-E	[Aqua satellite] Advanced Microwave Scanning Radiometer-E
AMU	Applied Meteorology Unit
ANG	Air National Guard
ANSP	Air Navigation Service Provider
AO	Announcement of Opportunity
AOC	[NOAA] Aircraft Operations Center
AOML	Atlantic Oceanographic and Meteorological Laboratory
AOR	Area of Responsibility
ARL	[NOAA] Air Resources Laboratory
ARM	Atmospheric Radiation Measurement Climate Research Facility
ARS	Agricultural Research Service
ARTCC	Air Route Traffic Control Center
ASCC	Army Service Component Commands
ASNE MSEA	[DOD] Air and Space Natural Environment Modeling and Simulation Executive Agent
ASOS	Automated Surface Observing System
ASR	Atmospheric System Research [activity in DOE/CESD]; Airport Surveillance Radar

ASR-11	Airport Surveillance Radar Model 11
ASR-9	Airport Surveillance Radar Model 9
ASWON	Automated Surface Weather Observation Network
ATCSCC	Air Traffic Control System Command Center
ATD	atmospheric transport and diffusion
ATDD	Atmospheric Turbulence and Diffusion Division [NOAA]
ATEC	U.S. Army Test and Evaluation Command
ATLAS	Autonomous Temperature Line Acquisition System
ATM	Air Traffic Management
ATO	Air Traffic Organization [FAA]
ATOP	Advanced Technologies and Oceanic Procedures
ATOS	Applications, Transactions, and Observations Subsystem
AWC	[NOAA/NECEP] Aviation Weather Center
AWG	Aviation Weather Group [FAA]
AWIPS	Advanced Weather Interactive Processing System
AWOS	Automated Weather Observing System
AWOS III-P/T	Automated Weather Observing System [variant of AWOS]
AWRP	Aviation Weather Research Program [FAA]
AWRT	Advanced Weather Radar Technique
AWSD	Aviation Weather Services Directorate
AWSS	Automated Weather Sensors Systems
BASC	Board on Atmospheric Sciences and Climate
BCTP	Battle Command Training Program
BIA	Bureau of Indian Affairs
BLM	Bureau of Land Management
BonD	Battlespace on Demand
CAC	[U.S. Army] Combined Arms Center
CAgM	[WMO] Commission for Agricultural Meteorology
CALIPSO	Cloud-Aerosol Lidar and Infrared Pathfinder Satellite Observations
CAP	Civil Air Patrol; Common Alerting Protocol
CASR	Committee for Aviation Services and Research
CBRNE	chemical, biological, radiological, nuclear, or explosive
CCAFS	Cape Canaveral Air Force Station
CCMC	Community Coordinated Modeling Center
CCSP	U.S. Climate Change Science Program
CDMP	Climate Database Modernization Program
CDR	climate data record
CEISC	Committee on Environmental Information Systems and Communications
CENR	[NSTC] Committee on Environment and Natural Resources
CENRS	[NSTC] Committee on Environment, Natural Resources, and Sustainability
CERIS	Coastal, Estuary Resource Information System
CESD	Climate and Environmental Sciences Division [DOE Office of Science]

CESM	Community Earth System Model
CESORN	Committee on Environmental Services, Operations, and Research Needs
CFC	chlorofluorocarbon
CFSR	Climate Forecast System Reanalysis
CHPS	Community Hydrologic Prediction System
CI-FLOW	Coastal-Inland Flood Observation and Warning
CICE	DOE/OS/CESD sea ice model
CICS	Cooperative Institute for Climate and Satellites
CICS-NC	Cooperative Institute for Climate and Satellites North Carolina
CIOS	Committee for Integrated Observing Systems
CIP	Current Icing Product
CISM	Community Ice Sheet Model (DOE/OS/CESD)
CJMTK	Commercial Joint Mapping Tool Kit
CLASS	Comprehensive Large-Array data Stewardship System
CLIVAR	Climate Variability and Predictability Experiment
CMAQ	Community Multi-scale Air Quality
CMAS	Commercial Mobile Alert System
CMD-P	Computer Meteorological Data-Profiler
CME	coronal mass ejection(s)
CMIP5	Coupled Model Intercomparison Project Phase 5
CMS	Carbon Monitoring System
CNMOC	Commander, Naval Meteorology and Oceanography Command
CNO	Chief of Naval Operations
COAMPS	Coupled Ocean/Atmosphere Mesoscale Prediction System
COASTAL	Coastal Oceanographic Applications and Services for Tides and Lakes
CoCoRaHS	Community Collaborative Rain, Hail, and Snow [network]
COLA	Center for Ocean-Land-Atmosphere Studies
COMNAVMETOCCOM	Naval Meteorology and Oceanography Command
CONPLAN	Concept of Operations Plan
COOP	Cooperative Observer Program
CO-OPS	Center for Operational Oceanographic Products and Services
COPC	Committee for Operational Processing Centers
CORMS	Continuous Operational Real-time Monitoring System
COSIM	Climate, Ocean and Sea Ice Modeling Project
COSMIC-2	Constellation Observing System for Meteorology Ionosphere and Climate-2
CPC	Climate Prediction Center
CPP	Command Post Platform
CSD	[NOAA/NWS] Climate Services Division
CSESMO	Committee for Space Environmental Sensor Mitigation Options
CVA	Ceiling and Visibility, Analysis [FAA/AWRP]
CVF	Ceiling and Visibility, Forecast [FAA/AWRP]
CWSU	Center Weather Service Unit
DAC	[AOML] Data Assembly Center; Department of the Army Civilian

DAPE	Data Acquisition, Processing, and Exchange
DASI	Digital Altimeter Setting Indicator
DATMS	Defense Information Switched Network Asynchronous Transfer Mode System
DEM	digital elevation model
DHS	U.S. Department of Homeland Security
DMCC	DOE Meteorological Coordinating Council
DMSP	Defense Meteorological Satellite Program
DOC	U.S. Department of Commerce
DOD	U.S. Department of Defense
DOE	U.S. Department of Energy
DOI	U.S. Department of the Interior
DOMSAT	domestic communication satellite
DOS	U.S. Department of State
DOT	U.S. Department of Transportation
DOTMLPF	doctrine, organization, training, materiel, leadership, education, personnel, and facilities
DPG	Dugway Proving Ground
DSCOVR	Deep Space Climate Observatory
DTC	Developmental Test Center; [U.S. Army] Developmental Test Command
DTRA	Defense Threat Reduction Agency
DTSS	Digital Topographic Support System
EAS	Emergency Alert System
EcoFOCI	Ecosystem-Fisheries Oceanography Coordinated Investigations
ECV	Essential Climate Variables
EdIWG	[CCSP] Education Interagency Working Group
ELV	Expendable Launch Vehicle(s)
EMC	[NOAA/NCEP] Environmental Modeling Center
EMI SIG	Emergency Management Issues Special Interest Group
EMSL	Environmental Molecular Sciences Laboratory
EOSDIS	Earth Observing System Data and Information System
EPA	U.S. Environmental Protection Agency
EPI	Enhanced Precipitation Identification
ERAM	En Route Automation Modernization
ERC	[hurricane] eyewall replacement cycles
ERDC	[USACE] Engineer Research and Development Center
EROS	[USGS] Earth Resources Observation and Science [center]
ESM	[DOE/OS/CESD] Earth System Models
ESRL	Earth System Research Laboratory
ESS	Environmental Sensor Station(s)
ESTP	[NASA] Earth Science Technology Program
ET	evapotranspiration
EUMETSAT	European Organisation for the Exploitation of Meteorological Satellites
FAA	Federal Aviation Administration

FAR	false alarm rate
FCAMMS	Fire Consortia for Advanced Modeling of Meteorology and Smoke
FCMSSR	Federal Committee for Meteorological Services and Supporting Research
FEC	Fire Executive Council
FEMA	Federal Emergency Management Agency
FFMP	Flash Flood Monitoring and Prediction
FHWA	Federal Highway Administration
FIP	Forecast Icing Product
FMF	Fleet Marine Force
FNMOC	Navy Fleet Numerical Meteorology and Oceanography Center
FOR	Flight Operations Review
FPAW	Friends/Partners in Aviation Weather
FRA	Federal Railroad Administration
FRD	[NOAA/ARL] Field Research Division
FSR	Forest Service Research
FTE	full-time equivalent
FWS	U.S. Fish and Wildlife Service
FY	fiscal year
GACC	Geographic Area Coordinating Center
GBS	Global Broadcast Service
GCOS	Global Climate Observing System
GEOSS	Global Earth Observation System of Systems
GFDL	Geophysical Fluid Dynamics Laboratory [NOAA-associated]
GFDN	Geophysical Fluid Dynamics Navy [model]
GHCN-M	Global Historical Climatology Network-Monthly
GIS	geographic information system
GLD	Global Lagrangian Drifters
GLERL	Great Lakes Environmental Research Laboratory
GLOBE	Global Learning and Observations to Benefit the Environment
GMD	[NOAA/OAR/ESRL] Global Monitoring Division
GODAE	Global Ocean Data Assimilation Experiment
GOES	Geostationary Operational Environmental Satellite
GOES-R	Geostationary Operational Environmental Satellite R
GOOS	Global Ocean Observing System
GOSIC	Global Observing Systems Information Center
GPS	Global Positioning System
GPS-Met	GPS-Meteorology
GRA	GOOS Regional Alliances
GRIP	Genesis and Rapid Intensification Processes [NASA project]
GSD	[ESRL] Global Systems Division
GTGN	Graphical Turbulence Guidance Nowcast
GTOS	Global Terrestrial Observing System
GTS	Global Telecommunications System

HALE	high altitude, long-endurance [UAS]
HAZUS	Multi-Hazard Loss Estimation Methodology
HCFC	hydrochlorofluorocarbon
HEL	high-energy laser
HELSTF	High Energy Laser Systems Test Facility
HF	high frequency
HFIP	Hurricane Forecast Improvement Project
HFPP	HRD Field Program Plan
HHWWS	Heat Health Watch Warning Systems
HMR	[Nuclear Regulatory Commission] hydrometeorological report(s)
HMT	Hydrometeorological Testbed
HPC	[NPAA/NCEP] Hydrometeorological Prediction Center
HPCMP	[DOD] High Performance Computing Modernization Program
HQDA	Headquarters, Department of the Army
HRD	[NOAA/OMAO] Hurricane Research Division
HSPD	Homeland Security Presidential Directive
HWRF	Hurricane Weather Research and Forecasting
HYPOP	Hybrid Coordinate Parallel Ocean Program [DOE model]
HYSPLIT	Hybrid Single Particle Lagrangian Integrated Trajectory [ATD model]
IBTrACS	International Best Track Archive for Climate Stewardship
ICAO	International Civil Aviation Organization
ICESCAPE	Impacts of Climate on Ecosystems and Chemistry of the Arctic Pacific Environment
ICMSSR	Interdepartmental Committee for Meteorological Services and Supporting Research
ICOADS	International Comprehensive Ocean-Atmosphere Data Set
IFEX	Intensity Forecast Experiment
IHC	Interdepartmental Hurricane Conference
IIP	International Ice Patrol
IMAAC	Interagency Modeling and Atmospheric Assessment Center
IMET	Incident Meteorologist
IMETS	Integrated Meteorological System
IMPROVE	Interagency Monitoring of Protected Visual Environments [program]
INL	Idaho National Laboratory
IPAWS	Integrated Public Alert and Warning System
IPB	intelligence preparation of the battlespace
IPCC	Intergovernmental Panel on Climate Change
IPE	intelligence preparation of the environment
IRTSS	Infrared Target Scene Simulator
ISCS	International Satellite Communications System
ISES	International Space Environment Service
ISMS	[DOE] Integrated Safety Management System
ISR	intelligence, surveillance, and reconnaissance
ISS	International Space Station

IT	information technology
IT R&D	Information Technology Research and Development [Program]
ITS	Intelligent Transportation Systems
ITWS	Integrated Terminal Weather System
IWEDA	Integrated Weather Decision Aid
IWGCCST	Interagency Working Group on Climate Change Science and Technology
IWRSS	Integrated Water Resources Science and Services
JAG	Joint Action Group
JAG/ADM	Joint Action Group on Architecture and Data Management (
JAG/CCM	Joint Action Group for Centralized Communications Management
JAG/JUTB	Joint Action Group for Joint Urban Test Beds
JAG/MD	Joint Action Group on Metadata
JAG/OCM	Joint Action Group for Operational Community Modeling
JAG/ODAA	Joint Action Group for Operational Data Acquisition for Assimilation
JAWF	Joint Agricultural Weather Facility
JAWS	Juneau Airport Wind System
JCIDS	Joint Capabilities Integration and Development System
JCSDA	Joint Center for Satellite Data Assimilation
JHT	Joint Hurricane Testbed
JPDO	Joint Planning and Development Office
JPSS	Joint Polar Satellite System
JSC	[NASA] Johnson Space Center
JTWC	Joint Typhoon Warning Center
KDP	Key Decision Point
KSC	[NASA] Kennedy Space Center
KSCWO	SOMD Weather Office at NASA Kennedy Space Center
LAN	local area network
LAP	Lightning Advisory Panel
LBS	Littoral Battlespace Sensing
LDCM	Landsat Data Continuity Mission
LIDAR	light detection and ranging
LLCC	lightning launch commit criteria
LLWAS	Low Level Wind shear Alerting System
LLWAS-NE	LLWAS Network Expansion
LRGS	[USGS] local readout ground station(s)
M&O	management and operating
MACPEX	Mid-latitude Airborne Cirrus Properties Experiment
MADA	Monsoon Area Drought Atlas
MADIS	Meteorological Assimilation Data Ingest System
MAGTF	Marine Air Ground Task Force

MALE	medium altitude, long endurance [UAS]
MASPS	Minimum Aviation Safety Performance Standards
MAW	Marine Aircraft Wing
MEF	Marine Expeditionary Force
MET	meteorological; Meteorology and Oceanography
MetMF(R)	Meteorological Mobile Facility-Replacement
METOC	meteorological and oceanographic
METOP	Meteorological Operational Polar
METSAT	meteorological satellite
MM5	Mesoscale Model Version 5
MMS-P	Meteorological Measurement Set-Profiler
MOC	[NextGen] Mid-term Operational Capability
MODIS	Moderate Resolution Imaging Spectroradiometer
MOPS	Minimum Operations Standards
MOU	memorandum of understanding
MPAR	multifunction phased array radar
MPCV	Multipurpose Crew Vehicle
MRMS	[FAA/AWRP] Multi-Radar Multi-Sensor [capability]
MSFC	Marshall Space Flight Center
MST	METOC Support Team
MTOE	Modified Table of Organization and Equipment
MTSAT	[Japanese] Multifunctional Transport Satellite
MWPI	Microburst Windspeed Potential Index
MWSG	Marine Wing Support Group
MWSS	Marine Wing Support Squadron
NADM	North American Drought Monitor
NADP	National Atmospheric Deposition Program
NAO	NOAA Administrative Order
NAS	National Airspace System
NASA	National Aeronautics and Space Administration
NASS	National Agricultural Statistics Service
NAVOCEANO	Naval Oceanographic Office
NAVOCEANOPSCOM	Naval Oceanography Operations Command
NBC	nuclear, biological, and chemical
NCA	National Climate Assessment
NCAR	National Center for Atmospheric Research
NCDC	National Climatic Data Center
NCEP	National Centers for Environmental Prediction
NCOM	Navy Coastal Ocean Model
NCV	National Ceiling and Visibility [FAA/AWRP Product Team]
NDBC	National Data Buoy Center
NDFD	National Digital Forecast Database
NDMC	National Drought Mitigation Center

NEB	Natural Environments Branch [MSFC]
NESDIS	[NOAA] National Environmental Satellite, Data, and Information Service
NEXION	NEXt-generation IONosonde
NEXRAD	Next-Generation Weather Radar
NextGen	Next Generation Air Transportation System
NFDRS	National Fire Danger Rating System
NFIP	National Flood Insurance Program
NGA	National Geospatial-Intelligence Agency
NGDC	National Geophysical Date Center
NGEE	Next-Generation Ecosystem Experiments
NHC	[NCEP] National Hurricane Center
NICC	National Interagency Coordination Center
NIDIS	National Integrated Drought Information System
NIFA	National Institute for Food and Agriculture
NIFC	National Interagency Fire Center
NITES	Navy Integrated Tactical Environmental System
NNEW	NextGen Network Enabled Weathe
NNSA	National Nuclear Security Administration
NOAA	National Oceanic and Atmospheric Administration
NOCMP	National Operational Coastal Modeling Program
NODC	National Oceanographic Data Center
NOGAPS	Navy Operational Global Atmospheric Prediction System
NOHRSC	National Operational Hydrologic Remote Sensing Center
NOMADS	National Operational Model Archive and Distribution System
NOP	Naval Oceanography Program
NOPC	National Operational Processing Centers Program Council
NORAD	North American Aerospace Defense Command
NOS	National Ocean Service
NOWCON	Network of Weather and Climate Observing Networks
NPDIA	National Plan for Disaster Impact Assessments: Weather and Water Data
NPOESS	National Polar-orbiting Operational Environmental Satellite System
NPP	NPOESS Preparatory Project
NPRB	North Pacific Research Board
NPS	National Park Service
NRC	Nuclear Regulatory Commission; National Research Council
NRCC	[FEMA] National Response Coordination Center
NRCS	Natural Resources Conservation Service
NRL	Naval Research Laboratory
NRL/MRY	Marine Meteorology Division of the Naval Research Laboratory [NRL Monterey]
NSF	National Science Foundation
NSIP	National Streamflow Information Program
NSIR	Office of Nuclear Security and Incident Response
NSPD	National Security Presidential Directive
NSSL	[NOAA] National Severe Storm Laboratory

NSWP	National Space Weather Program
NSWRC	NextGen Surveillance and Weather Radar Capability
NTAS	Northwest Tropical Atlantic Station
NTHMP	National Tsunami Hazard Mitigation Program
NUOPC	National Unified Operational Prediction Capability
NWCG	National Wildfire Coordinating Group
NWIS	National Water Information System
NWLON	National Water Level Observation Network
NWP	numerical weather prediction
NWR	NOAA Weather Radio
NWRT	National Weather Radar Testbed
NWS	[NOAA] National Weather Service
OAR	[NOAA] Office of Oceanic and Atmospheric Research
OCE	(USDA) Office of the Chief Economist
ODCS	[U.S. Army] Office of the Deputy Chief of Staff
OEP	Operational Evolution Partnership [FAA airport designation]
OFCM	Office of the Federal Coordinator for Meteorological Services and Supporting Research
OGC	Open Geospatial Consortium
OMAO	[NOAA] Office of Marine and Aviation Operations
OMB	Office of Management and Budget
ONR	Office of Naval Research
OPC	Ocean Prediction Center
OSSE	observing system simulation experiment
OSTEP	Ocean Systems Test and Evaluation Program
OSTM	Ocean Surface Topography Mission
OSTP	Office of Science and Technology Policy
OTN	[Defense Information Systems Agency] Optical Transport Network
OTSR	Optimum Track Ship Routing
OWS	[USAF] Operational Weather Squadron
P3I	Pre-Planned Product Improvement
PACAF	Pacific Air Forces
PARISE	Phased Array Radar Innovative Sensing Experiment
PCMDI	Program for Climate Model Diagnosis and Intercomparison
PLAN	Personal Localized Alert Network
PMEL	Pacific Marine Environmental Laboratory
PNE	PIRATA Northeast Extension [project]
POD	probability of detection
POES	Polar-orbiting Operational Environmental Satellite
POP	Parallel Ocean Program [DOE/OS/CESD model]
POPS	Primary Oceanographic Prediction System
PORTS®	Physical Oceanographic Real-Time System
PREDICT	Pre-Depression Investigation of Cloud-systems in the Tropics

PSD	[ESRL] Physical Sciences Division
PTWC	Pacific Tsunami Warning Center
QPE	quantitative precipitation estimation(s)
QPF	quantitative precipitation forecast
R&D	research and development
R2O	Research to Operations
RASCAL	Radiological Assessment System for Consequence Analysis
RAWS	Remote Automated Weather Stations (network)
RBSP	Ring Current, Radiation Belts, Solar Particles [satellite]
RDT&E	research development, test, and evaluation
RES	Office of Nuclear Regulatory Research
RFC	[NWS] River Forecast Center
RGCM	Regional and Global Climate Modeling [in DOE/OS/CESD]
RIMS	Radio Interference Measuring Set
ROMANS	Rocky Mountain Atmospheric Nitrogen and Sulfur
ROSES	[NASA] Research Opportunities in Space and Earth Sciences
RPA	Remotely Piloted Aircraft
RSMC	Regional Specialized Meteorological Center
RSTN	Radio Solar Telescope Network
RTC	Redstone Test Center
RTS	Ronald Reagan Ballistic Missile Defense Test Site
RTVS	Real-Time Verification System
RUC	Rapid Update Cycle [NWS forecast model]
RVR	Runway Visual Range
RWI	Reduce Weather Impact
RWMP	Road Weather Management Program
SAFETEA-LU	Safe, Accountable, Flexible, Efficient Transportation Equity Act: A Legacy for Users
SAS	[NextGen] Single Authoritative Source
SAWS	Stand Alone Weather Sensors [FAA]
SCAN	Soil Climate Analysis Network
SCAP	Security Certification and Accreditation Package
SCAPA	Subcommittee for Consequence Assessment and Protective Actions
SDO	Solar Dynamics Observatory
SDR	[CENRS] Subcommittee on Disaster Reduction
SEBN	Surface Energy Budget Network
SEES	Science, Engineering and Education for Sustainability
SEON	Solar Electro-optical Observing Network
SERVIR	Sistema Regional de Visualizacion y Monitoreo [NASA network]
SESAR	Single European Sky ATM Research
SFMR	stepped frequency microwave radiometer
SGOT	Strike Group Oceanography Team

SICPS	Standard Integrated Command Post Shelter
SIP	Societal Impacts Program
SIR	System Integration Review
SLEP	Service Life Extension Program
SLOSH	Sea, Lake and Overland Surges from Hurricanes [storm surge model]
SMAP	[NASA] Soil Moisture Active-Passive [Satellite Mission]
SMD	[NASA] Science Mission Directorate
SMDC	[U.S. Army] Space and Missile Defense Command
SMG	[NWS] Spaceflight Meteorology Group
SNOTEL	SNOw pack TELemetry
SOF	Special Operations Forces
SOHO	Solar and Heliospheric Observatory
SOMD	[NASA] Space Operations Mission Directorate
SOON	Solar Observing Optical Network
SOOP	Ship of Opportunity Program
SORD	[NOAA/ARL] Special Operations and Research Division
SPC	[NCEP] Storm Prediction Center
SRS	Solar Radio Spectrograph
SST	sea surface temperature
SSWSF	Snow Survey and Water Supply Forecasting Program
STAR	Center for Satellite Applications and Research
STEM	Science, Technology, Engineering, and Mathematics
STEREO	Solar Terrestrial Relations Observatory
STIWG	Satellite Telemetry Interagency Working Group
SWEF	Space Weather Enterprise Forum
SWIM	[NextGen] System Wide Information Management
SWPC	Space Weather Prediction Center
SWSI	State Surface Water Supply Index(es)
TAF	Terminal Aerodrome Forecast
TAO	Tropical Atmosphere-Ocean [Project]
TAO/TRITON	Tropical Atmosphere-Ocean/TRIangle Trans-Ocean buoy Network
TAWS	Target Acquisition Weapons Software
TDA	Tactical Decision Aid
TDWR	Terminal Doppler Weather Radar
TFCC	Task Force Climate Change
T-IWEDA	Tri-Service Integrated Weather Effects Decision Aid
TMC	traffic management center
TOC	Tactical Operations Center
TOE	Table of Organization and Equipment
TRADOC	U.S. Army Training and Doctrine Command
TRMM	Tropical Rainfall Measuring Mission
TSG	thermosalinograph
TSO	Technical Standard Order [FAA]

UAS	unmanned aircraft systems; unmanned aerial systems
UHF	ultrahigh frequency
UNEP	United Nations Environment Program
UNFCCC	United Nations Framework Convention on Climate Change
UNOLS	University-National Oceanographic Laboratory System
UNSWC	Unified National Space Weather Capability
URI	University of Rhode Island
USA	U.S. Army
USACE	U.S. Army Corps of Engineers
USAF	U.S. Air Force
USAICoE	U.S. Army Intelligence Center of Excellence
USARNORTH	U. S. Army North
USASMDC	U.S. Army Space and Missile Defense Command
USCG	U.S. Coast Guard
USCRN	U.S. Climate Reference Network
USDA	U.S. Department of Agriculture
USDM	U.S. Drought Monitor
USFF	U.S. Fleet Forces Command
USFS	U.S. Forest Service
USGCRP	U.S. Global Change Research Program
USGS	U.S. Geological Survey
USHCN	U.S. Historical Climatology Network
USNO	U.S. Naval Observatory
USNORTHCOM	U.S. Northern Command
USWRP	U.S. Weather Research Program
UV	ultraviolet
VAAC	Volcanic Ash Advisory Center
VAFB	Vandenberg Air Force Base
VFR	visual flight rules
VORTEX2	Verification of the Origins of Rotation in Tornadoes Experiment 2
VOS	Volunteer Observing System
WAFS	World Area Forecast System
WAMIS	World AgroMeteorological Information Service
WAOB	World Agricultural Outlook Board
WARP	Weather And Radar Processor [FAA]
WASDE	*World Agricultural Supply and Demand Estimates* [report]
WC/ATWC	West Coast/Alaska Tsunami Warning Center
WEBB	Water, Energy, and Biogeochemical Budgets
WFDSS	Wildland Fire Decision Support System
WFIP	Wind Forecast Improvement Project
WFO	National Weather Service Forecast Office

WG/DIAP	Working Group for Disaster Impact Assessments and Plans: Weather and Water Data (
WG/TBC	Working Group for Test Bed Coordination
WG/UM	Working Group for Urban Meteorology
WG/VA	Working Group on Volcanic Ash
WG/WIST	Working Group on Weather Information for Surface Transportation
WGA	Western Governors' Association
WHDE	Wind Hazard Detection Equipment
WIFS	WAFS Internet File Service
WIMS	weather information management system
WIS	WMO Information Service
WMO	World Meteorological Organization
WoF	Warn on Forecast [Program]
WP	Work Product
WRAP	Wildfire Research and Applications Partnership
WRD	[USGS] Water Resources Discipline
WRE-N	Weather Running Estimate-Nowcasts
WRF	Weather Research and Forecasting
WRTM	Weather-Responsive Traffic Management
WS	[USAF] Weather Squadron
WSDS	Wind Shear Detection Services
WSP	[FAA] Weather Systems Processor
WSR-88D	Weather Surveillance Radar-1988 Doppler
WTIC	Weather Technology in the Cockpit
WUI	Wildland-Urban Interface
WW3	Wave Watch III [model]
WWCB	*Weekly Weather and Crop Bulletin*
WWLLN	World Wide Lightning Locator Network
WXG	[USAF] Weather Group
XBT	Expendable BathyThermograph [Program]

Agency Points of Contact

U.S. Department of Agriculture   Mr. Mark Brusberg
                                                            Dr. Harlan Shannon

Department of Commerce
National Oceanic and Atmospheric Administration   Ms. Aria Remondi
    National Environmental Satellite Data and Information Service   Ms. Kelly Turner
                                                            Ms. Lisa Heilmeier
    National Ocean Service   Mr. Darren Wright
                                                            Mr. Jason Shadid
    National Weather Service   Mr. Patrick Nield
                                                            Ms. Sreela Nandi
    Office of Oceanic and Atmospheric Research   Mr. Michael Bettwy
                                                            Ms. Melissa Pratt-Zossoungbo
    Office of Marine and Aviation Operations   LCDR Bradley Fritzler
                                                            Mr. Jeffrey Weir

Department of Defense   Lt Col Daniel Weekley
                                                            Ms. Marsha Korose
Joint Staff   Lt Col Kyle Bellue
    U.S. Air Force   Mr. Ralph Stoffler
    U.S. Navy   Mr. Philip Vinson
    U.S. Marine Corps   Lt. Col. Jeff Wooldridge
    U.S. Army   Mr. William Spendley

Department of Energy   Mr. Rick Petty
                                                            Mr. Vaughn Standley
                                                            Mr. Joel Cline

Department of Homeland Security
Federal Emergency Management Agency   Mr. Daniel Porter (NOAA Liaison)
U.S. Coast Guard   Dr. Jonathan Berkson
IMAAC   Mr. Chad Gorman
                                                            Mr. Sean Crawford

Department of the Interior
U.S. Geological Survey   Mr. Robert Mason
National Park Service   Mr. John Vimont
Bureau of Land Management   Mr. Louis Brueggeman
    Office of Fire and Aviation   Mr. Grant Beebe
    Office of Soil Water and Air   Dr. Larisa Ford

Department of State   Dr. David Reidmiller

Department of Transportation
Federal Aviation Administration   Mr. Everette C. Whitfield
Federal Highway Administration   Mr. Paul Pisano
Federal Railroad Administration   Mr. Jared Withers

Environmental Protection Agency   Dr. Rohit Mathur
                                                            Mr. Thomas Pierce

National Aeronautics and Space Administration   Dr. Ramesh Kakar
                                                            Ms. Elizabeth Yoseph

National Science Foundation   Mr. Nicholas Anderson

Nuclear Regulatory Commission   Mr. Jason White

Smithsonian Institution   Dr. Richard Wunderman

www.ingramcontent.com/pod-product-compliance
Lightning Source LLC
Chambersburg PA
CBHW080801180526
45168CB00006B/2286